FROM INTERSTELLAR CLOUDS TO STAR-FORMING GALAXIES: UNIVERSAL PROCESSES?

IAU SYMPOSIUM 315

COVER ILLUSTRATION:

Overview of the star-forming structures which were discussed in Symposium 315. From top right to bottom left: High-redshift star-forming galaxies as observed by Herschel in the GOODS-N deep field (HerMES project); Andromeda galaxy as imaged by Herschel (HELGA project); structure of the Cygnus X Giant Molecular Cloud as revealed by Herschel in our own galaxy (HOBYS project); detailed substructures unveiled by ALMA in the protoplanetary disk surrounding the young star HL Tauri.

IAU SYMPOSIUM PROCEEDINGS SERIES

INTERNATIONAL ASTRONOMICAL UNION

UNION ASTRONOMIQUE INTERNATIONALE

International Astronomical Union

IAU

FROM INTERSTELLAR CLOUDS TO STAR-FORMING GALAXIES: UNIVERSAL PROCESSES?

PROCEEDINGS OF THE 315th SYMPOSIUM OF THE INTERNATIONAL ASTRONOMICAL UNION HELD IN HONOLULU, HAWAII, UNITED STATES AUGUST 3–7, 2015

Edited by

PASCALE JABLONKA
EPFL, Switzerland

PHILIPPE ANDRÉ
CEA - AIM Paris-Saclay, France

and

FLORIS VAN DER TAK
SRON & University of Groningen, The Netherlands

CAMBRIDGE
UNIVERSITY PRESS

CAMBRIDGE UNIVERSITY PRESS
University Printing House, Cambridge CB2 8BS, United Kingdom
32 Avenue of the Americas, New York, NY 10013, USA
10 Stamford Road, Oakleigh, Melbourne 3166, Australia

First published 2016

Printed in the UK by Bell & Bain, Glasgow, UK

Typeset in System LaTeX 2ε

A catalogue record for this book is available from the British Library Library of Congress Cataloguing in Publication data

This journal issue has been printed on FSC™-certified paper and cover board. FSC is an independent, non-governmental, not-for-profit organization established to promote the responsible management of the world's forests. Please see www.fsc.org for information.

ISBN 9781107135208 hardback
ISSN 1743-9213

Table of Contents

Comparison of low and high redshift star formation

Preface

The link between the structure of the interstellar medium in galaxies and the star formation process on both local and global scales is one of the fastest growing areas of astrophysical research. The availability of wide-field far-infrared and submillimeter surveys with, e.g., the *Spitzer*, *Herschel*, *Planck*, *Wise*, and *Akari* space observatories, coupled with the much improved capabilities of ground-based millimeter and submillimeter interferometers, have led to spectacular and decisive steps forward in our understanding of star formation modes from solar system scales (tens of AUs) to global (kpc) scales in galaxies.

The goal of Symposium 315 was to start building up a coherent picture of how star formation is fuelled on a wide range of scales. Our ambition was to bring together researchers working on star formation throughout the Universe from nearby clouds to local galaxies to the first star-forming galaxies at high redshift, and to make connections between the most recent observations and the most advanced numerical simulations. We wished to foster discussions around the fundamental question of whether the dominant mode of star formation is quasi-universal or environment-dependent, we encouraged debates on a number of critical issues such as the origin and universality of the stellar initial mass function, the nature of star formation 'laws', and the role of feedback.

We are grateful to the SOC members for their help, exquisite expertise and generous share of their time, and we warmly thank all the participants for the high quality of their talks and posters. May these proceedings reflect the enthousiasm we put in the organisation of this symposium that was held during the 2015 IAU general assembly.

Lausanne / Saclay / Groningen, Spring 2016,

Pascale Jablonka, Philippe André, and Floris van der Tak, co-chairs of the SOC

THE ORGANIZING COMMITTEE

The Chairs

Pascale Jablonka (EPFL, Switzerland)
Philippe André (CEA - AIM Paris-Saclay, France)
Floris van der Tak (SRON & University of Groningen, The Netherlands)

The SOC

Susanne Aalto (Chalmers University, Onsala, Sweden)
Yuri Aikawa (Kobe University, Japan)
Frank Bigiel (University of Heidelberg, Germany)
Alberto Bolatto (University of Maryland, USA)
Françoise Combes (Observatoire de Paris, France)
Steve Eales (Cardiff University, United Kingdom)
Neal Evans (University of Texas, Austin, USA)
Daisuke Iono (National Astronomical Observatory of Japan, Tokyo, Japan)
Jeremy Lim (University of Hong Kong, Hong-Kong)
Eve Ostriker (Princeton University, USA)
Monica Rubio (Universidad de Chile, Santiago, Chile)
Kazushi Sakamoto (Academia Sinica, Taipei, Taiwan)
Jonathan Williams (University of Hawaii, Honolulu, USA)

Acknowledgements

This symposium was sponsored by the Ecole Polythechnique Fédérale de Lausanne (EPFL, Switzerland), the Netherlands Institute for Space Research (SRON, The Netherlands), and Commissariat à l'énergie atomique et aux énergies alternatives (CEA, France). It was supported by the IAU Divisions J (Galaxies and Cosmology) and H (Interstellar Matter and Local Universe)

From Interstellar Clouds to Star-Forming Galaxies: Universal Processes?
Proceedings IAU Symposium No. 315, 2015
P. Jablonka, Ph. André & F. van der Tak, eds.

doi:10.1017/S174392131600716X

Atomic and Molecular Phases of the Interstellar Medium

Mordecai-Mark Mac Low

Dept. of Astrophysics, American Museum of Natural History, Central Park West at 79th Street, New York, NY 10024-5192, USA and Institut für Theoretische Astrophysik, Zentrum für Astronomie der Universität Heidelberg, Albert-Ueberle-Str. 2, 69121 Heidelberg, Germany
email: `mordecai@amnh.org`

Abstract. This review covers four current questions in the behavior of the atomic and molecular interstellar medium. These include whether the atomic gas originates primarily in cold streams or hot flows onto galaxies; what the filling factor of cold gas actually is in galactic regions observationally determined to be completely molecular; whether molecular hydrogen determines or merely traces star formation; and whether gravity or turbulence drives the dynamical motions observed in interstellar clouds, with implications on their star formation properties.

Keywords. astrochemistry, hydrodynamics, MHD, turbulence, shock waves, ISM: general, ISM: structure, ISM: molecules, ISM: kinematics and dynamics

1. Introduction

Star formation occurs primarily in atomic and molecular interstellar gas, although Keto (2002) emphasizes that around a massive enough star, even ionized gas can be accreted. In this review, I discuss a number of recent advances in our understanding of the origin and dynamics of that gas and its relation to star formation.

2. Cold Accretion and its Limits

The origin of this gas lies in gas accreted onto galaxies during their formation and subsequent evolution. Perspectives on how that gas arrives have changed several times recently. Originally, it was thought that all gas accreted into the halo of a galaxy the size of the Milky Way would be heated to the virial temperature of over 10^6 K, followed by gradual cooling and accretion. However, a decade ago, numerical simulations started to suggest that a major, possibly even dominant alternative was for dense streams to accrete from the cosmic web, dense enough to cool quickly when compressed and thus to never heat to the virial temperature (e.g. Fardal *et al.* 2001; Birnboim & Dekel 2003; Kereš *et al.* 2005; 2009; Ocvirk *et al.* 2008; Dekel & Birnboim 2006; Dekel *et al.* 2009; 2013; Goerdt *et al.* 2015). However, Nelson *et al.* (2013), using the AREPO Voronoi mesh code (Springel 2010), argued that cold streams are likely only important for small galaxies at high redshift (see Fig. 1), while their appearance at lower redshifts and for larger galaxies is a numerical artifact of the smoothed particle hydrodynamics (SPH) method, as exemplified in that paper by GADGET (Springel 2005).

Evidence for continuing accretion of gas in the modern Milky Way can be gained from observations of high velocity clouds of neutral gas falling towards the galactic plane at velocities over 70 km s^{-1}. Material falling in at slower velocities has been shown to have Galactic chemical abundances, while the high velocity clouds have far lower abundances, suggesting extragalactic origin, as reviewed by Putman *et al.* (2012). Simple analytic

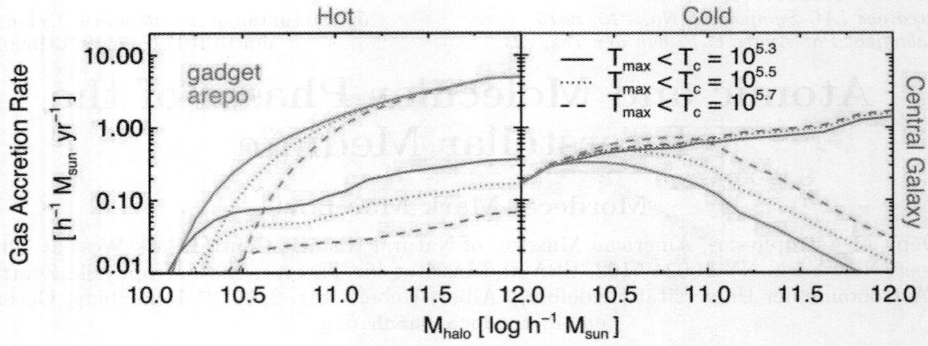

Figure 1. Comparison of gas accretion rates for hot gas (left, with temperature above T_c as shown in legend) and cold gas (right) as a function of halo mass, for SPH models (brown) and Voronoi mesh models (green). The two methods agree for small masses typical of high redshifts, but the Voronoi mesh code predicts most gas will reach high temperatures for large halo sizes, disagreeing with the SPH results. Adapted from Nelson *et al.* (2013).

models modeled on the flow of water into a bathtub and down the drain can guide our understanding of the cycle of gas into the atomic phase from infalling ionized gas, and then out of it again through star formation or expulsion of the gas in galactic outflows (Bouché *et al.* 2010; Dekel *et al.* 2013; Dekel & Mandelker 2014).

3. Why a Cold, Dense, ISM is Mostly Hot

The thermodynamic state of the atomic and diffuse molecular gas is controlled by the balance between optically-thin radiative cooling and photoelectric heating by far ultraviolet radiation. Between around 100 K and 10^4 K, the cooling curve is a sufficiently steep function of temperature to produce a region of thermal instability (Field *et al.* 1969, Wolfire *et al.* 1995, 2003). In the case of pressures typical of the Milky Way, this leads to a two-phase medium (solid line in Fig. 2). When pressures increase (dotted line), only the cold, dense phase remains thermally stable.

Ostriker *et al.* (2010) argued that, in fact, this transition leads to self-regulation of star formation. The pressure range within which a warm phase will be present is determined by the far ultraviolet radiation field produced by stars. If the vertical hydrostatic pressure exceeds that range, it will drive nearly all the gas into the cold phase, promoting star formation, and thus increasing the far ultraviolet field until a warm medium can be reestablished. Thus, the combination of thermal and hydrostatic equilibrium controls, in the terms of the bathtub model, how fast the bathtub drains.

While the point that high enough pressures produce a predominantly cold temperature distribution is widely understood, what is perhaps less appreciated is the resulting spatial distribution of the gas. Gatto et al. (2015) and Li *et al.* (2015) emphasize that the filling factor of the cold, dense gas will be small in such a high-pressure environment. In regions with supernova explosions (either from active star formation or even Type Ia supernovae from older populations) the space in between the cold, dense clumps is filled with hot, tenuous gas (see Fig. 3), the third phase of McKee & Ostriker (1977).

Observational tracers of atomic and molecular gas will show that the mass of the interstellar gas is almost entirely in cold, molecular material, leading to the description of the region as a fully molecular interstellar medium. The hot gas, on the other hand, is difficult to observe, as its density is low and temperature high enough that it emits primarily in bremsstrahlung. However, it still completely dominates the volume filling

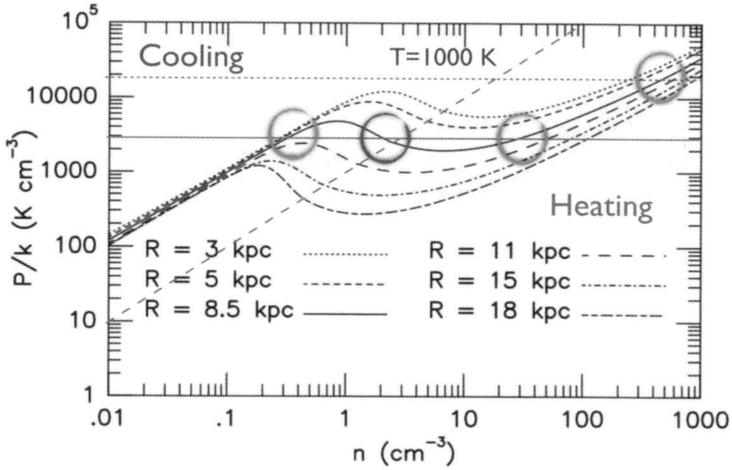

Figure 2. Thermal equilibrium curves at different radii in the Milky Way, compared to a typical pressure near the Solar circle (solid orange line), and compared to the pressure in the molecular ring or starburst galaxies (dashed orange line). Orange circled points represent stable equilibria, where small perturbations of temperature (motions perpendicular to the green-dashed line, with increasing temperature towards the left) result in restoring increases in cooling or heating, as labeled. Purple circled point is unstable. Adapted from Wolfire *et al.* (2003).

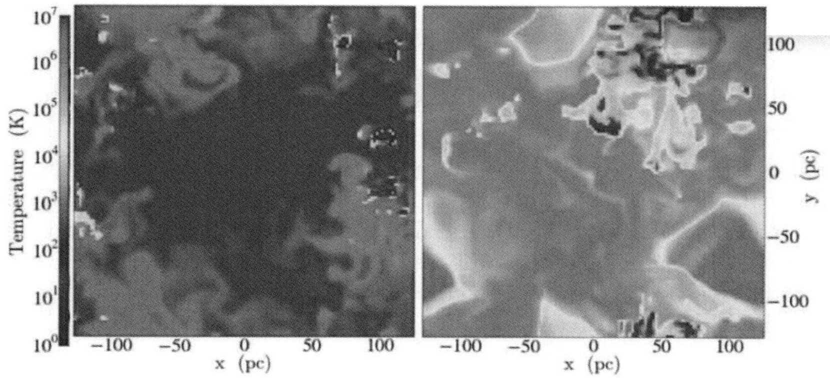

Figure 3. Temperature distributions on slices through periodic boxes of supernova-driven turbulence. On the left, supernovae explode in random positions, while on the right, half the supernovae explode in density peaks, substantially increasing the loss of energy to radiative cooling. On the left, pressures have reached high enough levels to eliminate the warm stable phase. Adapted from Gatto *et al.* (2015).

factor, and needs to be taken into account in any model of high-pressure, fully molecular, interstellar gas. Joung *et al.* (2009) demonstrated that following the Kennicutt-Schmidt law for star formation in stratified disks results in a three-phase medium, while Walch *et al.* (2015) showed that increasing the star formation rate by a factor of three already leads to a cold interstellar medium filled with hot gas.

4. Molecules are a Consequence of Star Formation

Observations have demonstrated a linear correlation between the surface density of star formation and that of molecular hydrogen (Rownd & Young 1999; Wong & Blitz 2002;

Bigiel *et al.* 2008; Leroy *et al.* 2008). This raises the question of whether this correlation implies a causal relationship. Is molecular hydrogen formation required for star formation to occur?

It has more recently become evident that many tracers of high density gas correlate just as well with the star formation rate as does molecular hydrogen. For example, the tracer molecule HCN was shown by Gao & Solomon (2004) to also correlate with star formation rate in both normal spiral galaxies and luminous and ultraluminous infrared galaxies. In the solar neighborhood, the number of young stellar objects in a cloud correlates with the mass of the cloud contained in regions with a K-band infrared extinction $A_K > 0.8$ (Lada *et al.* 2010). Clark & Glover 2014 indeed found thresholds for star formation in numerical simulations that reproduce this observed threshold. However, the physical process producing this threshold is *not* molecule formation, but rather dust shielding from heating by the background far ultraviolet radiation.

Indeed Krumholz *et al.* (2011) and Glover & Clark (2012) have demonstrated that atomic line cooling is as effective as molecular cooling in bringing gas down to star forming temperatures of 10-20 K. Figure 4 demonstrates that the same cold, star-forming gas appears in numerical simulations with full molecular and atomic cooling; with only atomic cooling, but still including molecular hydrogen formation; and with all molecular physics suppressed, but still including dust shielding from far ultraviolet heating. Only the removal of dust shielding from the simulation suppresses star formation.

Thus, it appears that the formation of molecular hydrogen does not control star formation. Rather, it seems, star formation, or more generally gravitational collapse of atomic gas, controls molecular hydrogen formation.

5. Molecular Cloud Dynamics Driven by Gravity Rather Than Turbulence

Molecular cloud dynamics offer a useful window to their origins. The argument has been made that they form by the sweeping up of smaller clouds in superbubbles or spiral arms (e.g. Tasker & Tan 2009). An alternative proposal is that clouds form from large scale gravitational collapse, as proposed by Hartmann *et al.* (2001) to explain the apparent correlation of star formation across entire molecular clouds.

Models of the supernova-driven, stratified interstellar medium can be used to examine these scenarios. One example of such a model derives from work by Joung & Mac Low (2005), Joung *et al.* (2009), and Hill *et al.* (2012). It includes both clustered and random supernovae, a background gravitational potential from the stellar disk and the halo, photoelectric heating and equilibrium ionization cooling, and magnetic fields. These models swept up gas into filaments that appeared similar to large molecular clouds. An example of such a filament is shown in top panel of Figure 5.

Ibáñez-Mejía *et al.* (2015) added the self-gravity of the gas to these models, turning it on after a full vertical galactic fountain had been set up over the course of over 200 Myr. This long preparation period also allowed plenty of time for the dense regions to come into equilibrium with the supernova-driven turbulence in the surrounding diffuse medium. However, when the self-gravity was turned on, immediate collapse ensued, as is shown in the example of Figure 5. One immediate and notable result was the enhancement of the filamentary morphology to the extreme levels characteristic of observed infrared dark clouds (e.g., Simon *et al.* 2006).

Within 5 Myr, this collapse leads to the formation of unresolved dense cores within which rapid star formation appears inevitable. As these first models have no mechanism to simulate the local prompt feedback in ionizing radiation, stellar winds and so forth

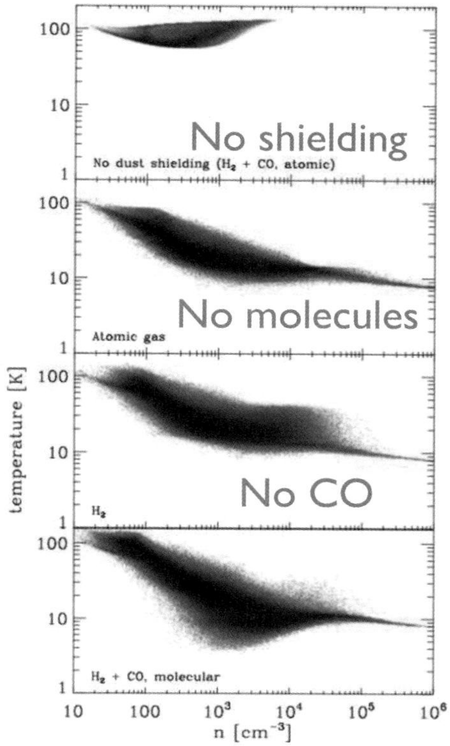

Figure 4. Distribution of temperature and pressure in models of turbulent, atomic gas including both atomic and molecular chemistry and cooling processes, as well as photoelectric heating modulated by dust shielding. The bottom panel shows the full model, while the upper panels have successively less physics. The first panel from the bottom neglects molecular line cooling, the second also neglects molecular hydrogen formation, and the top also neglects dust shielding. Gas reaches 10 K and high densities in all but the last of these models. Adapted from Glover & Clark (2012).

from this star formation, we can no longer follow its evolution after this point. However it is clear already that without this local feedback, star formation would reach unrealistic rates, just as first argued by Zuckerman & Palmer (1974).

In order to understand this rapid collapse and fragmentation, it is necessary to examine the dynamics of these dense regions. Ibáñez-Mejía *et al.* (2015) identified clouds using density contours and demonstrated that these clouds quite uniformly had velocity dispersions $\sigma < 1$ km s^{-1}, even at radii $R > 30$ pc, quite different from the Larson (1981) size-velocity dispersion relation. This is shown by the yellow dots in all three panels of Figure 6, in contrast to the observed relation presented as the dashed line, which is a fit to the subset of the Galactic Ring Survey (Jackson *et al.* 2006) presented in Heyer *et al.* (2009). These low velocity dispersions cannot resist gravitational collapse, and indeed the clouds do collapse, forming the same sort of filamentary structures familiar from the free gravitational collapse of the cosmic web. Although magnetic fields are present in the simulation, they are unable to effectively resist this collapse. As they collapse, their observed velocity dispersions increase to the observed values.

Ballesteros-Paredes *et al.* (2011) emphasize that the relationship between surface density, velocity dispersion, and radius expected from freely falling material differs by only

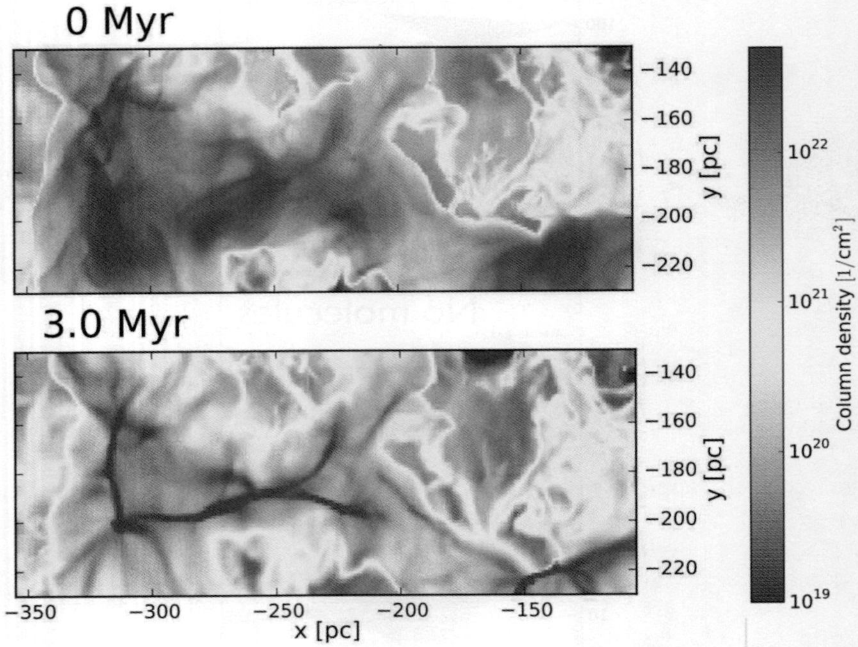

Figure 5. Vertical projection through a sub-region of a supernova-driven, stratified, magnetized model of the interstellar medium at times of 0 and 3 Myr after self-gravity has been turned on. (The full domain is a $1 \times 1 \times 20$ kpc vertical cut through the galactic disk.) Prompt collapse occurs, forming filaments strongly resembling observed IRDCs, with star-forming cores strung along them. Adapted from Ibáñez-Mejía *et al.* (2015).

a factor of $\sqrt{2}$ from that of material in virial equilibrium, and both appear consistent with the observations as summarized by Heyer *et al.* (2009).

It is worth noting that in a periodic box simulation of supernova-driven turbulence, neglecting stratification, but run at higher resolution, Padoan *et al.* (2015) find velocity dispersions after 6 Myr that agree with our results. However Padoan (priv. comm., 2015) finds similar velocity dispersions in their model even prior to the onset of self-gravity, in contradiction to our results. Further investigation will be needed to establish if the difference is due to their higher resolution or their neglect of stratification and confinement within the periodic box.

6. Summary

In this review, I have addressed a few current topics in the theory of the atomic and molecular interstellar medium. I have focused on the following questions, with my current best guesses as to the answers.

• Do cold streams contribute to gas accretion onto galaxies only at early times and for small galaxies, or far more generally as was originally claimed? The former appears to be the result of the numerical simulations most likely to be reliable.

• How small is the actual filling factor of dense, molecular gas in an interstellar medium when it is observationally determined to be completely molecular? Quite small, with the rest of the volume occupied by low-density, hot gas with temperatures exceeding 10^6 K.

• Does molecular hydrogen drive star formation, or rather does star formation produce molecular hydrogen and other high density tracers as a by-product? Molecular hydrogen

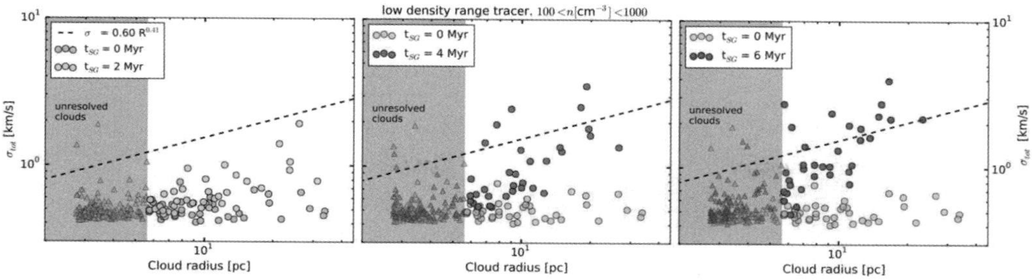

Figure 6. Radius-velocity dispersion relation for clouds drawn from the models prior to the action of self-gravity (yellow dots in all three panels), and for clouds at times of 2 Myr (left), 4 Myr (center) and 6 Myr (right) after self-gravity begins to act. For comparison, the observed relation derived from the data presented by Heyer *et al.* (2009) is shown by the dashed line. The shaded region shows clouds under 10 zones whose velocity dispersion is likely to be strongly suppressed by numerical viscosity. Adapted from Ibáñez-Mejía *et al.* (2015).

and other molecules appear to function far more as a tracer than a cause of star formation, particularly since atomic line cooling is nearly as effective as molecular line cooling.

- Are the observed velocity dispersions in molecular clouds caused by supernova-driven turbulence, or by gravitational collapse? Gravitational collapse seems more likely, but this question does not yet appear to be settled.

References

Ballesteros-Paredes, J., Hartmann, L. W., Vázquez-Semadeni, E., Heitsch, F., & Zamora-Avilés, M. A. 2011, *MNRAS*, 411, 65

Bigiel, F., Leroy, A., Walter, F., *et al.* 2008, *AJ*, 136, 2846

Birnboim, Y. & Dekel, A. 2003, *MNRAS*, 345, 349

Bouché, N., Dekel, A., Genzel, R., *et al.* 2010, *ApJ*, 718, 1001

Clark, P. C. & Glover, S. C. O. 2014, *MNRAS*, 444, 2396

Dekel, A. & Birnboim, Y. 2006, *MNRAS*, 368, 2

Dekel, A., Birnboim, Y., Engel, G., *et al.* 2009, *Nature*, 457, 451

Dekel, A. & Mandelker, N. 2014, *MNRAS*, 444, 2071

Dekel, A., Zolotov, A., Tweed, D., *et al.* 2013, *MNRAS*, 435, 999

Fardal, M. A., Katz, N., Gardner, J. P., *et al.* 2001, *ApJ*, 562, 605

Field, G. B., Goldsmith, D. W., & Habing, H. J. 1969, *ApJ (Letters)*, 155, L149

Gao, Y. & Solomon, P. M. 2004, *ApJ*, 606, 271

Gatto, A., Walch, S., Mac Low, M.-M. *et al.* 2015, *MNRAS*, 449, 1057

Glover, S. C. O. & Clark, P. C. 2012, *MNRAS*, 421, 9

Goerdt, T., Ceverino, D., Dekel, A., & Teyssier, R. 2015, *MNRAS*, 454, 637

Hartmann, L., Ballesteros-Paredes, J., & Bergin, E. A. 2001, *ApJ*, 562, 852

Heyer, M., Krawczyk, C., Duval, J., & Jackson, J. M. 2009, *ApJ*, 699, 1092

Hill, A. S., Joung, M. R., Mac Low, M.-M., *et al.* 2012, *ApJ*, 761, 189

Ibáñez-Mejía, J. C., Mac Low, M.-M., Klessen, R. S., & Baczynski, C. 2015, *ApJ*, submitted, arXiv:1511.05602

Jackson, J. M., Rathborne, J. M., Shah, R. Y., *et al.* 2006, *ApJ (Suppl.)*, 163, 145

Joung, M. K. R. & Mac Low, M.-M. 2006, *ApJ*, 653, 1266

Joung, M. R., Mac Low, M.-M., & Bryan, G. L. 2009, *ApJ*, 704, 137

Kereš, D., Katz, N., Weinberg, D. H., & Davé, R. 2005, *MNRAS*, 363, 2

Kereš, D., Katz, N., Fardal, M., Davé, R., & Weinberg, D. H. 2009, *MNRAS*, 395, 160

Keto, E. 2002, *ApJ*, 580, 980

Krumholz, M. R., Leroy, A. K., & McKee, C. F. 2011, *ApJ*, 731, 25

Lada, C. J., Lombardi, M., & Alves, J. F. 2010, *ApJ*, 724, 687

Leroy, A. K., Walter, F., Brinks, E., *et al.* 2008, *AJ*, 136, 2782
Li, M., Ostriker, J. P., Cen, R., Bryan, G. L., & Naab, T. 2015, *ApJ*, 814, 4
Nelson, D., Vogelsberger, M., Genel, S., *et al.* 2013, *MNRAS*, 429, 3353
Ocvirk, P., Pichon, C., & Teyssier, R. 2008, *MNRAS*, 390, 1326
Ostriker, E. C., McKee, C. F., & Leroy, A. K. 2010, *ApJ*, 721, 975
Padoan, P., Pan, L., Haugbølle, T., & Nordlund, Å. 2015, *ApJ*, submitted, arXiv:1509.04663
Putman, M. E., Peek, J. E. G., & Joung, M. R. 2012, *ARAA*, 50, 491
Rownd, B. K. & Young, J. S. 1999, *AJ*, 118, 670
Simon, R., Rathborne, J. M., Shah, R. Y., Jackson, J. M., & Chambers, E. T. 2006, *ApJ*, 653, 1325
Springel, V. 2005, *MNRAS*, 364, 1105
Springel, V. 2010, *MNRAS*, 401, 791
Tasker, E. J. & Tan, J. C. 2009, *ApJ*, 700, 358
Walch, S., Girichidis, P., Naab, T., *et al.* 2015, *MNRAS*, 454, 238
Wolfire, M. G., McKee, C. F., Hollenbach, D., & Tielens, A. G. G. M. 1995, *ApJ*, 453, 673
Wolfire, M. G., McKee, C. F., Hollenbach, D., & Tielens, A. G. G. M. 2003, *ApJ*, 587, 278
Wong, T. & Blitz, L. 2002, *ApJ*, 569, 157

From Interstellar Clouds to Star-Forming Galaxies: Universal Processes?
Proceedings IAU Symposium No. 315, 2015
P. Jablonka, Ph. André & F. van der Tak, eds.

doi:10.1017/S1743921316007171

Accretion Onto the Milky Way: The Smith Cloud

Felix J. Lockman

National Radio Astronomy Observatory†
P.O. Box 2
Green Bank, WV 24944 USA
email: jlockman@nrao.edu

Abstract. Active gas accretion onto the Milky Way is observed in an object called the Smith Cloud, which contains several million solar masses of neutral and warm ionized gas and is currently losing material to the Milky Way, adding angular momentum to the disk. It is several kpc in size and its tip lies 2 kpc below the Galactic plane. It appears to have no stellar counterpart, but could contain a stellar population like that of the dwarf galaxy Leo P. There are suggestions that its existence and survival require that it be embedded in a dark matter halo of a few 10^8 solar masses.

Keywords. Galaxy: evolution, Galaxy: formation, Galaxy: halo, galaxies: ISM

1. Introduction

The puzzle of the origin of high velocity clouds (HVCs) is still with us more than 50 years after their initial discovery (Muller *et al.* 1963), and despite the enormous progress made in the past decades in understanding their distances, size, mass, metallicity, covering fraction, and ionization state (e.g., Wakker & van Woerden 1997; Lockman *et al.* 2002; Fox *et al.* 2006; Wakker *et al.* 2008; Thom *et al.* 2008; Shull *et al.* 2009; Putman *et al.* 2012; Lehner *et al.* 2012). HVCs are now known to exist around nearby galaxies in a population that extends at least 50 kpc from the host system (Thilker *et al.* 2004; Grossi *et al.* 2008; Westmeier *et al.* 2008; Putman *et al.* 2009). The possibility that *some* HVCs may be intra-group or even intergalactic, or associated with dark matter halos, or tracers of faint dwarf galaxies, cannot be excluded (e.g., Blitz *et al.* 1999; Braun & Burton 1999; Adams *et al.* 2013). Except for the Magellanic Stream, the origin of even the most well-studied HVC is completely unknown.

2. The Smith Cloud

Perhaps the most exceptional HVC is the Smith Cloud (Smith 1963), first identified in observations at $35'$ resolution where it shows some head-tail structure, as do about 10% of Milky Way HVCs (Brüns *et al.* 2000). Observations with the 100 meter Green Bank Telescope (GBT) at $9'$ resolution reveal a spectacular cometary cloud, showing extensive evidence of interaction between the cloud and the gaseous halo of the Milky Way (Lockman *et al.* 2008). Figure 1 shows an HI image from new unpublished GBT observations (Lockman *et al.* 2016). The new data reveal some previously unknown cloud components symetrically placed aside the main body of the cloud, as well as a very extended "tail".

† The National Radio Astronomy Observatory is a facility of the National Science Foundation, operated under cooperative agreement by Associated Universities, Inc.

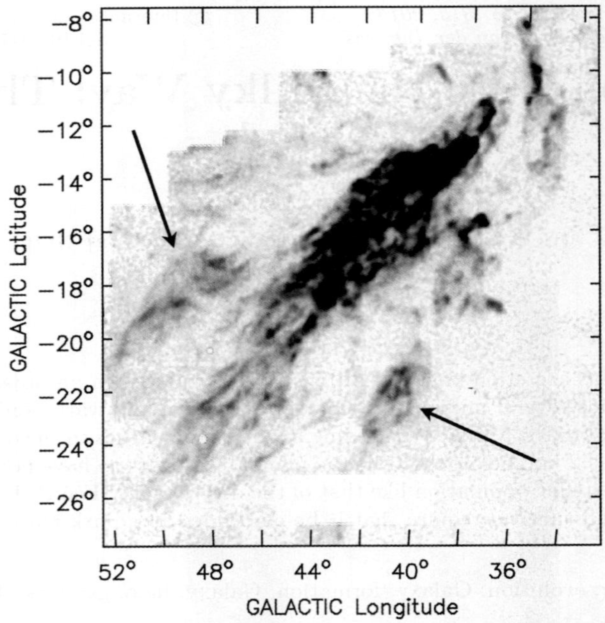

Figure 1. The Smith Cloud in HI from new unpublished observations with the GBT. The two features marked with arrows have the appearance of having been shed by the cloud as it moves toward the Galactic plane; each contains $\sim 5 \times 10^4$ $M_\odot$.

Much is now known about the Smith Cloud. Its distance has been estimated from three independent techniques, which agree quite well (Lockman *et al.* 2008). Measurement of interstellar absorption lines against stars at varying distance provide the strongest constraint (Wakker *et al.* 2008). Basic properties of the cloud are listed in Table 1.

Faint Hα emission detected from the cloud suggests that it contains as much ionized as neutral gas (Bland-Hawthorn *et al.* 1998; Putman *et al.* 2003; Hill *et al.* 2009). From a study of the rotation measure of background AGN, Hill *et al.* (2013) deduce the presence of a magnetic field with an amplitude 8 μG in the decelerated parts of the cloud. New data, currently being analyzed, should improve both of these estimates.

There is no obvious stellar counterpart to the cloud. Stark *et al.* (2015) looked for young stars but did not find a statistically significant excess toward the cloud compared to adjacent areas. Contamination by the Milky Way at the low latitude of the cloud makes this work difficult, however, and Stark *et al.* emphasize that they cannot rule out a stellar population like that of Leo P, a dwarf galaxy with a similar HI mass and $M_V = -9.4$ (McQuinn *et al.* 2013).

3. Interaction

Evidence of interaction with the Milky Way abounds. The overall morphology of the Smith Cloud suggests that it is moving towards the Galactic plane and enountering the Galactic halo, shedding pieces of itself as it goes. The tip of the cloud at $z \approx -2$ kpc is closer to the Galactic plane than the top of some HI superbubbles (Pidopryhora *et al.* 2007). The densest portions of the cloud lie closest to the Galactic plane while the "tail", 10° further from the plane, is much more diffuse. There is kinematic evidence of the interaction as well. Figure 2 shows a velocity-position cut across the center of the cloud. The main body of the cloud is distinct from Galactic HI, but there are kinematic

Table 1. Smith Cloud Properties

Distance[1,2]	12.4 ± 1.3 kpc
R_{gal}	7.6 ± 1.0 kpc
z	-2.2 kpc
size	$\approx 3 \times 1$ kpc
M_{HI} [2]	2.0×10^6 $M_\odot$
M_{H+} [3]	$> 1.0 \times 10^6$ $M_\odot$
$B_{\|\|}$ [4]	8 μG

Notes: 1) Wakker *et al.* 2008. 2) Lockman *et al.* 2008, 2016. 3) Hill *et al.* 2009. 4) Hill *et al.* 2013.

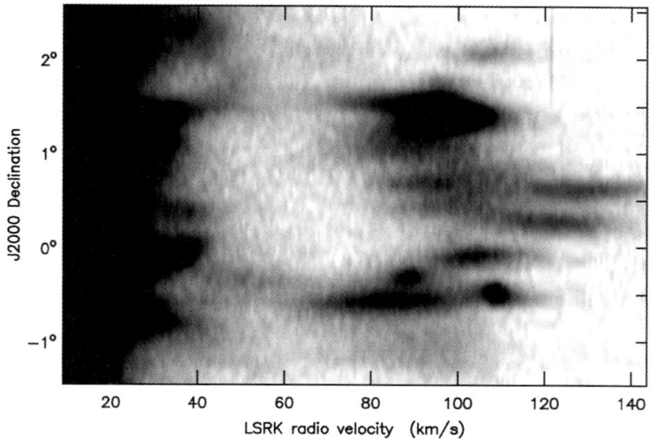

Figure 2. Position-velocity cut across the minor axis of the Smith Cloud. The main part of the cloud lies at $V_{LSR} \approx 100$ km s^{-1}, but its edges show kinematic "bridges" to the lower-velocity Galactic gas, indicating that portions of the cloud are decelerating and being accreted by the Galaxy.

"bridges" between the cloud edges and lower-velocity disk/halo gas. Here we see direct evidence that portions of the cloud have been decelerated and are now blending with the Galaxy. There is also a ridge of decelerated gas conforming to the upper edge of the cloud (see Figure 4 of Lockman *et al.* 2008). Several holes in the main body of the cloud at $V_{LSR} \approx 100$ km s^{-1} correspond to ~ 100 $M_\odot$ HI clumps at ~ 50 km s^{-1} lower V_{LSR}, suggesting that pieces of the cloud have broken away and been decelerated by the medium it is encountering. There seems to be no doubt that the Smith Cloud is being destroyed by its interaction with the Milky Way and is losing gas to the Galaxy. A similar pattern of accretion is observed in the HVC Complex H (Lockman 2003).

4. Trajectory and Progenitor

Lockman *et al.* (2008) used the change of V_{LSR} along the cloud's major axis to estimate a trajectory and found that the cloud has a total space velocity ≈ 300 km s^{-1}, well below the escape velocity of the Milky Way. While this analysis was based on early data and does not include the full extent of the cloud in Fig. 1, there seems little doubt that the cloud's largest velocity component is in the direction of Galactic rotation – as it interacts with the Milky Way it therefore adds angular momentum to the disk.

It is likely that the main body of the cloud was never more than 4 kpc from the Galactic plane and its "tail" is considerably closer to us than its head, though its exact orientation with respect to the plane of the sky is unknown. There is even evidence for a extended component of the cloud ahead of the main body (Lockman 2012). Clarifying these points is clearly important to understanding the origin and fate of the cloud.

Simulations suggest that clouds falling into the Milky Way halo are easily disrupted (Putman *et al.* 2012 and references therein), so given its striking appearance, evidence for interaction, and rather modest total space velocity, a natural question is why the Smith Cloud has survived as long as it has. One answer, first proposed by Nichols & Bland-Hawthorn (2009) and presented in more detail by Nichols *et al.* (2014) is that the Smith Cloud is the baryonic component of a dark matter sub-halo. A dark matter halo $\sim 3 \times 10^8$ $M_\odot$ is sufficient to stabilize the cloud core while not massive enough to induce significant star formation, which is not observed. This model may be entirely consistent with the Smith Cloud being a dwarf galaxy similar to Leo P, but a dwarf whose accretion and disruption by the Milky Way are now being witnessed at close range.

The Smith Cloud is an extraordinary object whose study may illuminate many aspects of the growth and evolution of galaxies.

References

Adams, E. A. K., Giovanelli, R., & Haynes, M. P. 2013, *ApJ*, 768, 77
Bland-Hawthorn, J., Veilleux, S., Cecil, G. N., *et al.* 1998, *MNRAS*, 299, 611
Blitz, L., Spergel, D. N., Teuben, P. J., Hartmann, D., & Burton, W. B. 1999, *ApJ*, 514, 818
Braun, R. & Burton, W. B. 1999, *A&A*, 341, 437
Brüns, C., Kerp, J., Kalberla, P. M. W., & Mebold, U. 2000, *A&A*, 357, 120
Fox, A. J., Savage, B. D., & Wakker, B. P. 2006, *ApJS*, 165, 229
Grossi, M., Giovanardi, C., Corbelli, E., *et al.* 2008, *A&A*, 487, 161
Hill, A. S., Haffner, L. M., & Reynolds, R. J. 2009, *ApJ*, 703, 1832
Hill, A. S., Mao, S. A., Benjamin, R. A., Lockman, F. J., & McClure-Griffiths, N. M. 2013, *ApJ*, 777, 55
Lehner, N., Howk, J. C., Thom, C., *et al.* 2012, *MNRAS*, 424, 2896
Lockman, F. J., Murphy, E. M., Petty-Powell, S., & Urick, V. J. 2002, *ApJS*, 140, 331
Lockman, F. J. 2003, *ApJ*, 591, L33
Lockman, F. J., Benjamin, R. A., Heroux, A. J., & Langston, G. I. 2008, *ApJ*, 679, L21
Lockman, F. J. 2012, *EAS Publications Series*, 56, 189
McQuinn, K. B. W., Skillman, E. D., Berg, D., *et al.* 2013, *AJ*, 146, 145
Muller, C. A., Oort, J. H., & Raimond, E. 1963, *Acad. Sci. Paris Comptes Rendus*, 257, 1661
Nichols, M. & Bland-Hawthorn, J. 2009, *ApJ*, 707, 1642
Nichols, M., Mirabal, N., Agertz, O., Lockman, F. J., & Bland-Hawthorn, J. 2014, *MNRAS*, 442, 2883
Pidopryhora, Y., Lockman, F. J., & Shields, J. C. 2007, *ApJ*, 656, 928
Putman, M. E., Bland-Hawthorn, J., Veilleux, S., *et al.* 2003, *ApJ*, 597, 948
Putman, M. E., Peek, J. E. G., Muratov, A., *et al.* 2009, *ApJ*, 703, 1486
Putman, M. E., Peek, J. E. G., & Joung, M. R. 2012, *Ann. Rev. Astr. Ap.*, 50, 491
Shull, J. M., Jones, J. R., Danforth, C. W., & Collins, J. A. 2009, *ApJ*, 699, 754
Smith, G. P. 1963, *BAN*, 17, 203
Stark, D. V., Baker, A. D., & Kannappan, S. J. 2015, *MNRAS*, 446, 1855
Thilker, D. A., Braun, R., Walterbos, R. A. M., *et al.* 2004, *ApJ*, 601, L39
Thom, C., Peek, J. E. G., Putman, M. E., *et al.* 2008, *ApJ*, 684, 364
Wakker, B. P. & van Woerden, H. 1997, *Ann. Rev. Astr. Ap.*, 35, 217
Wakker, B. P., York, D. G., Wilhelm, R., *et al.* 2008, *ApJ*, 672, 298
Westmeier, T., Brüns, C., & Kerp, J. 2008, *MNRAS*, 390, 1691

From Interstellar Clouds to Star-Forming Galaxies: Universal Processes?
Proceedings IAU Symposium No. 315, 2015
P. Jablonka, Ph. André & F. van der Tak, eds.

doi:10.1017/S1743921316007183

Resolving the Transition from Molecular to Atomic at 1/5 Solar Metallicity in the Small Magellanic Cloud

Katherine E. Jameson[1], Alberto D. Bolatto[1], Mark Wolfire[1], Monica Rubio[2], Rodrigo Herrera Camus[1,3] and the HS[3] Collaboration

[1]Astronomy Department and Laboratory for Millimeter-wave Astronomy, University of Maryland, College Park, MD, 20742, USA
email: kjameson@astro.umd.edu

[2]Departamento de Astronomía, Universidad de Chile, Camino El Observatorio 1515, Casilla 36-D, Santiago, Chile

[3]Max-Planck-Institut für Extraterrestrische Physik, Giessenbachstrasse, D-85748 Garching, Germany

Abstract. At a distance of 61 kpc, the Small Magellanic Cloud (SMC) affords an absolutely unique view of the low metallicity star-forming interstellar medium, providing the nearest laboratory to study processes relevant to star formation at high redshifts. We present new ALMA 7m-array maps of ^{12}CO and ^{13}CO $(2-1)$ for one of the four observed regions in the Southwest Bar of the SMC. These maps are the first high-resolution ($\sim 6''$ $\sim$ 1.7 pc) images of CO in a molecular cloud at 1/5 Solar metallicity. We show the structure of photodissociation regions for the first time at 1/5 Solar metallicity by combining the new ALMA data with Herschel maps of [C II] and [O I]. We present preliminary evidence that there is extended, faint ^{12}CO $(2-1)$ emission near where we expect the H I-to-H_2 transition. We also compare our data to the low metallicity 3D simulations by Glover & Mac Low (2011) and Shetty *et al.* (2011).

Keywords. ISM: clouds, ISM: structure, galaxies: ISM, Magellanic Clouds

1. Introduction

Due to its proximity and our ability to achieve high spatial resolution, the SMC provides the ideal laboratory to study detailed properties of the molecular gas in star-forming regions in a low mass, $M_{*,\rm SMC} = 3 \times 10^8$ $M_\odot$ (Skibba *et al.* 2012), and low metallicity, $Z_{\rm SMC} \sim 1/5$ $Z_\odot$ (Pagel 2003), environment. In low metallicity environments ^{12}CO, the most common tracer of the total amount of molecular gas (dominated by H_2), emits weakly and often goes undetected. The SMC has been studied extensively in ^{12}CO with the earliest surveys completed using the Columbia 1.2m (Rubio *et al.* 1991). The early survey of both Clouds (Israel *et al.* 1993) and of the Southwest Bar of the SMC (Rubio *et al.* 1993) using the Swedish-ESO Submillimetre Telescope (SEST) showed the CO emission to be under-luminous compared to the Milky Way by a factor of a few at small scales ($\sim$10 pc) and by $\sim 10-20$ at large scales ($\sim$200 pc) in the SMC. A "CO-faint" molecular phase is predicted by analytical and numerical models that show the CO abundance drops off below a mean visual extinction of $A_V \sim 2$ mag (Wolfire *et al.* 2010; Glover & Mac Low 2011; Shetty *et al.* 2011) where most carbon exists as C^+ while H_2 remains molecular because it is strongly self-shielding (Tielens & Hollenbach 1985; Wolfire *et al.* 1989; Kaufman *et al.* 1999). Studies of the SMC find this phase to encompass $80\%-90\%$ of all the H_2 (Israel 1997; Pak *et al.* 1998; Leroy *et al.* 2007; Leroy *et al.* 2011; Bolatto *et al.* 2011), likely dominating the molecular reservoir available to star formation.

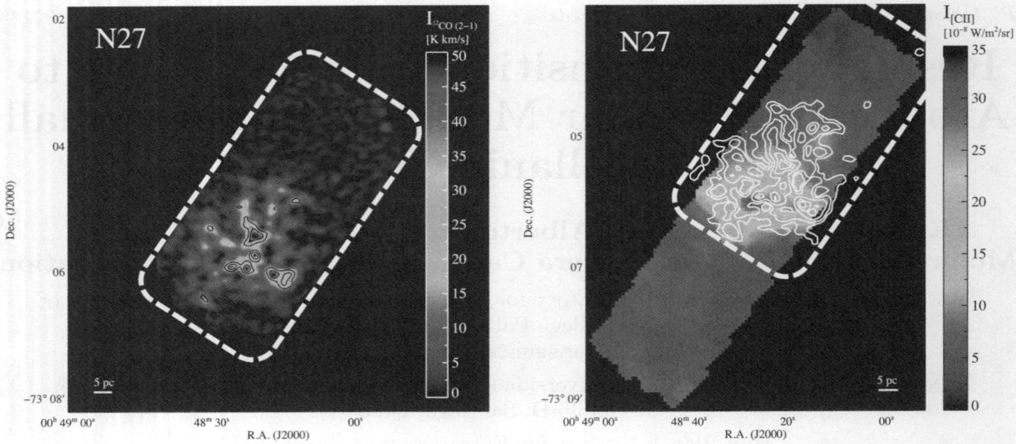

Figure 1. Left: Integrated intensity map of the combined ALMA 7m-array and APEX single-dish ^{12}CO $(2-1)$ image of the N27 region with black contours showing the ^{13}CO $(2-1)$ integrated intensity map at 2, 4, and 6 K km s^{-1} (beam size $\sim 7'' \times 5''$). In both panels the dashed white lines shows the approximate coverage of the ALMA 7-m array map. Right: Image of the [C II] integrated intensity for the N27 region with white contours showing the combined CO integrated intensity map at 3, 6, 15, 30, 40, and 50 K km s^{-1}. We see evidence of faint ^{12}CO emission out to low A_V and close to the expected atomic-to-molecular transition, and there is similarity between the [C II] and CO structure, which suggests that CO is tracing the molecular cloud structure.

The "CO-faint" molecular gas can be traced using diffuse γ-rays, dust continuum emission, and [C II] and [O I] line emission. Large-scale maps of the total molecular gas in the SMC have been made using dust continuum emission using *Spitzer* data (Leroy *et al.* 2007; Leroy *et al.* 2009; Bolatto *et al.* 2011), and more recently *Herschel* data (Jameson *et al.* 2015a), that show more extended molecular gas than what is traced by the large-scale NANTEN ^{12}CO map (Mizuno *et al.* 2001). Using the dust-based molecular gas estimate, Jameson *et al.* (2015a) finds no strong evidence that low metallicity affects the large-scale relationship between star formation and molecular gas, but metallicity is expected to affect the small-scale structure of the molecular gas and their surrounding photodissociation regions. The emission from [C II] can also trace the "CO-faint" molecular phase (e.g., Madden *et al.* 1997; Pineda *et al.* 2013), and observations from *Herschel* allow us to attain high resolution ($\sim 10''$) [C II] maps. When combined with new high resolution ^{12}CO maps from ALMA ($\sim 6''$ resolution), we can study the structure of the molecular gas and resolve the H I-to-H_2 transition. We present new *Herschel* [C II] and ALMA CO data for one of the targeted regions in the Southwest Bar of the SMC and show that there appears to be faint CO emission near the H I-to-H_2 transition.

2. Observations

Our results are based on our new ALMA 7m-array CO observations and *Herschel* [C II] and [O I] data. The *Herschel* data come from the *Herschel* Spectroscopic Survey of the SMC (HS^3; Jameson *et al.* 2015b). The survey consists of four $1.5' \times 6'$ strips covering star-forming regions and one $2' \times 2'$ map targeted at a peak in the dust-based molecular gas estimate not detected in the NANTEN ^{12}CO map. The regions sample a range of radiation field strength, star formation, and detected molecular gas in the Southwest Bar and in N83/N84 of the SMC. We mapped four approximately $1.5' \times 3'$

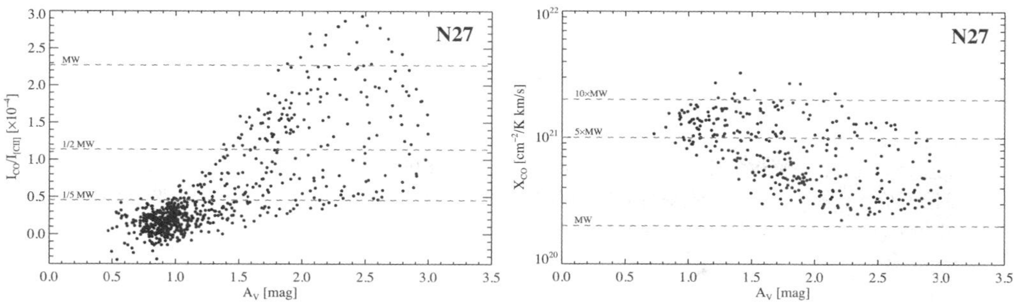

Figure 2. Left: The ratio of the integrated intensity of ^{12}CO (using the APEX ^{12}CO $(2-1)$ data and assuming thermalized emission to convert to $(1-0)$) to that of [C II] as a function of A_V for the N27 region in the SMC. The red dashed lines indicate different scalings of the canonical value for the Milky Way of $\sim 1/4400$ (Stacey *et al.* 1991). Right: The CO-to-H_2 conversion factor (X_{CO}) using [C II]-based estimate of N_{H_2} (assuming all [C II] emission comes from molecular gas) with $\geqslant 3\sigma$ ^{12}CO $(2-1)$ detections for the N27 region. The red dashed lines show different scalings of X_{CO} for the Milky Way of 2×10^{20} cm^{-2} (K km s^{-1})$^{-1}$ (Bolatto *et al.* 2013).

regions overlapping the HS^3 regions in ^{12}CO, ^{13}CO, and C^{18}O $(2-1)$ in the Southwest Bar with the ALMA 7m-array (resolution $\sim 6''$). The ALMA 7m-array maps have been combined with single-dish data from APEX ($\theta = 29.5''$; Rubio *et al.* 2015). For the preliminary results, we focus on the N27 region.

3. Preliminary Results

The [C II] and [O I] lines were detected ($S/N \gtrsim 3$) throughout all of the HS^3 regions. The ALMA 7m-array data shows clear detections of ^{12}CO and ^{13}CO $(2-1)$ emission in all of the regions, but C^{18}O is not detected (see Figure 1). In N27, we find ^{12}CO/^{13}CO ratios of $\sim 10-15$ in the peaks of the ^{12}CO map, which is consistent with previous measurements (Israel *et al.* 2003) and the typical ratios observed in the Milky Way. We see good correspondence between the ^{12}CO and [C II] emission and a fairly flat [O I]/[C II] ratio of ~ 0.3 across the regions. We investigate the structure of the photodissociation region and molecular cloud by using the visual extinction (A_V) as an indicator of depth within the cloud. The optical depth at 160 μm (τ_{160}) map from Lee *et al.* (2015) is converted to A_V using $A_V \sim 2200\tau_{160}$. The left panel of Figure 2 shows that the CO/[C II] ratio in N27 tends to increase at higher A_V deeper into the cloud and reach ratios similar to the value found in the Milky Way. Finding CO/[C II] and ^{12}CO/^{13}CO ratios similar to the Milky Way is consistent with theoretical expectations that once enough dust shielding is built up, the structure at low metallicity will resemble that of a high metallicity cloud.

An $A_V \sim 1$ corresponds to the typical extinction where ^{12}CO becomes optically thick. Our observation of faint ^{12}CO emission at $A_V < 1$ indicates that we are tracing optically thin ^{12}CO emission that corresponds to the "CO-faint" molecular gas (Wolfire *et al.* 2010). Figure 2 shows [C II] emission from the cloud depths we would expect to see a significant amount of "CO-faint" molecular gas and we have some detections of ^{12}CO $(2-1)$ down to an $A_V \sim 0.5$, which is close to the expected range for the H I-to-H_2 ($A_V \sim 0.2-0.4$). This suggests that this translucent molecular phase is "CO-faint" as opposed to the terminology that refers to it as "CO-dark" gas. The lower CO/[C II]ratios relative to the Milky Way value is likely indicating the presence of this this "CO-faint" component where the carbon primarily exists as C$^+$ and not as CO.

The extended, low level [C II] emission can potentially emanate from ionized gas, warm neutral medium (WNM), cold neutral medium (CNM), a translucent molecular phase, or some combination. The HS^3 data includes [N II] observations, but there are no detections in the SMC regions, which indicates that the ionized gas is not contributing to the [C II] emission. The [O I]/[C II] ratio is a good indicator of the density of the gas, and comparing the ratio values to new heating and cooling curves for the SMC from Wolfire *et al.* (2015) show that the values we observe indicate densities of $n \sim 100 - 300$ cm^{-3}. These densities correspond to the CNM and potentially molecular gas. Combined with the observation that the extended [C II] and ^{12}CO emission structure is similar to that seen for the mid-IR H_2 rotational lines, it is likely that this gas contains a molecular component.

We then make the simplifying assumption that all of the [C II] emission originates from H_2 gas, and convert the observed [C II] intensity into an H_2 column density (N_{H_2}) using the [C II] cooling curve for the SMC from Wolfire *et al.* (2015). This produces [C II]-based estimates of N_{H_2} $1 - 2 \times 10^{21}$ cm^{-2}. Using these values of N_{H_2}, we estimate the ^{12}CO-to-H_2 conversion factor X_{CO} (see Figure 2, right panel). The general trend of decreasing X_{CO} with increasing A_V is consistent with the results for a LMC/SMC-type molecular cloud from Shetty *et al.* (2011) and can be explained by the enhanced photodissociation of CO in the outskirts of the cloud at lower metallicity.

References

Bolatto, A. D., Leroy, A. K., Jameson, K., *et al.* 2011, *ApJ*, 741, 12
Bolatto, A. D., Wolfire, M., & Leroy, A. K. 2013, *ARA&A*, 51, 207
Glover, S. C. O. & Mac Low, M.-M. 2011, *MNRAS*, 412, 337
Israel, F. P., Johansson, L. E. B., Lequeux, J., *et al.* 1993, *A&A*, 276, 25
Israel, F. P. 1997, *A&A*, 328, 471
Israel, F. P., Johansson, L. E. B., Rubio, M., *et al.* 2003, *A&A*, 406, 817
Jameson, K. E., Bolatto, A. D., Leroy, A. K., *et al.* 2015, *ApJ*, submitted (arXiv:1510.08084)
Jameson, K. E., Bolatto, A. D., Wolfire, M., *et al.* 2015, *in prep*
Kaufman, M. J., Wolfire, M. G., Hollenbach, D. J., & Luhman, M. L. 1999, *ApJ*, 527, 795
Lee, C., Leroy, A. K., Schnee, S., *et al.* 2015, *MNRAS*, 450, 2708
Leroy, A., Bolatto, A., Stanimirović, S., *et al.* 2007, *ApJ*, 658, 1027
Leroy, A. K., Bolatto, A., Bot, C., *et al.* 2009, *ApJ*, 702, 352
Leroy, A. K., Bolatto, A., Gordon, K., *et al.* 2011, *ApJ*, 737, 12
Madden, S. C., Poglitsch, A., Geis, N., Stacey, G. J., & Townes, C. H. 1997, *ApJ*, 483, 200
Mizuno, N., Yamaguchi, R., Mizuno, A., *et al.* 2001, *PASJ*, 53, 971
Pagel, B. E. J. 2003, in: Charbonnel, C. and Schaerer, D. and Meynet, G. (eds.), *CNO in the Universe*, ASP Conference Series Volume 304 (San Francisco: ASP,) p. 187
Pak, S., Jaffe, D. T., van Dishoeck, E. F., Johansson, L. E. B., *et al.* 1998, *ApJ*, 498, 735
Pineda, J. L., Langer, W. D., Velusamy, T., & Goldsmith, P. F. 2013, *A&A*, 554, A103
Rubio, M., Garay, G., Montani, J., & Thaddeus, P. 1991, *ApJ*, 368, 173

Rubio, M., Garay, G., Montani, J., & Thaddeus, P. 1991, *ApJ*, 368, 173

Rubio, M., Lequeux, J., & Boulanger, F. 1993, *A&A*, 271, 9
Rubio, M., Bolatto, A. D., Jameson, K. E., *et al.* 2015, *in prep*
Shetty, R., Glover, S. C., Dullemond, C. P., *et al.* 2011, *MNRAS* 415, 3253
Skibba, R. A., Engelbracht, C. W., Aniano, G., *et al.* 2012, *ApJ* 761, 42
Stacey, G. J., Geis, N., Genzel, R., *et al.* 1991, *ApJ*, 373, 423
Tielens, A. G. G. M. & Hollenbach, D. 1985, *ApJ*, 291, 722
Wolfire, M. G., Hollenbach, D., & Tielens, A. G. G. M. 1989, *ApJ* 344, 770
Wolfire, M. G., Hollenbach, D., & McKee, C. F. 2010, *ApJ* 716, 1191
Wolfire, M. G., *et al.* 2015, *in prep*

From Interstellar Clouds to Star-Forming Galaxies: Universal Processes?
Proceedings IAU Symposium No. 315, 2015
P. Jablonka, Ph. André & F. van der Tak, eds.

doi:10.1017/S1743921316007195

Astrochemistry in external galaxies: how to use molecules as probes of their physical conditions

Serena Viti
Department of Physics and Astronomy, University College London, Gower St., London, WC1E 6BT, UK
email: serena.viti@ucl.ac.uk

Abstract. It is now well established that chemistry in external galaxies is rich and complex. In this review I will explore whether one can use molecular emissions to determine their physical conditions. There are several considerations to bear in mind when using molecular emission, and in particular molecular ratios, to determine the densities, temperatures and energetics of a galaxy, which I will briefly summarise here. I will then present an example of a study that uses multiple chemical and radiative transfer analyses in order to tackle the too often neglected 'degeneracies' implicit in the interpretation of molecular ratios and show that only via such analyses combined with multi-species and multi-lines high spatial resolution data one can truly make molecules into powerful diagnostics of the evolution and distribution of molecular gas.

Keywords. astrochemistry, galaxies: ISM, galaxies: nuclei, galaxies: individual (NGC 1068)

1. Introduction

The first detections of molecular emission in external galaxies were made in the 1970s. Since then it has become clear that chemistry in external galaxies can be complex and that, although most molecular emission is probably extended, it is certainly not all co-existent. CO emission is routinely used to trace the molecular mass of a galaxy, as it traces H_2 at large scale (see review by Bolatto *et al.* (2013)), although one notes that the X-factor used in the Milky Way to convert CO integrated line intensities to H_2 column densities may give misleading results if applied to other galaxies (e.g. Bell *et al.* 2006). However, it is the observations of molecules with a high critical density that allow us to trace the reservoir for star formation or describe the process of star formation itself, including determining the influence of newly formed stars on their environments.

The objective of this review is to establish how one can use molecular emissions of different species to explore the physical conditions of galaxies and their likely evolutionary status, as we routinely do for our own Galaxy. In other words can astrochemistry provide the tools to explore the physical conditions and energetics of external galaxies?

There are two main challenges for the astrochemist. The first challenge is that even for the closest galaxies (e.g. M82) with the most powerful interferferometer to date (ALMA), the highest spatial resolution that one can achieve is ~0.2 pc which is still larger than an individual star forming core. In fact, for most galaxies a much worse resolution is achieved, with entire star-forming regions being unresolved (e.g. for NGC 1068, at a distance of 14 Mpc, the highest spatial resolution achieved is 30 pc, Garcia-Burillo *et al.* 2014). Hence the telescope beam will usually encompass emissions from many types of region and the molecular emission will be an 'ensemble' of multi-phase gas where the spatial and temporal effects are diluted in the beam. This will challenge the interpretation of molecular line emission, especially for galaxies very different from our own. The range of physical parameters (gas densities, UV fields, cosmic ray ionization rates, dust properties, etc.) will differ across galaxies (Williams & Viti 2013). Hence we can not necessarily use information about the Milky Way as a reliable guide to the properties of other galaxies.

The second challenge for the astrochemist is the fact that molecular spectra of nearby galaxies are now proving to be almost as rich as those of the Milky Way (Aladro *et al.* 2011, Aladro *et al.* 2013, Costagliola *et al.* 2015). For example, an unbiased molecular line survey in the wavelength range 1.7–2.3 mm towards the nuclear region of the starburst galaxy NGC 253 shows the presence of more than 20 different molecular species (Martin *et al.* 2006), all with relatively high abundances (see Table 6.2 in Williams & Viti 2013). The concomitant detection of, say, CO, SiO, and CH_3OH emissions in a spatially unresolved galaxy does not necessarily mean that they are emitted from the same gas: the first molecule may indicate the presence of cold diffuse clouds, the second strong shocks, and the third dense star-forming cores. Hence such chemical complexity cannot be explained by a one-component model.

In order to stand up to these challenges the interpretaation of molecular emission must take into consideration the chemical routes to the formaton and destruction of each species, as well as their sensitivity to the physical and chemical conditions of the gas.

2. How to interpret molecular observations in external galaxies

Molecular observations are an ideal tool to trace a wide range of densities and energetics in the interstellar medium of a galaxy, largely because of the wide range of critical densities across the different molecular species and across the transitions of the same molecule, as well as the dependendencies of chemical reactions on the energy available to the system i.e the kinetic temperature of the gas. In fact star formation as well as AGN activity are both affected by the molecular gas available to the system.

Many studies have shown that molecular ratios, such as HCO^+/HCN or HCN/CO differ across different types of galaxies expecially between AGN-dominated galaxies and starburst-dominated galaxies. However the derived abundance ratios for an individual galaxy also highly differ across studies depending on the transitions observed, the available resolution, and the method used in deriving the column densities.

The differences between molecular ratios in starburst or AGN galaxies as compared to our own Galaxy have been interpreted as anomalous at times and an indication either of high ionisation rates or of high star formation rates. On the other hand, it is likely that multiple components, each with different physical conditions, are contributing to the emission (e.g. Giant Molecular Clouds, where star formation take place, will have average number densities of the order of 10^3 cm^{-3} and T = 50–100 K, while the star-forming clumps may have number densities up to $\sim 10^7$ cm^3 and temperatures as low as 10 K), so that a knowledge of the individual components is required to interpret such ratios. This will also explain why the use of different transitions in deriving the abundance ratios as well as different beam sizes lead to substantial differences within the same galaxy. An understanding of the chemistry behind each molecule and its dependencies on the density and temperature of the gas will lead to a more coordinated approach to understanding the nature of galaxies by using molecular line emissions.

In fact it is well known that the degeneracy between density and temperature implies that single transitions of one molecule should not be used to determine quantitatively the gas density or temperature, especially in extragalactic environments. Chemical models show that almost any density can be representative of a large range of environments; for example, a density of $\leqslant 10^4$ cm^{-3} may indicate an ensemble of dark/quiescent gas, potential site of future star foramtion; on the other hand these are also typical densities of PDRs. An average density of 10^5 cm^{-3} may indicate that star forming gas is dominant, but it may also be a signature of shocked-compressed gas. Finally a high density of $\sim 10^6$ cm^{-3} may be typical of left over gas from episodes of starbursts or highly shocked gas driven by AGN (Krips *et al.* 2011; Viti *et al.* 2014). In Figure 1 two chemical models showing the fractional abundance of selected species as a function of cosmic ray ionization rate are plotted: the top one represent a steady-state calculation using a PDR gas-phase code (Bisbas *et al.* 2012) at a density of 10^5 cm^{-3}; the bottom figure shows the results from a time dependent gas-grain chemical model (Viti *et al.* 2004) which is run in two phases: the first phase follows the collapse from a diffuse gas to a density of 10^5 cm^{-3}. The temperature is kept constant at 10 K and atoms and molecules freeze on to the dust and are involved in surface reactions. The second phase follows the gas as its temperature is increased to 100 K due to star formation taking place, with subsequent sublimation of the icy mantles formed during the first phase. The two models are clearly very different and simulate physically different environments; and yet for the same cosmic ray ionization rate, some of the abundances are remarkably similar: take for example the HCN/HCO$^+$ at a cosmic ray 100 times that of our own Galaxy. Clearly, this ratio alone would not be able to inform the observer whether it is tracing star forming gas, AGN or PDR dominated gas without any prior information or more molecular observations.

2.1. *Why are molecular ratios 'degenerate'?*

Most molecular species have more than one route of formation and destruction. For example, let us consider a collapsing core undergoing collapse from a diffuse state, e.g. 100 cm^{-3}, to a high density of, say, 10^7 cm^{-3}, under standard interstellar conditions and let us look at the chemistry of HCO$^+$ and HCN, the two most commonly observed extragalactic species, after CO. At very low densities the following routes of formation are important:

$$\mathrm{H + HCN^+ \rightarrow HCN + H^+} \tag{2.1}$$

$$\mathrm{C^+ + H_2O \rightarrow HCO^+ + H^+} \tag{2.2}$$

while at intermediate densities:

$$\mathrm{N + CH_2 \rightarrow HCN + H} \tag{2.3}$$

$$\mathrm{C + NH_2 \rightarrow HCN + N} \tag{2.4}$$

and

$$\mathrm{H_3^+ + CO \rightarrow HCO^+ + H_2} \tag{2.5}$$

$$\mathrm{CH + O \rightarrow HCO^+ + e^-} \tag{2.6}$$

contribute ~30% each to the formation of HCN and HCO$^+$ respectively. Finally, at high densities:

$$\mathrm{NH_3 + CN \rightarrow HCN + NH_2} \tag{2.7}$$

and Reaction 2.5, are the two dominant routes for HCN and HCO$^+$ respectively.

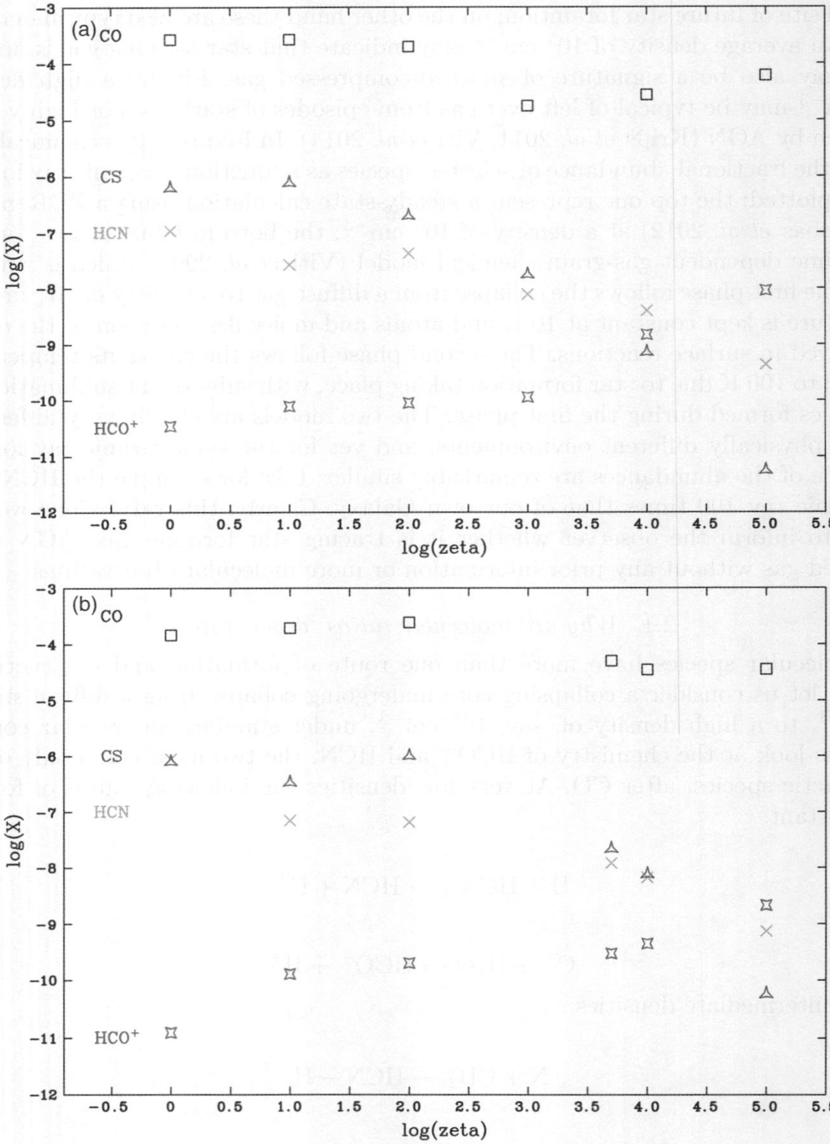

Figure 1. Fractional abundances (with respect to the total number of hydrogen nuclei) of selected species as a function of cosmic ray ionization rate for two types of chemical models. See text.

The formation and destruction routes are not just dependent on the density of course. For constant density, variation in cosmic ray ionization rate can lead to different routes of formation and destruction. For example, CS, another routinely observed molecule in extragalactic environemts, is very sensitive to cosmic ray ionization rates. At $\zeta > 100\ \zeta_0$ the main destruction mechaniscm for CS is: CS + O → CO + S, while at low ionization rates H_3O^+ + CS → HCS^+ + H_2O is dominant.

From an observational point of view the variations in the chemistry are of course reflected in the observed line intensities. Figure 2 shows theoretical line intensities for CS for a very small grid of chemical and radiative transfer models.

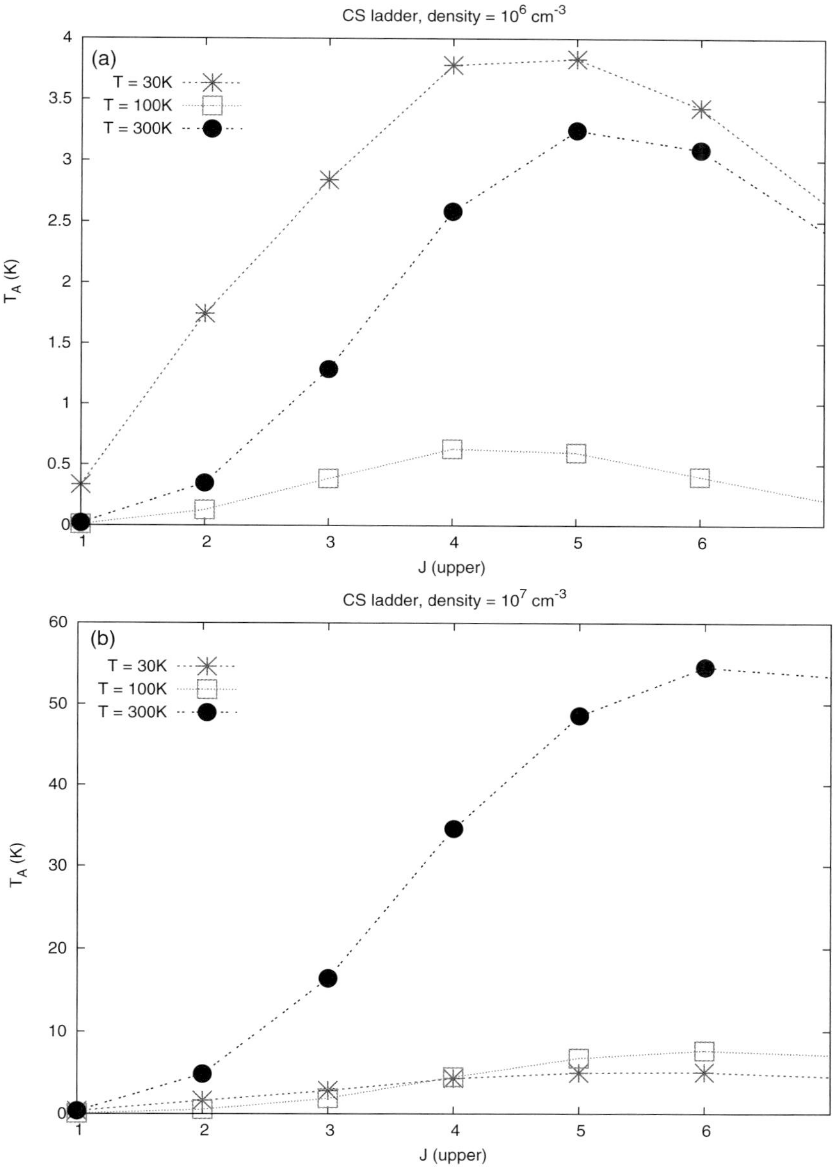

Figure 2. Theoretical CS ladders calculated by coupling chemical (UCL_CHEM, Viti *et al.* 2011) and line radiative transfer models (SMMOL, Kelly *et al.* 2015).

Already from such a small grid (where we only used two gas densities and three gas temperatures, keeping constant the cosmic ray ionisation rate and not including shocks, for example) one can see how sensitive the intensities of the CS transitions are; it is especially interesting to see how the ladder peaks at different J depending on the gas temperatures and densities.

When one considers that hundreds of species involved in thousands of reactions may be present in the gas, then disentangling the chemistry from the physics is not a trivial task, especially as we must not forget the role of dust and surface reactions. For example, methanol is routinely observed in extragalactic environments, and yet there is no efficient

Table 1. Table 10 from Viti *et al.* (2014). Grid of chemical models. The cosmic ray ionization rate, the radiation field, the gas temperature, and the gas density are listed in columns 2–5.

Model n^o	ζ (s^{-1})	G_0 (Draine)	T (K)	$n_{final/preshock}$ (cm^{-3})	Shock
1	1	1	100	10^4	N
2	1	1	100	10^5	N
3	1	1	100	10^6	N
4	10	1	100	10^4	N
5	10	1	100	10^5	N
6	10	1	100	10^6	N
7	1	10	100	10^4	N
8	1	10	100	10^5	N
9	1	10	100	10^6	N
10	10	10	100	10^4	N
11	1	500	100	10^4	N
12	500	1	100	10^4	N
13	5000	1	100	10^4	N
14	10^5	1	100	10^5	N
15	10	1	100	2×10^6	N
16	1	1	100	5×10^6	N
17	1	1	100	2×10^6	N
18	1	1	200	10^5	N
19	1	1	200	10^4	N
20	1	1	200	10^6	N
21	10	1	200	10^6	N
22	1	10	200	10^6	N
23	10	1	200	10^5	N
24	1	10	200	10^5	N
25	1	1	–	10^4	Y
26	10	1	–	10^5	Y
27	1	1	–	10^5	Y

way of forming it in the gas at low temperatures (T < 100 K). It is likely that most of the CH_3OH observed in extragalactic environments has formed on dust grains and then sublimated back in the gas phase.

In summary, it is important to emphasize that chemistry leads to a large range of ratios, heavily dependent on small changes in the physical and chemical (initial) conditions of the gas. Hence one *cannot* use molecular ratios without considering of the chemistry behind them and without paying attention to individual abundances. Performing multiple, independent analyses of the observational data is therefore essential as we shall demonstrate in the next section.

3. NGC 1068: a case study

The unprecedented resolution of ALMA has finally allowed us to resolve structures inside nearby galaxies. NGC 1068, a composite starburst/AGN galaxy, is a Seyfert 2 galaxy, at a distance of $\sim$ 14 Mpc. Many studies have focused on this galaxy (e.g. Usero *et al.* 2004, Israel *et al.* 2009, Kamenetzky *et al.* 2011, Hailey-Dunsheath *et al.* 2012, Aladro *et al.* 2013, Schinnerer, *et al.* 2000, Garcia-Burillo *et al.* 2010, Krips *et al.* 2011, Garcia-Burillo *et al.* 2014, Viti *et al.* 2014) and have shown that molecular tracers of dense gas are essential to spatially resolve the distribution, kinematics, and excitation of the circumnuclear gas of NGC 1068, as well as to study the relationship between the $r \sim 1-1.5$ kpc starburst ring and the $r \sim 200$ pc circumnuclear disc (CND) located around the AGN. In Garcia-Burillo *et al.* (2014) ALMA Cycle 0 observations in Bands 7 and 9 of several molecular transitions, namely CO $(3-2)$ and $(6-5)$, HCO^+ $(4-3)$, HCN $(4-3)$, and CS $(7-6)$ within the $r \sim 200$ pc CND, were presented (see Figure 3).

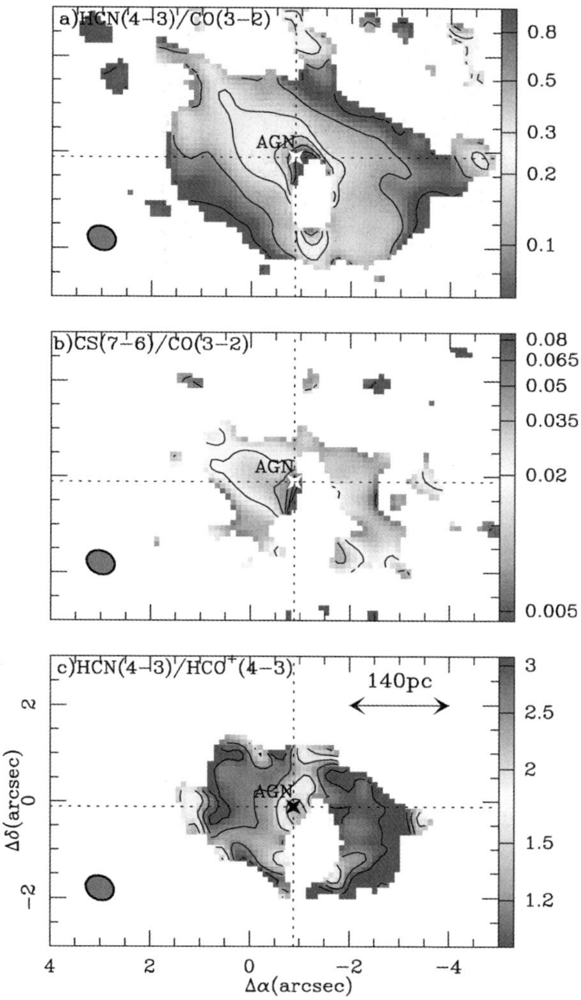

Figure 3. Figure 5 from Viti *et al.* (2014). HCN$(4-3)$/CO$(3-2)$, CS$(7-6)$/CO$(3-2)$ and HCN$(4-3)$/HCO$^+(4-3)$ velocity-integrated intensity ratios in the CND. The filled ellipses represent the spatial resolutions used to derive the line ratio maps: $0.6'' \times 0.5''$ PA $= 60^0$, for all panels.

Viti *et al.* (2014) performed a chemical analysis of the gas within the CND with the aim of quantifying the chemical differentiation and of determining the chemical origin of such differentiation: the ALMA data were used together with lower resolution IRAM Plateau de Bure Interferometer data covering the lower-J transitions of each molecule. Five different analyses were performed: an LTE analysis, including a rotation diagram; three different analyses using RADEX (van der Tak *et al.* 2007) differing in the number of ratios and species used; and finally a purely *ab initio* analysis using a time dependent gas-grain chemical mode. While each analysis led to its own detailed conclusions, all five indicated that *more than one gas-phase component is necessary to fit all the available molecular ratios uniquely.* Chemically this is certainly not unexpected, as discussed in the section above. Of particular interest to this review is Figure 12 in Viti *et al.* (2014), which we include here together with their Table 10, in Figure 4 and Table 1 respectively.

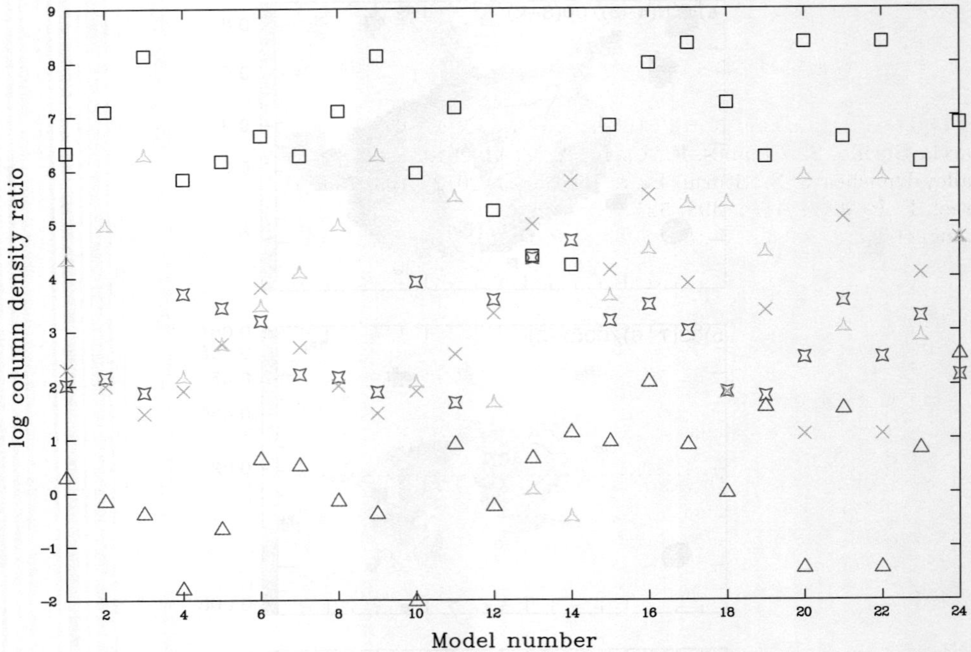

Figure 4. Theoretical fractional abundances ratios for the grid of UCL_CHEM models at high A_V (10 mags). The black squares are CO/HCO^+ ratios; red stars = CO/HCN; green crosses = CO/CS; cyan triangles = HCN/CS; blue triangles = HCN/HCO^+. The numbering in the x-axis refers to the Model number (see Table 1).

One notes first of all that the y-axis shows logarithmic values of column density ratios, and yet a very large range of ratios are displayed. None of them are anomalous in that the grid of models performed covers reasonable values of densities, cosmic ray ionozation rate, temperatures and radiation field. Viti *et al.* (2014) found that each subregion of the CND could in fact be fitted by more than one set of column densities and, more importantly, that there is not one single model that can explain *all* ratios: multiple gas components even within each sub region are required. In fact, the multi-analyses performed in Viti *et al.* (2014) seem to indicate that there has to be a pronounced chemical differentiation across the CND and that each subregion could be characterized by a 3-phase component ISM, with two gas phases at different densities but both subjected to a high ionization rate, and one gas component comprising of shocked gas where most likely the CS arises from.

4. Concluding remarks

Recent years have witnessed a plethora of extragalactic molecular surveys which brought detailed information about the amount and distribution of the molecular gas in different types of galaxies. High resolution multi-species as well as multi-line molecular observations coupled with multiple independent analyses take us a step closer towards disentangling the gas components and energetics of nearby galaxies.

References

Aladro, R., Martín, S., Martín-Pintado, J., *et al.* 2011, *A&A*, 535, A84

Aladro, R., Viti, S., Bayet, E., *et al.* 2013, *A&A*, 549, 39
Bolatto, A., Wolfire, M., & Leroy, A. K. 2013, *ARA&A*, 51, 207
Bell, T. A., Roueff, E., Viti, S., & Williams, D. A. 2006, *MNRAS*, 371, 1865
Bisbas, T. G., Bell, T. A., Viti, S., Yates, J., & Barlow, M. J. 2012, *MNRAS*, 427, 2100
Costagliola, F., Sakamoto, K., Muller, S., *et al.* 2015, *A&A*, accepted (arXiv:1506.09027)
García-Burillo, S., Combes, F., Usero, A., *et al.* 2014, *A&A*, 567, 125
Hailey-Dunsheath, S., Sturm, E., & Fischer, J. 2012, *ApJ*, 755, 57
Israel, F. P. 2009, *A&A*, 493, 525
Kamenetzky, J., Glenn, J., & Maloney, P. R. 2011, *ApJ*, 731, 83
Kelly, G., Viti, S., Bayet, E., Aladro, R., & Yates, J. 2015, *A&A*, 578, 70
Krips, M., Martín, S., Eckart, A., *et al.* 2011, *ApJ*, 736, 37
Martín, S., Mauersberger, R., Martín-Pintado, J., Henkel, C., & García-Burillo, S. 2006, *ApJS*, 164, 450
Usero, A., García-Burillo, S., Fuente, A., Martín-Pintado, J., & Rodríguez-Fernández, N. J. 2004, *A&A*, 419, 897
Van der Tak, F. F. S., Black, J. H., Schöier, F. L., Jansen, D. J., & van Dishoeck, E. F. 2007, *A&A*, 468, 627
Viti, S., Collings, M. P., Dever, J. W., McCoustra, M. R. S., & Williams, D. A. 2004, *MNRAS*, 354, 1141
Viti, S., Jimenez-Serra, I., Yates, J. A., *et al.* 2011, *ApJ*, 740, 3L
Viti, S., García-Burillo, S., Fuente, A., *et al.* 2014, *A&A*, 570, 28
Williams, D. A. & Viti, S. 2013, *Cambridge University Press*

From Interstellar Clouds to Star-Forming Galaxies: Universal Processes?
Proceedings IAU Symposium No. 315, 2015
P. Jablonka, Ph. André & F. van der Tak, eds.

doi:10.1017/S1743921316007201

High-J CO Intensity Measurements for Galaxies Observed by the Herschel FTS

Julia Kamenetzky[1], Naseem Rangwala[2], Jason Glenn[3], Philip Maloney[3] and Alex Conley[3]

[1]Steward Observatory, University of Arizona, Tucson, AZ, U.S.
email: jkamenetzky@as.arizona.edu
[2]NASA Ames Research Center, Mountain View, CA, U.S.
[3]Center for Astrophysics and Space Astronomy, University of Colorado, Boulder, CO, U.S.

Abstract. Molecular gas is the raw material for star formation and is commonly traced by the carbon monoxide (CO) molecule. The atmosphere blocks all but the lowest-J transitions of CO for observatories on the ground, but the launch of the Herschel Space Observatory revealed the CO emission of nearby galaxies from J = 4−3 to J = 13−12. Herschel showed that mid- and high-J CO lines in nearby galaxies are emitted from warm gas, accounting for approximately 10% of the molecular mass, but the majority of the CO luminosity. The energy budget of this warm, highly-excited gas is a significant window into the feedback interactions among molecular gas, star formation, and galaxy evolution. Likely, mechanical heating is required to explain the excitation. Such gas has also been observed in star forming regions within our galaxy.

We have examined all ∼300 spectra of galaxies from the Herschel Fourier Transform Spectrometer and measured line fluxes or upper limits for the CO J = 4 − 3 to J = 13 − 12, [CI], and [NII] 205 micron lines in ∼200 galaxies, taking systematic effects of the FTS into account. We will present our line fitting method, illustrate trends available so far in this large sample, and preview the full 2-component radiative transfer likelihood modeling of the CO emission using an illustrative sample of 20 galaxies, including comparisons to well-resolved galactic regions. This work is a comprehensive study of mid- and high-J CO emission among a variety of galaxy types, and can be used as a resource for future (sub)millimeter studies of galaxies with ground-based instruments.

Keywords. galaxies: ISM – ISM: molecules – submillimeter

1. Introduction

Molecular gas is the raw material for star formation, and it is also influenced by the ongoing processes of star formation, life, and death. This work focuses on CO as a tracer of molecular gas, specifically the new view of the high-J lines (up to J = 13−12) afforded to us by the *Herschel* Fourier Transform Spectrometer. The atmosphere blocks all but the lowest-J transitions of CO for observatories on the ground, but the launch of the Herschel Space Observatory revealed the CO emission of nearby galaxies from J = 4−3 to J = 13−12.

The early science results from the *Herschel* demonstrated that the high-J emission from CO was significantly higher than predicted by standard, $T_{kin} < 100$ K molecular gas described by the low-J lines in the nearby starburst galaxy M82 (Panuzzo *et al.* 2010). Subsequent studies, such as Kamenetzky *et al.* (2014) and references therein, found this was true for a variety of galaxies, shown in Figure 1.

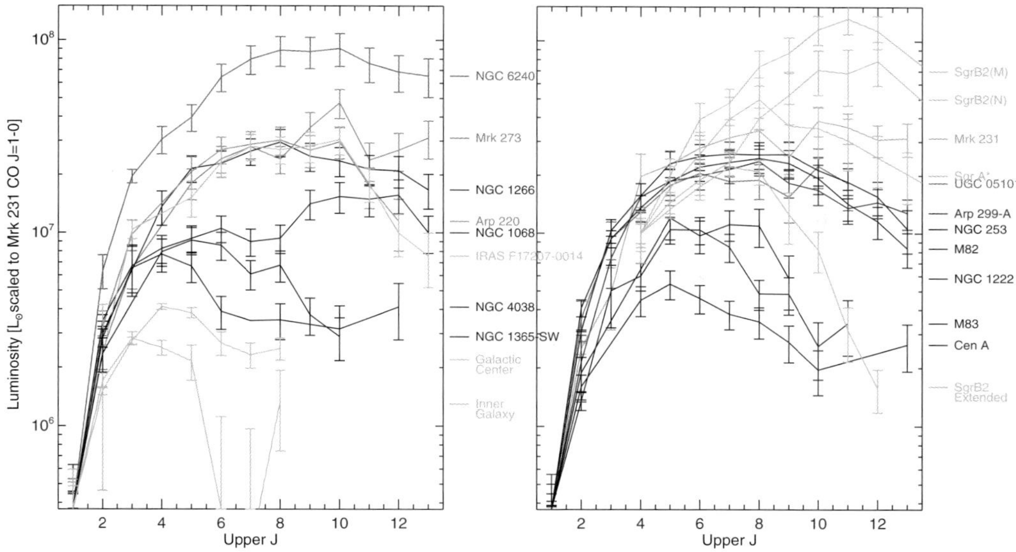

Figure 1. Normalized CO SLEDS, figure and caption reproduced from Kamenetzky *et al.* (2014) All $J = 1{-}0$ luminosities are scaled to match that of Mrk 231 (3.7×10^5 $L_{\odot}$). SLEDs are colored to indicate increasing $L_{\rm FIR}$ with increasing lightness. Placement in the left or right panel is for clarity only. On the left panel, the SLEDs of the Galactic center ($|l| < 2\overset{\circ}{.}5$) and the Inner Galaxy ($2\overset{\circ}{.}5 < |l| < 32\overset{\circ}{.}5$), also normalized, are shown in green for comparison (Fixsen *et al.* 1999). On the right panel, we show the SLEDs of two star-forming cores and the extended envelope of Sgr B2 (Etxaluze *et al.* 2013) and that of Sgr A* (Goicoechea *et al.* 2013). Because these SLEDs begin at $J = 4{-}3$, we scale $J = 4{-}3$ to 10^7 $L_{\odot}$ for visual comparison of the shapes. None of the SLEDs are corrected for dust extinction.

2. New Survey

This new work is a compilation of all extragalactic SPIRE FTS spectra from the *Herschel* mission. These proceedings summarize the findings of Kamenetzky *et al.* 2015, submitted to The Astrophysical Journal. To briefly summarize the processing methodology for all $\sim$ 300 spectra from the archive, we first corrected for source/beam coupling using the SPIRE PSW photometry maps. We fit the CO $J = 4{-}3$ through $J = 13{-}12$ lines, two [CI] lines, and [NII] 205 μm line using an approach which better takes into account the systematic, line-profile-like ringing of the SPIRE spectrum better than least squares minimization. Additional low-J ($J = 1{-}0$, $J = 2{-}1$, $J = 3{-}2$) lines were collected from the literature and obtained at the the Arizona Radio Observatory.

3. Findings

As shown in Figure 2, we find a superlinear slope in CO $J = 1{-}0$ for the relation $\mathrm{Log}(L_{\rm FIR}) = \mathrm{a}\ \mathrm{Log}(L'_{\rm CO}) + \mathrm{b}$. As one moves to higher-J lines, the relation flattens and stays linear by about $J = 5{-}4$. Individual populations (e.g. only (U)LIRGS) show some variation in the line fits.

The consequence of the aforementioned linear fits is that the ratio of high-J to low-J CO line luminosities only varies by about a factor of 10 throughout this whole sample (Figure 3), which covers many orders of magnitude of far infrared luminosity (and star formation rate). We found that many models predict steep drop-offs in high-J to low-J ratios with lower star formation rates, such as Narayanan & Krumholz (2014). The models

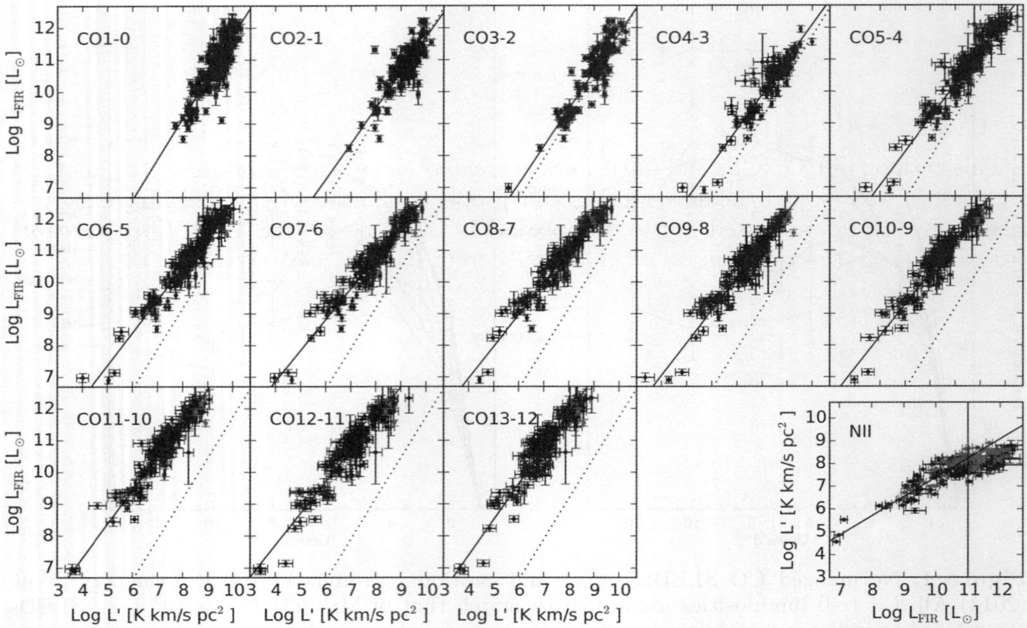

Figure 2. CO vs. L_{FIR}, figure and caption reproduced from Kamenetzky *et al.* (2015, submitted). The y-axis is the L_{FIR} in the beam for comparison to the CO measurement. Low-J lines may include multiple measurements for the same galaxy if available in the literature. Blue data points indicate resolved (Sinc-Gaussian) measurements from the FTS. The CO $J = 1{-}0$ line fit is shown as a dotted line on each other CO plot for comparison. The [NII] line is shown with the x- and y-axes flipped compared to CO as is more common. It is only fit up to Log(L_{FIR}) = 11 due to the well-known line deficit above this value.

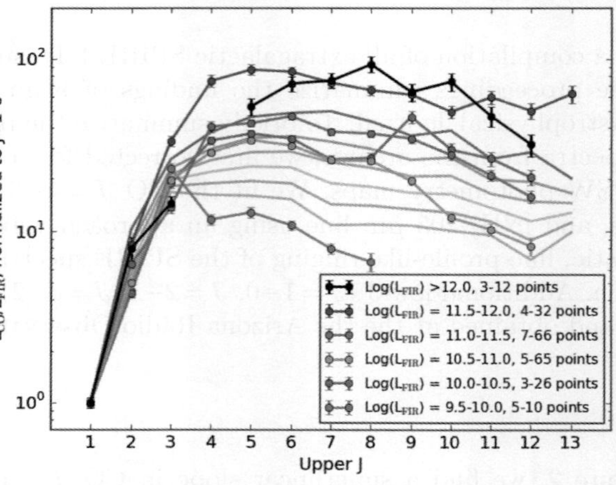

Figure 3. Average SLEDs by L_{FIR} Ranges, figure and caption reproduced from Kamenetzky *et al.* (2015, submitted). For L_{FIR} bin ranges shown in the legend, the value of $L_{\mathrm{CO}}/L_{\mathrm{FIR}}$ was averaged if measurements existed for at least 3 galaxies. All SLEDs were then divided by the value of the CO $J = 1{-}0$ line to demonstrate the difference in relative excitation of the CO ladder. The number of data points used in each SLED may change with each line, which is why a range is given in the legend. The highest luminosity bin is dominated by more distant galaxies, where the CO $J = 4{-}3$ line is likely to be redshifted out of the FTS band, which is why that black data point is missing. The lighter lines (in both plots) with no data markers indicate the predictions from the slopes in Figure 2 for the center of each log bin.

of Kazandjian *et al.* (2015) include mechanical heating, which can produce flatter SLEDs, but only at much higher ratios than we see here.

4. Conclusions

This catalog of CO, [CI], and [NII] lines is created using a uniform processing method for dealing with source/beam coupling and proper estimates of the line flux measurements, given the correlated nature of spectral bins in the SPIRE FTS. Over multiple orders of magnitude, the mid- to high-J CO lines show linear slopes for the relation $\text{Log}(L_{\text{FIR}}) = \text{a Log}(L'_{\text{CO}}) + \text{b}$. As a result, the average SLEDs show increasing mid- to high-J CO luminosity with increasing L_{FIR}, but only in the range from a few to ~ 100. Current theoretical models do not reproduce flat SLEDs with appropriate ratios. More details are available in Kamenetzky *et al.* (2015, submitted).

References

Etxaluze, M. *et al.* 2013, *Astron. Astrophys.*, 556, A137

Fixsen, D. J., Bennett, C. L., & Mather, J. C. 1999, *Astrophys. J.*, 526, 207

Goicoechea, J. R. *et al.* 2013, *Astrophys. J.*, 769, L13

Kamenetzky, J., Rangwala, N., Glenn, J., Maloney, P. R., & Conley, A. 2014, *Astrophys. J.*, 795, 174

Kamenetzky, J., Rangwala, N., Glenn, J., Maloney, P. R., & Conley, A. 2015, submitted to *Astrophys. J.*, arXiv:1508.05102

Kazandjian, M. V., Meijerink, R., Pelupessy, I., Israel, F. P., & Spaans, M. 2015, *Astron. Astrophys.*, 574, A127

Narayanan, D. & Krumholz, M. R. 2014, *Mon. Not. R. Astron. Soc*, 442, 1411

Panuzzo, P. *et al.* 2010, *Astron. Astrophys.*, 518, L37

From Interstellar Clouds to Star-Forming Galaxies: Universal Processes?
Proceedings IAU Symposium No. 315, 2015
P. Jablonka, Ph. André & F. van der Tak, eds.

doi:10.1017/S1743921316007213

Giant Molecular Cloud Populations in Nearby Galaxies

Annie Hughes[1], Sharon Meidt[2], Dario Colombo[3], Andreas Schruba[4], Eva Schinnerer[2], Adam Leroy[5], Tony Wong[6] and the PAWS, MAGMA, CANON and M31-CARMA survey teams

[1]Institute for Research in Astrophysics & Planetology, CNRS
email: `Annie.Hughes@irap.omp.eu`

[2]Max Planck Institute for Astronomy

[3]Department of Physics, University of Alberta

[4]Max Planck Institute for Extraterrestrial Physics

[5]Department of Astronomy, The Ohio State University

[6]Astronomy Department, University of Illinois

Abstract. We present new results from a comparative analysis of the resolved giant molecular cloud (GMC) populations in six nearby galaxies. We show that the GMCs in denser environments – M51, the centre of NGC6946 – have greater CO surface brightness and higher velocity dispersions relative to their size than GMCs in less dense environments. We find systematic differences in the GMC mass distribution among galaxies, such that more of the molecular gas in the low-mass galaxies (M33, the Large Magellanic Cloud) and the outer disk of M31 is located in low mass clouds. Using the number density of GMCs in the interarm regions of M51, we argue that GMC destruction in this region is regulated by shear, and that cloud lifetimes there are finite and short, ∼20 to 30 Myr. Our results indicate the importance of galactic environment on the evolution of GMCs, and on a galaxy's global pattern of star formation.

Keywords. ISM: clouds, galaxies: ISM, (galaxies:) Magellanic Clouds, (galaxies:) Local Group

---

1. Introduction

Most star formation occurs within giant molecular clouds (GMCs): cold, dense ($T \sim 10\,\mathrm{K}$, $n \sim 10^3\,\mathrm{cm}^{-3}$) structures with a typical mass of $\sim 10^5\,\mathrm{M}_\odot$ and a characteristic size of $\sim 40\,\mathrm{pc}$. Studies of the Milky Way's GMC population have shown that they are in approximate virial equilibrium, with scaling relations between their size, velocity dispersion and CO luminosity (e.g. Solomon *et al.* 1987; henceforth S87). Investigating GMCs in external galaxies presents the opportunity to learn how the distribution, density structure and dynamical state of star-forming clouds depends on galactic environment, and how galactic-scale processes influence GMC formation and destruction. Until recently, however, the observations required to resolve extragalactic GMCs were technically demanding. In this contribution, we compare the properties of GMCs identified in wide-field, high resolution CO surveys of six galaxies: M51 (Pety *et al.* 2013; Schinnerer *et al.* 2013), NGC6946 (Donovan Meyer *et al.* 2013), NGC628 (Schinnerer *et al.*, in prep.), M31 (Schruba *et al.*, in prep.), M33 (Rosolowsky *et al.* 2007) and the Large Magellanic Cloud (LMC, Wong *et al.* 2011). We highlight results that connect the properties and timescales of individual clouds to their galactic context. The results in Sections 1 and 2 are presented in Hughes *et al.* (in prep.); the results in Section 3 are presented in Meidt *et al.* 2015.

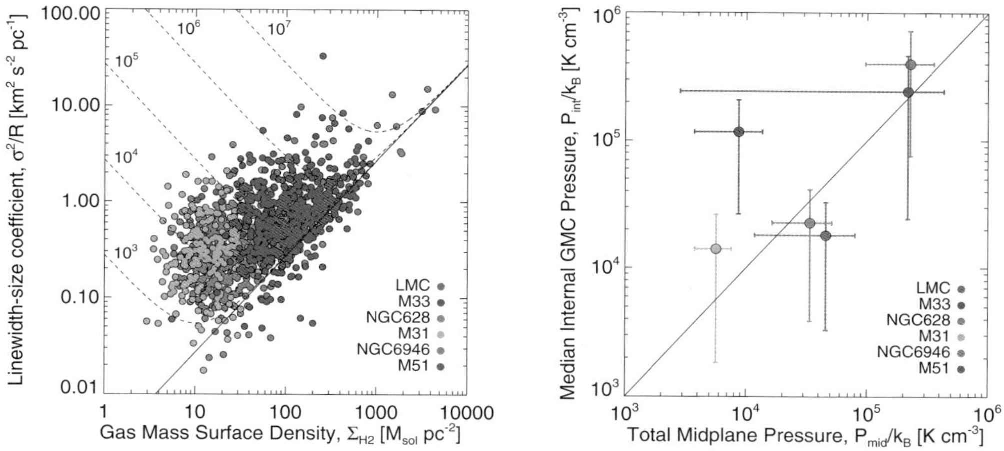

Figure 1. Left. The coefficient of the linewidth-size relation, $a^2 = \sigma_v^2/R$, versus the molecular gas mass surface density Σ_{H_2} for GMCs in M51, NGC6946, NGC628, M31, M33 and the LMC. The solid diagonal line shows the solution for simple gravitational equilibrium. The V-shaped curves represent solutions to Equation 2.1, for different values of the external pressure between $P_{ext}/k_B = 10^3$ and 10^7 K cm^{-3}. In contrast to the Larson scaling relations, the Σ_{H_2} values of GMCs vary by more than two orders of magnitude, and the linewidth-size coefficient increases with increasing Σ_{H_2}. Right. Median internal pressure of the GMCs versus an estimate of the external pressure within the region of each galaxy covered by the CO observations. The internal pressures are estimated from GMC properties and the external pressures are estimated using an expression derived by Koyama & Ostriker (2009) for the total pressure at the disk midplane (see text). The solid diagonal line indicates equality. The vertical error bars indicate the dispersion of the internal pressure estimates for individual GMCs, while the horizontal bars indicate the range of external pressures across the surveyed region.

2. Physical Properties of Extragalactic GMCs

From a theoretical perspective, it might be expected that GMC properties depend on their galactic environment (e.g. Elmegreen 1989), but this has been difficult to establish observationally. The first comparative studies of GMC populations in nearby (mostly Local Group) galaxies instead found that the physical properties of extragalactic GMCs are very similar to the properties of GMCs in the inner Milky Way (e.g. Blitz *et al.* 2007; Bolatto *et al.* 2008). There are several caveats to the universality of this conclusion, primarily that the first extragalactic surveys could only parameterise the brightest, most massive GMCs in external galaxies, and that the diversity of galactic environments in Local Group galaxies is rather limited. Moreover, a re-analysis of the GMCs in the S87 catalogue has shown that these GMCs are inconsistent with the Larson scaling relations that are regularly used as the basic empirical benchmark of cloud dynamics (Larson 1981) (Heyer *et al.* 2009). Several recent works targeting individual galaxies have reported variations in the typical mass and dynamical state of GMCs (e.g. Hirota *et al.* 2011; Utomo *et al.* 2015), but differences in methodology, as well as the heterogeneity of the original survey datasets, have prevented a conclusive answer emerging from literature measurements.

In the left panel of Figure 1, we plot a^2 versus Σ for the GMC populations in our six galaxies. We emphasize that the datapoints in these plots represent our best effort to compare 'like with like', rather than extracting the most precise information possible from each datacube. The GMC properties are estimated using the same CPROPSTOO algorithms for identifying cloud structures and measuring their properties (Leroy *et al.* in prep), and the algorithms are applied to datacubes that have been brought to the same spectral ($5\,\mathrm{km\,s^{-1}}$) and spatial (53 pc) resolution. Note that this can yield quite different results to the properties determined from the original, higher resolution datasets. The sensitivity (per $5\,\mathrm{km\,s^{-1}}$ channel) of the homogenized datasets is 0.5 K for NGC6946, $\sim$0.2 K for M51 and M33, and $\sim$0.05 K for the LMC, NGC628 and M31.

If extragalactic GMCs strictly obeyed Larson's scaling relations, the datapoints in the left panel of Figure 1 would cluster around $\Sigma \sim 100\,\mathrm{M_\odot\,pc^{-2}}$, and $a^2 = \sigma_\mathrm{v}^2/R = (\pi G\Sigma/5)$. Instead, a^2 increases with Σ for extragalactic GMCs, both among and within individual galaxies, and that typical mass surface density of GMCs varies between $\sim$10 and $10^3\,\mathrm{M_\odot\,pc^{-2}}$. The solid diagonal line in the left panel of Figure 1 represents $a^2 = (\pi G\Sigma/5)$, which is where GMCs in simple virial equilibrium should lie. The majority of extragalactic GMCs follow a trend that is roughly parallel to but displaced upwards from the relationship for virialized clouds, i.e. they have an excess of kinetic energy relative to simple virial equilibrium. Following the formulation of Field *et al.* (2011), the dashed curves in the left panel of Figure 1 indicate the solutions for clouds in a pressure-bounded equilibrium configuration,

$$a^2 = \frac{\sigma_{rmv}^2}{R} = \frac{1}{3}\left(\pi\Gamma G\Sigma + \frac{4P_{ext}}{\Sigma}\right) \tag{2.1}$$

for external ISM pressures P_{ext}/k_B between 10^3 and $10^7\,\mathrm{cm^{-3}\,K}$. Γ is a factor that depends on the internal density distribution of a cloud. We adopt $\Gamma = 0.6$, corresponding to a cloud with uniform density. The data points do not lie along the dashed curves, which means that the GMCs in each galaxy do not tend towards pressure equilibrium for a single value of P_{ext}. If pressure is important to the dynamical state of clouds, then the large scatter for each GMC population suggests that clouds within a single galaxy experience significant external pressure variations.

Nevertheless, global properties of the host galaxy do seem to influence the average GMC properties. The right panel of Figure 1 shows the average internal pressure of the GMC population in each galaxy versus the total ISM pressure at the disk midplane. We estimate the internal pressure $P_{int}/k_B \sim \rho\sigma_\mathrm{v}^2 = 1176\left(\frac{M}{R^3}\right)\sigma_\mathrm{v}^2\,\mathrm{cm^{-3}\,K}$, where M is the cloud mass in solar masses, R is the cloud radius in parsecs, and σ_v is the cloud velocity dispersion in $\mathrm{km\,s^{-1}}$. The midplane pressure P_{mid} for each galaxy is estimated by combining maps of the stellar and gas surface densities and the atomic gas velocity dispersion (each at 500 pc resolution) according to equation 8 of Koyama & Ostriker (2009). Defined in this way, P_{mid} approximates the mass-weighted mean pressure (including self-gravity) in the interstellar gas. It is a factor of $\sim$10 greater than the pressure in the diffuse non-self-gravitating ISM, and a factor of $\sim$4 greater than the formulation for the external pressure of Blitz & Rosolowsky (2006) and Elmegreen (1989) that have been used in several previous studies. The vertical error bars in the right panel of Figure 1 reflect the 1σ dispersion of the P_{int} values in each GMC catalogue; the horizontal error bars reflect the range in the per-pixel P_{mid} values across each survey's field-of-view. For both axes, measurement and systematic uncertainties are not taken into account.

Regardless of the precise formulation for P_{mid}, we obtain the robust result that the typical internal pressure of GMCs is greater in galaxies with higher midplane pressures. More work is required to understand the physical origin of the observed correlation. The average internal pressure of GMCs also increases with quantities such as the stellar surface density Σ_* and the molecular fraction of the neutral ISM $R_{mol} = \frac{\Sigma_{\mathrm{H}_2}}{\Sigma_{\mathrm{HI}}}$; the current data are insufficient to determine whether the ISM pressure is responsible for the variation in GMC properties, or whether the main driver is a quantity that contributes to P_{mid}. M33 appears to lie slightly away from the main trend. This may partly be due to the modest signal-to-noise of the BIMA+FCRAO datacube, in which we detect only ~ 70 clouds, presumably those with the highest CO surface brightness. A new survey by the IRAM 30 m telescope has identified > 500 clouds in M33 (Druard et al. 2014), and should yield a more reliable estimate for typical GMC properties in M33.

3. GMC Mass Distributions in Nearby Galaxies

The GMC mass distribution describes how a galaxy's molecular gas reservoir is distributed across structures of different mass. In the Milky Way, GMCs follow $dN \propto M^{\gamma} dM$, with $\gamma \sim -1.5$ in the inner disk (S87) and $\gamma \sim -1.8$ in the outer disk (Heyer, Carpenter & Snell 2001). In star formation theory and numerical modelling, a value of $\gamma \sim -1.5$ has been used as both a benchmark (e.g. Wada, Spaans & Kim 2000) and input (e.g. Krumholz & McKee 2005), although the universality of this γ value has not been empirically verified. New surveys of CO emission in M51 (Colombo *et al.* 2014) and M33 (Gratier *et al.* 2012) instead suggest that the shape of the GMC mass distribution varies with galactocentric radius and kinematic environment, although the physical agents that determine the shape of the distribution are yet to be securely identified.

While GMC mass distributions in several nearby galaxies have now been studied (e.g. Rosolowsky 2005; Fukui *et al.* 2008; Gratier *et al.* 2012; Colombo et al. 2014), a direct comparison of these results is not straightforward since the shape of the mass distribution depends on the resolution and sensitivity of the input datasets, as well as the decomposition algorithm and the completeness limit (e.g. Reid *et al.* 2010). In Figure 2, we show the cumulative mass distributions for the GMC populations of our six galaxies, using the same GMC catalogues as in Section 2. Strictly, the curves in Figure 2 represent the CO luminosity function of each GMC population. We convert the CO luminosity to mass using the simple scaling $M_{GMC} = \alpha_{\rm CO} L_{\rm CO}$, assuming a constant $\alpha_{CO} = 4.4\,{\rm M_{\odot}}\,({\rm K\,km\,s^{-1}\,pc^{2}})^{-1}$ for all galaxies. This is equivalent to $X_{\rm CO,MW} = 2\times 10^{20}\,{\rm cm^{-2}}\,({\rm K\,km\,s^{-1}})^{-1}$, and includes the mass contribution of helium. In the left panel, the curves are normalized by the deprojected area that was surveyed in each galaxy. In this case the vertical offsets between the curves reflect a combination of the different sensitivity limits of the input CO surveys and genuine physical variations in the number surface density of GMCs. The surveys of LMC, M31 and NGC628 have roughly equivalent sensitivity, so the relative amplitude of their mass distributions shows that NGC628 is much richer in molecular clouds than the surveyed regions of M31 and the LMC. M51 is likewise much more GMC-rich than M33.

In the right panel of Figure 2, the curves are normalized by the total number of clouds. This normalization is useful to compare the shape of the mass distribution in the different galaxies. Overall, the mass distribution shapes are quite similar, although the mass distribution continues to a much higher maximum cloud mass in M51 and NGC6946 than in the other galaxies. Across the mass range where the distributions are well-represented by a power-law, the distribution slope is also flatter (by ~ 0.5 dex) in M51 and NGC6946. Following Rosolowsky (2005), we model the cumulative GMC mass distribution as a truncated power-law with the form

$$N(M' > M) = N_0 \left[\left(\frac{M}{M_0} \right)^{\gamma+1} - 1 \right], \tag{3.1}$$

where γ is the slope of the differential mass spectrum, M_0 is the maximum mass in the distribution, and N_0 is the number of clouds more massive than the mass limit where

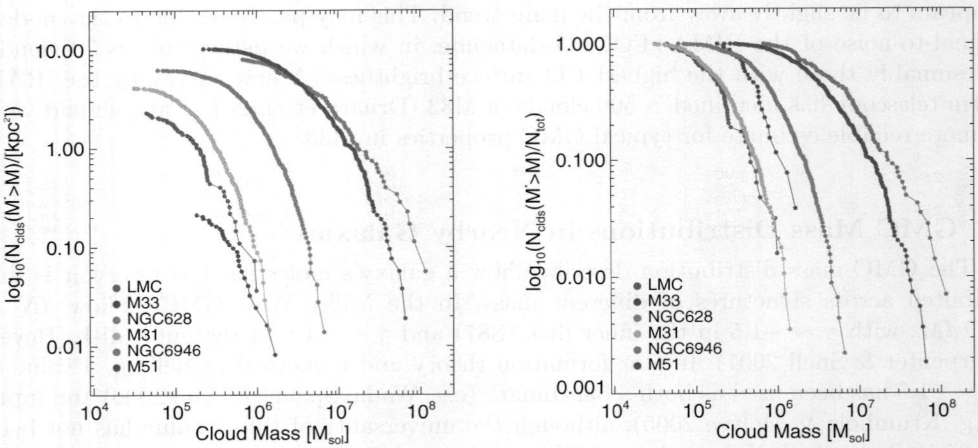

Figure 2. Cumulative mass distributions of GMCs in M51, NGC6946, NGC628, M31, M33 and the LMC, normalized by the area of the CO survey (left panel) and by the total number of identified clouds (right). The vertical offset between the curves in the left panel reflects variations in the number density of GMCs. The right panel shows that the slope of the GMC mass distribution is flatter in the denser, more massive systems (M51, NGC6946).

the mass distribution deviates from a simple power-law. A truncated power-law provides a better description of the mass distribution than a simple power-law for most of our galaxies. The exceptions are M33 and the LMC, where relatively few clouds are identified and a simple power-law fit cannot be excluded. As noted by previous studies, we find that the best-fitting values of γ, N_0 and M_0 are sensitive to the lower mass limit that we include in the fit. If we restrict the fit to clouds with masses above the median GMC mass, we find $\gamma \sim -1.8$ for M51 and NGC6946, $\gamma \sim -2.2$ for M31 and NGC628, and $\gamma < -2.5$ for M33 and the LMC. Using a lower mass limit of only 5 times the RMS sensitivity of each survey decreases the best-fitting values of γ by ~ 0.4 dex, but the general trend among galaxies is preserved. If the GMC mass distribution in M33 follows a truncated power-law, then the steep slope that we derive may stem from the modest sensitivity of the BIMA+FCRAO survey.

Is the shape of the GMC mass distribution universal, or does it depend on properties of the host galaxy? While more work is required to identify the physical mechanisms that regulate the growth and destruction of GMCs, our analysis tends to favour the idea that the shape GMC mass distribution depends on galactic environment. In Figure 3, we plot the slope of the GMC mass distribution versus several basic quantities that characterize the local physical conditions: Σ_* (left), $\Sigma_{\rm gas} = \Sigma_{\rm HI} + \Sigma_{\rm H_2}$ (middle) and R_{mol} (right). These are empirical estimates measured on 500 pc scales. These plots show that the slope of the GMC mass distribution tends to flatten as the galactic environment becomes denser and more molecule-rich. The observed trends roughly follow $\gamma \propto \Sigma_*^{0.6}$, $\gamma \propto \Sigma_{\rm gas}^{0.8}$, and $\gamma \propto 0.2R_{mol}$, but the data are insufficient to characterize the trends very precisely. Qualitatively similar trends are observed for the upper 10$^{\rm th}$ percentile GMC mass M_{10} and the number surface density of GMCs $\mathcal{N}$, but the absolute values of M_{10} and $\mathcal{N}$ are dependent on the sensitivity of the input datasets. A survey of a diverse sample of nearby galaxies with uniform sensitivity and spatial resolution would place these results on a more quantitative footing.

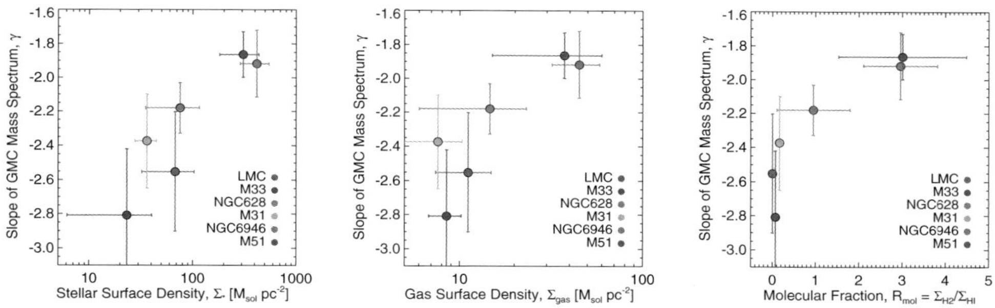

Figure 3. The power-law slope of the GMC mass distribution versus the local stellar surface density Σ_* (left), neutral gas surface density $\Sigma_{\rm gas} = \Sigma_{\rm HI} + \Sigma_{\rm H_2}$ (middle), and molecular fraction $R_{mol} = \Sigma_{\rm H_2}/\Sigma_{\rm HI}$. The quantities on the x-axis are measured on 500 pc scales. The vertical error bars represent the uncertainty in the fit. The horizontal error bars indicate the 1σ dispersion of values within the area of the CO surveys. Taken together, the plots show that the slope of the GMC mass distribution tends to flatten as the galactic environment becomes denser and more molecule-rich.

4. GMC Lifetimes: An Observational Estimate in M51

Resolved observations of GMC populations in nearby galaxies present a new opportunity to place empirical constraints on the timescale of GMC evolution. This timescale is important for star formation theory, as it sets a natural limit to the timescale over which gas can be transformed into stars. Previous efforts to reconstruct GMC lifetimes in external galaxies have used the association between GMCs and young stellar phenomena with better known evolutionary timescales, e.g. HII regions and young stellar clusters, to estimate the lifetime of the associated GMC (e.g. Leisawitz *et al.* 1989). In M33 and the Large Magellanic Cloud, this approach yields a typical GMC lifetime of ~ 30 Myr (e.g. Kawamura *et al.* 2009; Miura et al. 2012). The universality of this result – i.e. whether star-forming clouds in different galactic environment evolve over a similar timescale – is uncertain. Using the PAWS dataset for M51, we have developed a new method for estimating GMC lifetimes in disk galaxies that is independent of their relationship with young stellar phenomena (Meidt et al. 2015). The method requires wide-field, sensitive observations that capture a significant number of clouds (more than a few hundred), as well as detailed knowledge of gas dynamics in the galactic disk. The basic premise is that a short cloud lifetime (i.e. less than a galactic orbital period) should manifest itself as a decrease in the number of clouds across the inter-arm region between strong spiral arms. The observed decrease in cloud numbers constrains the ratio between the cloud lifetime and the travel time between the arms.

More formally, for an initial population of clouds N_0 at time t_0, we can estimate the characteristic cloud lifetime τ by measuring the cloud population N_1 at a later time t_1. Assuming the time evolution of GMCs is dominated by their destruction, then the cloud population will be reduced to zero after τ, i.e.

$$N_0 - \frac{N_0 - N_1}{t_1 - t_0}\tau = 0\,. \tag{4.1}$$

In this equation, the rate $-\frac{(N_0 - N_1)}{(t_1 - t_0)}$ represents the effective rate of change in the GMC population (i.e. the net result of both formation and destruction processes), assuming that the rates of cloud formation and destruction remain constant with time. When the cloud formation rate is low,

$$\frac{N_0 - N_1}{t_1 - t_0} = N_0 \left(\frac{1}{\tau_{\rm dest}} - \frac{1}{\tau_{\rm grow}} \right) \approx N_0 \left(\frac{1}{\tau} \right). \tag{4.2}$$

In this equation, $\tau_{\rm dest}$ is the required to reduce the initial cloud population to zero, and $\tau_{\rm grow}$ is the time required to increase the initial population by N_0.

Estimating τ from the observed spatial distribution of GMCs therefore involves measurements of N_1 and N_0 for any two adjacent regions between which there is a well-defined travel time $t_1 - t_0$. Counting the number of clouds at different spatial locations along a common trajectory serves as a proxy for the time evolution of a individual cloud. The method is well-suited to the interarm region of disk galaxies chiefly because: (1) in contrast to the spiral arms where there are significant non-circular motions, interarm clouds follow a roughly circular path and their angular velocity V_ϕ can be approximated by the circular velocity in the disk; and (2) cloud formation mechanisms are less efficient in the interarm region, hence τ is likely to provide a more direct measure of the true cloud destruction timescale $\tau_{\rm dest}$. With sufficient cloud numbers, multiple interarm zones could be considered, yielding several independent measures of τ. In practice, we apply the method to the PAWS observations of M51 by dividing the interarm into two halves (zone I and zone II) spanning the entire interarm region. In this case, counting the cloud numbers in the two zones $N_0 = N_I$ and $N_1 = N_{II}$ is equivalent to the time for an individual cloud to cross half the interarm distance $t_{travel}/2$, where t_{travel} is the full interarm distance divided by the angular velocity V_ϕ. Equation 4.1 can then be rewritten

$$\tau = \frac{t_{travel}}{2} \frac{N_I}{N_I - N_{II}}. \tag{4.3}$$

In M51, we have applied our method to six radial bins covering the interarm zone between galactocentric radii $R = 1.3$ and $R = 3.3$ kpc. Using the rotation curve model derived by Meidt et al. (2013) to estimate the circular velocity, the full interarm travel time t_{travel} across this radial range varies between 25 and 50 Myr. On average, we find that the cloud population decreases by $\sim 80\%$ from zone I to zone II, suggesting that a characteristic cloud lifetime that is shorter than t_{travel}. Inside $R \sim 2.5$ kpc, we find $\tau \sim 20$ to 30 Myr. This lifetime only applies to clouds with masses above the PAWS completeness limit, i.e. $M >$ a few times 10^5 $M_\odot$. The radial variation in the ratio $\frac{N_I}{N_I - N_{II}}$ is modest. This is inconsistent with a constant cloud lifetime since t_{travel} increases with galactocentric radius. The observed trend is qualitatively consistent with shear being the primary agent of cloud destruction in the interarm of M51's inner disk, since shear weakens with galactocentric radius. Quantitatively, our estimated cloud lifetimes are in good agreement with the shear timescale (as measured from the background Oort A) for $R < 2.5$ kpc. At larger radii, however, this good agreement breaks down. Maps of high-mass star formation tracers in M51 indicate that the presence of numerous giant HII regions at $R \sim 2.5$ kpc (e.g. Meidt *et al.* 2013, Schinnerer *et al.* 2013). Cloud destruction in these radial bins may therefore be governed by the interplay of shear and star formation feedback, rather than shear alone.

5. Summary and Outlook

The next decade promises to be a productive moment for connecting the cloud- and galactic-scale views of star formation. The first systematic comparisons of extragalactic GMC properties contained only ~ 100 clouds in total, mostly in the Local Group dwarfs; facilities such as ALMA and NOEMA will resolve thousands of GMCs across a diversity of systems at much larger distances. The results highlighted in this contribution provide an initial look at how galactic environment influences both the overall structure of the molecular ISM in galaxies, and the dynamical state and evolutionary timescale of individual clouds. With a new generation of extragalactic datasets that can be directly compared to cloud-scale models of star formation, we can expect good progress in understanding how galaxies regulate their gas density distribution, and how the physics of star formation and feedback on cloud scales leads to the observed global pattern of star formation in galaxies.

References

Blitz, L. & Rosolowsky, E. 2006, *ApJ* 650, 933
Blitz, L., Fukui, Y., Kawamura, A. *et al.* 2006, in: B. Reipurth, D. Jewitt, & K. Keil (eds.), *Protostars and Planets V* (Tucson: University of Arizona Press), p. 81
Bolatto, A., Leroy, A., Rosolowsky, E. *et al.* 2008, *ApJ*, 686, 948
Colombo, D., Hughes, A., Schinnerer, E. *et al.* 2014, *ApJ* 784, id. 3
Donovan Meyer, J., Koda, J., Momose, R. *et al.* 2013, *ApJ*, 772, id. 107
Druard, C., Braine, J., Schuster, K. *et al.* 2014, *A&A* 567, id. 118
Elmegreen, B. 1989, *ApJ*, 338, 178
Field, G., Blackman, E., & Keto, E. 2011, *MNRAS*, 416, 710
Fukui, Y., Kawamura, A., Minamidani, T. *et al.* 2008, *ApJS*, 178, 56
Gratier, P., Braine, J., Rodriguez-Fernandez, N. *et al.* 2012, *A&A* 542, id. 108
Heyer, M., Carpenter, J., & Snell, R. 2001, *ApJ*, 551, 852
Heyer, M., Krawczyk, C., Duval, J. *et al.* 2009, *ApJ*, 699, 1092
Hirota, A., Kuno, N., Sato, N., Nakanishi, H. *et al.* 2011, *ApJ*, 737, id. 40
Hughes, A., Meidt, S., Colombo, D. *et al.* 2013, *ApJ*, 779, id. 46
Kawamura, A., Mizuno, Y., Minamidani, T. *et al.* 2009, *ApJS*, 184, 1
Koyama, H. & Ostriker, E. 2009, *ApJ*, 693, 1346
Krumholz, M. & McKee, C. 2005, *ApJ*, 630, 250
Larson, R. 1981, *MNRAS*, 194, 809
Leisawitz, D., Bash, F., & Thaddeus, P. 1989, *ApJS*, 70, 731
Meidt, S., Hughes, A., Dobbs, C. *et al.* 2015, *ApJ*, 779, id. 45
Meidt, S., Schinnerer, E., García-Burillo, S. *et al.* 2013, *ApJ*, 779, id. 45
Miura, R., Kohno, K., Tosaki, T. *et al.* 2012, *ApJ*, 761, id. 37
Pety, J., Schinnerer, E., Leroy, A. *et al.* 2013, *ApJ* 779, id. 43
Reid, M.., Wadsley, J., Petitclerc, N. *et al.* 2010, *ApJ*, 719, 561
Rosolowsky, E 2005, *PASP*, 117, 1403
Rosolowsky, E., Keto, E., Matsushita, S. *et al.* 2007, *ApJ* 661, 830
Schinnerer, E., Meidt, S., Pety, J. *et al.* 2013, *ApJ*, 779, id. 42
Solomon, P., Rivolo, A., Barrett, J. *et al.* 1987, *ApJ*, 319, 730
Utomo, D., Blitz, L., Davis, T. *et al.* 2015, *ApJ*, 803, id. 16
Wada, K., Spaans, M., & Kim, S. 2000, *ApJ*, 540, 797
Wong, T., Hughes, A., Ott, J. *et al.* 2011, *ApJS*, 197, id. 16

From Interstellar Clouds to Star-Forming Galaxies: Universal Processes?
Proceedings IAU Symposium No. 315, 2015
P. Jablonka, Ph. André & F. van der Tak, eds.

doi:10.1017/S1743921316007225

Feedback Regulated Turbulence, Magnetic Fields, and Star Formation Rates in Galactic Disks

Chang-Goo Kim and Eve C. Ostriker

Department of Astrophysical Sciences, Princeton University
4 Ivy Lane, Princeton, New Jersey, USA 08544
email: cgkim@astro.princeton.edu, eco@astro.princeton.edu

Abstract. We use three-dimensional magnetohydrodynamic (MHD) simulations to investigate the quasi-equilibrium states of galactic disks regulated by star formation feedback. We incorporate effects from massive-star feedback via time-varying heating rates and supernova (SN) explosions. We find that the disks in our simulations rapidly approach a quasi-steady state that satisfies vertical dynamical equilibrium. The star formation rate (SFR) surface density self-adjusts to provide the total momentum flux (pressure) in the vertical direction that matches the weight of the gas. We quantify feedback efficiency by measuring feedback yields, $\eta_c \equiv P_c/\Sigma_{\rm SFR}$ (in suitable units), for each pressure component. The turbulent and thermal feedback yields are the same for HD and MHD simulations, $\eta_{\rm th} \sim 1$ and $\eta_{\rm turb} \sim 4$, consistent with the theoretical expectations. In MHD simulations, turbulent magnetic fields are rapidly generated by turbulence, and saturate at a level corresponding to $\eta_{\rm mag,t} \sim 1$. The presence of magnetic fields enhances the total feedback yield and therefore reduces the SFR, since the same vertical support can be supplied at a smaller SFR. We suggest further numerical calibrations and observational tests in terms of the feedback yields.

Keywords. galaxies: ISM, galaxies: star formation, galaxies: magnetic fields, turbulence, MHD, methods: numerical

---

1. Introduction

"What determines the SFR in galaxies?" In order to answer this long-standing, fundamental question, a correlation between the SFR and the gas content has been extensively explored. Among many studies since the pioneering work by Schmidt (1959), Kennicutt (1998) presents a well-defined power-law relationship between total gas surface density (Σ) and the SFR surface density ($\Sigma_{\rm SFR}$) for galaxies as a whole, $\Sigma_{\rm SFR} \propto \Sigma^{1+p}$ with $p = 0.4$. This observed correlation was soon widely accepted as the "Kennicutt-Schmidt law" (KS law) and used as a star formation recipe for large scale galaxy formation and cosmological simulations.

The observed power-law index with $p = 0.4$ of the KS law makes it tempting to infer a simple dimensional relationship, $\Sigma_{\rm SFR} = \Sigma/t_{\rm dep}$, with the gas depletion time related to the gas free-fall time, $t_{\rm dep} \propto t_{\rm ff} \sim (G\rho)^{-1/2}$. With a fixed gas scale height, this relation would imply $p = 0.5$, close to the observed value. Many theoretical studies based on this simple argument have been investigated, with low star formation efficiency per free-fall time $\epsilon_{\rm ff} \equiv t_{\rm ff}/t_{\rm dep} \sim 1\%$ (e.g., Krumholz *et al.* 2012). On scales of molecular clouds, the low efficiency has been attributed to the broad density probability distribution function generated by supersonic turbulence (e.g., Padoan *et al.* 2014 and references therein), but it is unclear whether this picture can be simply extended to large scales (>0.1-1 kpc).

Moreover, recent high-resolution observations of nearby galaxies reveal more complex correlations (Bigiel *et al.* 2008; Leroy *et al.* 2008). In particular, the simple power-law relation between $\Sigma_{\rm SFR}$ and Σ fails in the low surface density regime ($\Sigma < 10\ M_\odot\ {\rm pc}^{-2}$), where the gas is predominantly atomic. Rather, the power-law index becomes steeper and/or varies from one galaxy to another. The SFR shows a tighter correlation with the stellar surface density Σ_* or its combination with Σ (e.g., Leroy *et al.* 2008).

The increasing complexity of the observed KS law in the low-Σ regime implies that Σ is not the only control parameter of star formation. In this article, we describe a fundamental correlation based on physical causality between the SFR surface density and the total pressure ($P_{\rm tot}$). In § 2 and § 3, we respectively summarise the theory from Ostriker *et al.* (2010); Ostriker & Shetty (2011); Kim *et al.* (2011) and simulations from Kim *et al.* (2011, 2013); Kim & Ostriker (2015).

2. Theory

The interstellar medium (ISM) disk in an equilibrium state should satisfy vertical force balance between gravity and pressure gradients, which can be directly derived from the momentum equation of MHD.† In the integrated form, the vertical dynamical equilibrium can be written as $\mathcal{W} = \Delta P_{\rm tot}$, a balance between the total weight of gas and the momentum flux differences across the gas disk.

The other condition to satisfy is the energy/momentum equilibrium between gain from star formation feedback and loss in the dissipative ISM. Since cooling and turbulence dissipation time scales are typically short compared to dynamical time scales, continuous injection of energy and momentum is necessary to heat gas and drive turbulence. The far-UV radiation from massive young stars is the major heating source in the atomic ISM via the photoelectric effect onto grains. The momentum injection from SNe is the dominant source of the turbulence driving. Because the energy and momentum from stellar radiation and SNe fundamentally derive from nuclear processes, SF feedback is a highly efficient way to balance losses in the ISM.

Figure 1 shows a schematic diagram for (a) energy/momentum equilibrium and (b) vertical dynamical equilibrium as well as the connection between two equilibria. Since the ISM mostly gains energy/momentum through star formation feedback, the total pressure (momentum flux) is set by the SFR surface density. The total pressure determined by Figure 1(a) provides the vertical support in Figure 1(b), which should match with the dynamical equilibrium pressure $P_{\rm DE}$ (or the weight of the gas $\mathcal{W}$).

The equilibrium is naturally a stable one. For example, if the SFR gets higher than equilibrium, enhanced thermal heating and turbulence driving set the total pressure higher than the equilibrium level. The disk becomes thermally and dynamically hotter, vertically expanding and dispersing cold, dense clumps to quench further star formation. On the other hand, when the SFR drops below of the level of equilibrium, reduced feedback makes the disk thermally and dynamically cold and susceptible to gravitational collapse, forming more stars. Therefore, the SFR is self-regulated to satisfy both equilibria shown in Figure 1.

† Although the resulting equation would be essentially the same with so called *hydrostatic equilibrium* with an effective (total) sound speed, we prefer to term this *vertical dynamical equilibrium* since the ISM is highly dynamic and equilibrium holds only in an average sense.

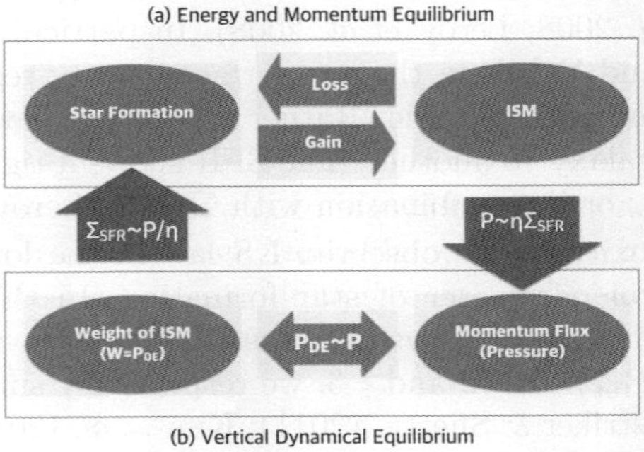

Figure 1. Schematic diagram of the equilibrium theory. (a) Energy/momentum equilibrium sets the total pressure in response to the SFR, $P_{\rm tot} \sim \eta\Sigma_{\rm SFR}$. (b) Vertical dynamical equilibrium constrains the total momentum flux (pressure), $P_{\rm tot} \sim P_{\rm DE}$, and hence the SFR, $\Sigma_{\rm SFR} \sim P_{\rm DE}/\eta$.

In order to quantify this process, we define the *feedback yield* of any pressure component "c" (= thermal, turbulent, or magnetic) in suitable units as

$$\eta_c \equiv \frac{P_{c,3}}{\Sigma_{\rm SFR,-3}}, \tag{2.1}$$

where $P_{c,3} \equiv P_c/(10^3 k_B \ {\rm cm^{-3}\ K})$ and $\Sigma_{\rm SFR,-3} \equiv \Sigma_{\rm SFR}/(10^{-3}\ M_\odot\ {\rm pc^{-2}\ Myr^{-1}})$. The feedback yield can be considered as a energy/momentum conversion efficiency of the star formation feedback, depending on the detailed thermal and dynamical processes in the ISM. To first order, we can simply connect the thermal and turbulent pressures with the SFR surface density linearly. Our adopted cooling and heating formalism gives $\eta_{\rm th} = 1.2$, and the momentum feedback prescription with specific momentum injection per star formation $(p_*/m_*) = 3000$ km/s gives $\eta_{\rm turb} = 3.6$.

3. Numerical Simulations

Utilizing the *Athena* code (Stone *et al.* 2008), we run three-dimensional simulations for the local patch of galactic disks including optically thin cooling, galactic differential rotation, self-gravity, vertical external gravity, and magnetic fields. We apply a spatially-constant, time-varying heating rate $\Gamma \propto \Sigma_{\rm SFR}$, and also momentum injection from SNe $\propto \Sigma_{\rm SFR}$. While in some other recent simulations, the SFR and SN rate is pre-specified and constant in time, for all of our models the time-dependent SFR and SN rate are self-consistently set by self-gravitating localized collapse.

Our simulations achieve a quasi-steady state after a few vertical oscillation times (less than one orbit time). We confirm vertical dynamical equilibrium using horizontally and temporally averaged vertical profiles that are converged for different initial and boundary conditions as well as numerical resolutions. For a wide range of disk conditions such that $0.1 < \Sigma_{\rm SFR,-3} < 10$, the equilibrium thermal and turbulent pressures give consistent feedback yields, $\eta_{\rm th} = 1.3\Sigma_{\rm SFR,-3}^{0.14}$ and $\eta_{\rm turb} = 4.3\Sigma_{\rm SFR,-3}^{0.11}$, respectively.

In MHD simulations, the time scales to reach a quasi-steady state depend on the initial magnetizations. For initial magnetic energy varying by two orders of magnitude, however, saturated states converge to the same asymptote for the turbulent magnetic fields, whose energy is about a half of the turbulent kinetic energy. The final turbulent magnetic fields provide additional vertical support that is directly related to the turbulent pressure and hence the SFR, giving rise to $\eta_{\rm mag,t} \sim 1$ for solar neighborhood models. Since we fix disk parameters for MHD models, further investigation is necessary to calibrate the detailed dependence of $\eta_{\rm mag,t}$ on $\Sigma_{\rm SFR}$.

4. Concluding Remarks

We have developed a theory for self-regulation of the SFR based on energy/momentum equilibrium and vertical dynamical equilibrium, and confirmed and calibrated this theory with numerical simulations. The former equilibrium gives rise to the correlation between $\Sigma_{\rm SFR}$ and $P_{\rm tot}$ owing to the physical causality; the higher/lower SFR causes higher/lower energy density and momentum flux ($P_{\rm tot} = \eta\Sigma_{\rm SFR}$). The latter equilibrium sets $\Sigma_{\rm SFR}$ based on the requirement $P_{\rm tot} = P_{\rm DE}$. Therefore, correlations between $\Sigma_{\rm SFR}$ and galactic properties are caused by dependences embedded in $P_{\rm DE}$ such that

$$P_{\rm DE} \equiv \mathcal{W} \approx \frac{\pi G \Sigma^2}{2} + \Sigma\sigma_z(2G\rho_{\rm sd})^{1/2} \tag{4.1}$$

where $\rho_{\rm sd}$ is the midplane density of stars plus dark matter and σ_z is total vertical velocity dispersion. The observed complexities in the KS law for the low-Σ regime naturally arise when the second term in RHS of Equation (4.1) dominates.

Lastly, we suggest further numerical calibrations and observational tests of the equilibrium theory. Theorists who include any form of star formation feedback can calibrate η_c from a "P_c-$\Sigma_{\rm SFR}$" plot for each measured pressure (turbulent, thermal, magnetic) in their simulations. It could be interesting to check consistency and/or to investigate differences among simulations with different setups, including comparing global *vs.* local models. Additional calibrations of other components such as radiation pressure and cosmic ray pressure would enable comparison of the relative importance of each component.

It is difficult to measure the total pressure directly from observations even in the solar neighborhood. However, the dynamical equilibrium pressure can be determined from direct observables (such as those from Leroy *et al.* 2008) with proper assumptions.† From the "$P_{\rm DE}$-$\Sigma_{\rm SFR}$" observed plot, one can measure total feedback yield $\eta = P_{\rm DE,3}/\Sigma_{\rm SFR,-3}$. This can be compared with the sum of the theoretical values η_c as a test of the self-regulation theory, also constraining the dominant sources of SF feedback.

References

Bigiel, F., Leroy, A. K., Walter, F., Brinks, E., de Blok, W. J. G., Madore, B., & Thornley, M. D. 2008, *AJ*, 136, 2846

Kennicutt, Jr., R. C 1998, *ApJ*, 498, 541

Kim, C.-G., Kim, W.-T., & Ostriker, E. C. 2011, *ApJ*, 743, 25

Kim, C.-G., Ostriker, E. C., & Kim, W.-T. 2013, *ApJ*, 776, 1

Kim, C.-G. & Ostriker, E. C. 2015, *ApJ submitted*

Krumholz, M. R., Dekel, A., & McKee, C. F. 2012, *ApJ*, 745, 69

Leroy, A. K., Walter, F., Brinks, E., Bigiel, F., de Blok, W. J. G., Madore, B., & Thornley, M. D. 2008, *AJ*, 136, 2782

Ostriker, E. C., McKee, C. F., & Leroy, A. K. 2010, *ApJ*, 721, 975

Ostriker, E. C. & Shetty, R. 2011, *ApJ*, 731, 41

Padoan, P., Federrath, C., Chabrier, G., Evans II, N. J., Johnstone, D., Jrgensen, J. K., McKee, C. F., & Nordlund,. 2014, *Protostars and Planets VI*, 77

Schmidt, M. 1959, *ApJ*, 129, 243

Stone, J. M., Gardiner, T. A., Teuben, P., Hawley, J. F., & Simon, J. B. 2008, *ApJS*, 178, 137

† Please refer to the following link for a practical guide: `http://www.astro.princeton.edu/~cgkim/to_observers.html`

From Interstellar Clouds to Star-Forming Galaxies: Universal Processes?
Proceedings IAU Symposium No. 315, 2015
P. Jablonka, Ph. André & F. van der Tak, eds.

doi:10.1017/S1743921316007237

Molecules in Action: Extragalactic ISM at high resolution

Sergio Martín[1,2]

[1]European Southern Observatory,
Alonso de Córdova 3107, Vitacura, Santiago, Chile
email: smartin@eso.org
[2]Joint ALMA Observatory,
Alonso de Córdova 3107, Vitacura, Santiago, Chile

Abstract. During the last two decade, observations have shown the potential of molecular tracers to get insights into the physical processes taking place in the central regions of active galaxies. However, observations were severely limited by both sensitivity and resolution. This resulted also in a limited sample of bright enough galaxies where molecular species other than carbon monoxide could be observed. Current instruments like ALMA and the upcoming NOEMA are already changing our view of the extragalactic ISM molecular observations. In fact, it is now possible to study the physical properties of individual spatially resolved star forming GMCs in external galaxies, as well as resolving the physical structure of the ISM in the surroundings of AGNs at scales of a few parsecs. Here I quickly review some of the most recent observational studies in the nuclear regions of galaxies which are setting new standards in the ways we can study the extragalactic ISM properties.

Keywords. ISM: abundances, ISM: molecules, galaxies: abundances, galaxies: active, galaxies: ISM, galaxies: seyfert, galaxies: starburst

---

1. Introduction: Craving high resolution

Almost a decade ago, the first unbiased spectral line survey was carried out towards the starburst galaxy NGC 253 (Martín *et al.* 2006) with the aim of exploring the limits of the observable chemical complexity in the extragalactic ISM. This work showed that increasing the number of observed species allowed direct comparison with Galactic environments which could provide clues on the overall leading heating mechanism in the central regions of galaxies. Also driven by the technical developments in available instantaneous bandwidth of mm and submm telescopes, a large number of spectral line surveys were carried out towards the brightest nearby active galaxies (see review by Martín 2011). The spectral line survey towards the starburst M 82 clearly showed that when compared to that of NGC 253, even at the large spatial scales sampled by single dish observations, the evident molecular abundance variations could be linked to the evolutionary state of the nuclear starburst (Aladro *et al.* 2011). However such molecular variations were not so evident when trying to establish a comparative chemical composition towards AGN dominated galaxies or towards luminous and ultra luminous galaxies through deep wide band spectral observations (Costagliola *et al.* 2011, 2015; Aladro *et al.* 2013, 2015; Martín *et al.* 2011).

Very deep single dish observations are able to increment our knowledge about the overall molecular composition of the extragalactic ISM. Such is the case of the recent tentative detection of HC_5N towards NGC 253 and NGC 4418 (Aladro *et al.* 2015; Costagliola *et al.* 2015). However, the sampled regions (larger than a few hundred parsecs) severely hinders the possibility of establishing a direct link between the enhancement of a given species and the type of galactic activity and therefore the diagnostic potential of molecular observations.

A number of pioneering interferometric studies resulted in compelling imaging of the extragalactic ISM in molecular lines other than carbon monoxide, the "universal" molecular gas tracer (García-Burillo *et al.* 2000, 2001, 2002, 2010; Usero *et al.* 2007; Meier & Turner 2005, 2012). Though this studies showed the potential of mm and submm spectral observations to dissect the ISM through the enhanced emission selected molecular species, the number of available species was limited by sensitivity. This sensitivity issue biased chemical studies in galaxies towards the brightest objects in the sky.

2. Starburst and AGN molecular observations at high resolution

As a result of the observations carried out during its Cycle 0, the kickoff of ALMA, even at a fraction of its full potential, has shown to be a game changer also in the field of extragalactic astrochemistry. The ALMA observations towards NGC 253 represents an excellent example. Though originally aiming at the mapping of CO and the dense gas tracers HCN and HCO^+ (Leroy *et al.* 2015) the high sensitivity and large available bandwidth resulted in the detection and identification of 24 extra molecular transitions as well as some unidentified features (Meier *et al.* 2015). Though the images were probing scales similar to previous extragalactic works (50–100 pc), the unprecedented sensitivity allowed the imaging of faint transitions such as the rarer isotopologue of carbon monoxide, $C^{17}O$. The multi molecular imaging allowed to sketch a map of the different heating mechanisms in play within the central molecular zone of NGC 253. SiO and HNCO tracing the shocked molecular gas orbiting the nuclear region, with the latter disappearing with increasing PDR contribution towards center as determined from the $CN/C^{17}O$ and $C_2H/H^{13}CN$ line ratios. The dense gas traced by HCN, HCO^+ and CN located within the galactic plane at the base of the molecular outflow observed in CO (Bolatto *et al.* 2013) and the hydrogen recombination lines tracing the star formation right at the very inner regions (Bendo *et al.* 2015).

However, an even more compelling case is that of the direct imaging of the circumnuclear disk around the supermassive black holes in the nuclei of AGN galaxies. Not only it has become possible the multi molecular observation towards the brightest nearby Seyfert 2 luminous AGN, the galaxy NGC 1068 (Takano *et al.* 2014; Nakajima *et al.* 2015) but also fainter objects such as the galaxy NGC 1097 (Izumi *et al.* 2013; Martín *et al.* 2015) which is more than an order of magnitude fainter in X-ray and IR luminosity, as well as in molecular gas. Despite the different luminosities of their AGNs, as shown in Fig. 1 the distribution of species in the CND of both galaxies is remarkably similar, which suggests that the chemistry that we are observing is not directly driven by the X-ray radiation from the AGN. Within the central $\sim$100 pc around the AGN species efficiently formed in the surface of dust grains, such as HNCO and CH_3OH, clearly delineate the circumnuclear disk. Other molecules like the shock tracer SiO or even more complex ones like cyanoacetylene (HC_3N) and acetonitrile (CH_3CN) are observed to be spatially unresolved at the current resolution towards the very nuclear region. The compactness of the emission of these species shows the existence of molecular gas clumps at very close distances from the supermassive black hole which must dense enough to be able to protect these complex organic molecules from the pervasive X-ray dissociating radiation. The presence of SiO at such close distance from the AGN lead to think of its possible enhancement due to X-ray chemistry rather than just tracing shock material (García-Burillo *et al.* 2010). Actually, the achieved resolution is already close or at least comparable to that of existing chemical models of the the ISM chemistry in molecular

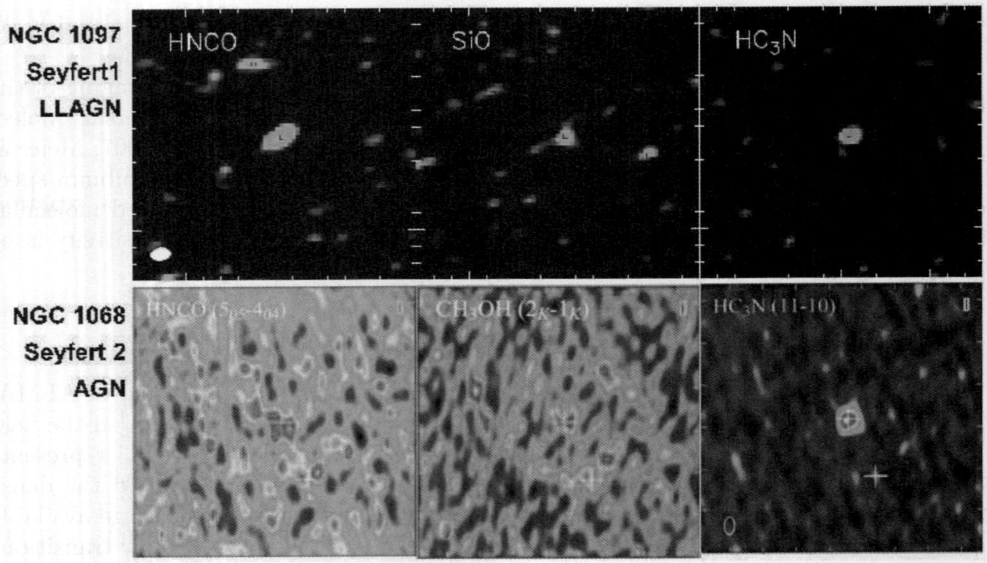

Figure 1. Multi-molecular observations within the inner ~100 pc of the low luminosity Sy 1 NGC 1097, and the bright Sy 2 nucleus in NGC 1068. Molecular observation show that even though there is an order of magnitude difference in luminosity between both nuclei, the ISM structure around the AGN show strong similarities. HNCO and CH_3OH trace well the dense CND while SiO and HC_3N are unresolved and emitting within the very nucleus. Dense gas clumps must exist within the central tens of parsec in order to shield species such as HC_3N and CH_3CN from the dissociation. Composite figure extracted from Martín *et al.* (2015) and Takano *et al.* (2014).

disks surrounding AGNs (Harada *et al.* 2013). Indeed, these models are able to reproduce such large abundances of HC_3N within 10 – 20 pc from the AGN. The comparison of such models with even higher resolution multi-molecule imaging will impose important constraints in the physical conditions of the ISM around the AGN.

3. HCN/HCO$^+$/CS as AGN-vs-SB diagnostic tools

The search for key molecular fingerprints which would allow the creation of AGN-vs-SB diagnostic diagrams has been around for more than a decade now. The pioneering work by Kohno *et al.* (2001) followed by multi-transition excitation studies by Krips *et al.* (2008) showed that the differences observed in the HCN/HCO$^+$ ratio was actually the result of abundance differences and therefore might be a good probe to differentiate between AGN and SB dominated galaxies. This differentiation was supported by chemical modeling which could reproduce the overproduction of HCN in X-ray dominated regions (i.e. Meijerink *et al.* 2007). However the support from models was marginal since such HCN/HCO$^+$ ratio enhancement was only above unity for relatively unattenuated gas in the surroundings of molecular clouds, which may not be representative of the dense gas medium around AGNs (Izumi *et al.* 2013).

Still, the differentiation based on this line ratio seems to be more prominent when using high resolution observations. The diagnostic diagram using the HCN/HCO$^+$ in the high-J 4 – 3 transitions together with the HCN/CS ratio (Izumi *et al.* 2013) shows such ratios in AGN dominated galaxies to be a factor ~ 4 higher than in SB dominated galaxies. The wider sample of observations explored by Izumi *et al.* (2015, submitted) shows clear evidence of the enhancement of these two ratios in the presence of an AGN. However, the high resolution observations of the HCN/HCO^+ towards both NGC 1068 and NGC 1097 (García-Burillo *et al.* 2014; Martín *et al.* 2015) proved that the highest HCN emission was not originated at the central few tens of parsec of the AGN but rather in the surrounding CND.

It has been claimed that the enhancement of HCN in the surroundings of the AGNs is actually related to the production of HCN via the endothermic hydrogenation reaction of CN

$$CN + H_2 \rightarrow HCN + H, \ \gamma = 820 \text{ K}$$

where γ is the reaction barrier. Modeling of the chemistry of molecular clouds in the vicinity of an AGN shows how, irrespective of variations in the X-ray ionization rate by several orders of magnitude, the HCN fractional abundance does not show strong variations. However, increasing the temperature of the gas between 100 and 800 K results in a dramatic HCN abundance enhancement (Harada *et al.* 2010, 2013; Izumi *et al.* 2013). Given that the heating due to the AGN would not be enough in low luminosity AGN such as the one in NGC 1097, hot gas chemistry likely induced by mechanical heating due to AGN jets and/or outflows could be the responsible for the observed abundances in the central regions of AGN dominated galaxies (Loenen *et al.* 2008; Aalto *et al.* 2012; Izumi *et al.* 2013; Martín *et al.* 2015). This mechanical heating could also be linked to the higher abundances of SiO at the very central region around the supermassive black hole.

4. Molecular probes of compact obscured nuclei

Molecular emission in the mm and submm has been claimed to be an excellent probe of the nuclei of dust obscured galaxies as these wavelengths are barely affected by obscuration. However, that might not apply to the central regions of compact obscured galaxies (CON).

CONs could be defined as luminous and ultra-luminous infrared galaxies, radiating most of their energy as thermal dust emission in the IR, with very compact (few tens of parsec) Compton thick nuclei. These objects might be the dominant population among the most luminous extragalactic objects, but still, reaching their inner regions may be an observational challenge even at millimeter wavelengths.

Some examples of this kind of objects would be the LIRG NGC 4418 (i.e. Costagliola *et al.* 2015) or the nearest ULIRG, Arp 220. Though the archetypical Arp 220 has been studied for decades at all possible wavelengths, its two nuclei keep surprising us when going to the highest resolutions. Recent interferometric observations with the SMA and PdBI resulted in the detection of P-cygni profiles in the high density tracers HCO^+ and SiO towards both nuclei (Sakamoto *et al.* 2009; Tunnard *et al.* 2015) which has been interpreted as $\sim$100 km s^{-1} molecular gas outward motions. During the ALMA Cycle 0 of observations, a number of projects have also aimed to peer into the nuclei of Arp 220. Wilson *et al.* (2014) imaged the dust continuum at 690 GHz deriving dust temperatures of 80 and 200 K for the eastern and western nucleus, respectively. The kinematics of the molecular gas around the nuclei was studied via a fit of a simple disk model to the dense gas emission traced by HCN (Scoville *et al.* 2015). However opacity towards the nuclei does play a crucial role in the observed line profiles and needs to be taken into account for any study of Arp 220. Such is the case of the work presented by Rangwala *et al.* (2015) where they modeled the CO $6-5$ kinematics of the eastern nucleus with a highly turbulent dis with high line center opacity and a large temperature gradient to account

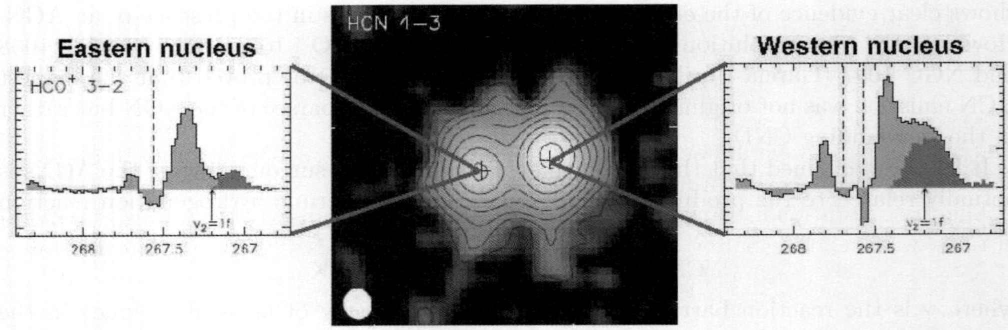

Figure 2. Integrated emission of HCN $J = 4 - 3$ transition towards the two nuclei of Arp 220 at a resolution of $\sim 0.5''$. The HCO^+ $J = 3 - 2$ spectra extracted from both nuclei are shown. Heavy absorption systems account for ~70% of the emission preventing the direct observation of the central few tens of parsecs around the nuclei. Dark grey histograms show the strong contribution from the vibrationally excited ($v_2 = 1$) transitions of HCN to the HCO^+ profiles. This vibrationally excited emission is likely the best probe of the inner hot regions around their nuclear power sources. At the current resolution this source might be either an AGN or a hot compact starburst.

for the central observed absorption. The various absorption system towards the Western nucleus (Martín *et al.* 2015 submitted) makes it more difficult to accurately model its kinematics.

Arp 220 showed an extremely prolific spectrum in the frequency scan carried out with the SMA at 1.3 mm (Martín *et al.* 2011). In these observations most of the emission lines showed a double peak profile which had been classically considered as two molecular components resulting from the distinct kinematics of the two nuclei. However, recent observations have shown that absorption close to the systemic velocity appears to be severely affecting the molecular emission in Arp 220 towards the central few tens of parsec. The recently completed spectral line scan at $0.65''$ resolution with ALMA toward Arp 220, covering frequencies between 210 and 360 GHz, results again in a mind-blowing molecular complexity towards both nuclei with dozens of molecules identified and hundreds of transitions blended in a confusion limited spectrum.

All the standard tracers of molecular gas, namely CO, HCN, HCO^+, HNC and CS, show absorbed profiles in the direction of both Arp 220 nuclei. Either due to self-absorption or continuum absorption, these tracers do not allow us to peer into the central molecular zone around the two nuclei. Fig. 2 shows how approximately 70% of the line is absorbed towards both nuclei. This fact, which is likely extended to most CONs, does prevent an accurate measurement of either the overall molecular gas content through CO or the dense gas content via the classical dense gas tracers. This absorption effect has implications in the conclusions from KennicuttSchmidt laws where the molecular gas content is derived from absorbed profiles. The study by Graciá-Carpio *et al.* (2008) in a sample of LIRGs and ULIRGs resulted in an enhancement of the star formation efficiency (SFE) of the dense gas towards the higher end of luminosities of the sampled galaxies. However, the dense gas content based on HCN $1 - 0$ observations might also affected by absorption, similar to what is observed towards Arp 220. Indeed, some of the higher signal to noise detections in this work showed double peak profiles which would support this absorption scenario. The estimate of the dense gas content in this galaxies would therefore be underestimated, leading to an overestimate of the SFE.

5. The ultimate tracer: Vibrationally excited emission lines

Apart from the difficulty of estimating the amount of both low and high density molecular gas in compact obscured nuclei, it still remains the question on how to actually observe their very inner regions affected by both obscuration and absorption. It appears that the answer might be actually in the vibrational levels of the commonly observed dense gas tracers, as shown in Fig. 2.

The first vibrational transition of HCN was detected in absorption towards Arp 220 in radio wavelengths (Salter *et al.* 2008). Only during the last five years that a number of detections of vibrationally excited HCN and HC_3N have been reported towards the LIRG NGC 4418 (Sakamoto *et al.* 2010; Costagliola *et al.* 2013, 2015), and the ULIRGs Arp 220 (Martín *et al.* 2011) and IRAS 20551-4250 (Imanishi & Nakanishi 2013). Indeed vibrationally excited emission has only been detected towards luminous infrared galaxies in the extragalactic ISM. In the case of HC_3N, the vibrational emission is one of the main polluters in the mm and submillimeter spectrum of NGC 4418 and Arp 220 (Martín *et al.* 2011; Costagliola *et al.* 2015), with groups of transitions every $\sim$9 GHz.

The vibrational transitions of HCN are separated by $\sim$1000 K from the ground level and cannot be collisionally excited at such high observed intensity (Sakamoto *et al.* 2010; Imanishi & Nakanishi 2013). They can however be efficiently excited through infrared pumping provided a bright source of IR emission ($T(14\ \mu\mathrm{m}) > 100\,\mathrm{K}$: Aalto *et al.* 2015) in the nuclei of these galaxies. Thus the detection of vibrationally excited emission requires the presence of a very hot nucleus with a temperature gradient out to the cooler surrounding envelope which would be responsible for the observed absorption in all tracers. This idea is supported by the study of the excitation of the vibrational transitions of HC_3N which yields high excitation temperatures $>$ 300 K in both NGC 4418 and Arp 220.

The largest sample of galaxies with vibrationally excited HCN emission detected, consisting on 8 luminous IR galaxies, was recently compiled by Aalto *et al.* (2015). This study tentatively showed an correlation between the $L_{\mathrm{vib}}/\mathrm{L_{FIR}}$ ratio and the velocity of the observed outflows in the sampled galaxies, with the highest vibrational emission luminosity in sources where no outflow was previously detected. This fact has been claimed as a possible trace of an early stage of evolution of their nuclear power source. Whether this powering source is an AGN or a hot compact starburst will require higher resolution observations.

In the case of Arp 220, evidences are put forth that suggest the presence of an obscured AGN in its Western nucleus (see Downes & Eckart 2007, and references therein). The heating due to the supermassive black hole might be at the origin of the vibrationally excited emission. Higher resolution observations of these transitions will finally reveal whether an AGN is the source of the huge IR luminosity of Arp 220 and whether both nuclei harbor such hidden accreting supermassive black holes.

References

Aalto, S., Garcia-Burillo, S., Muller, S., *et al.* 2012, *A&A*, 537, A44
Aalto, S., Martin, S., Costagliola, F., *et al.* 2015, arXiv1504.06824
Aladro, R., Martín, S., Martín-Pintado, J., *et al.* 2011, *A&A*, 535, A84
Aladro, R., Martín, S., Riquelme, D., *et al.* 2015, arXiv1504.03743
Aladro, R., Viti, S., Bayet, E., *et al.* 2013, *A&A*, 549, A39
Bendo, G. J., Beswick, R. J., D'Cruze, M. J., *et al.* 2015, *MNRAS*, 450, L80
Bolatto, A. D., Warren, S. R., Leroy, A. K., *et al.* 2013, *Nature*, 499, 450
Costagliola, F., Aalto, S., Rodriguez, M. I., *et al.* 2011, *A&A*, 528, A30

Costagliola, F., Aalto, S., Sakamoto, K., *et al.* 2013, *A&A*, 556, A66
Costagliola, F., Sakamoto, K., Muller, S., *et al.* 2015, arXiv1506.09027
Downes, D. & Eckart, A. 2007, *A&A*, 468, L57
García-Burillo, S., Combes, F., Usero, A., *et al.* 2014, *A&A*, 567, A125
García-Burillo, S., Martín-Pintado, J., Fuente, A., & Neri, R. 2000, *A&A*, 355, 499
García-Burillo, S., Martín-Pintado, J., Fuente, A., & Neri, R. 2001, *ApJ*, 563, L27
García-Burillo, S., Martín-Pintado, J., Fuente, A., Usero, A., & Neri, R. 2002, *ApJ*, 575, L55
García-Burillo, S., Usero, A., Fuente, A., *et al.* 2010, *A&A*, 519, A2
Graciá-Carpio, J., García-Burillo, S., Planesas, P., Fuente, A., & Usero, A. 2008, *A&A*, 479, 703
Harada, N., Herbst, E., & Wakelam, V. 2010, *ApJ*, 721, 1570
Harada, N., Thompson, T. A., & Herbst, E. 2013, *ApJ*, 765, 108
Imanishi, M. & Nakanishi, K. 2013, *AJ*, 146, 91
Izumi, T., Kohno, K., Martín, S., *et al.* 2013, *PASJ*, 65, 100
Kohno, K., Matsushita, S., Vila-Vilaró, B., *et al.* 2001, in J. H. Knapen, J. E. Beckman, I. Shlosman, & T. J. Mahoney (eds.), *The Central Kiloparsec of Starbursts and AGN: The La Palma Connection*, *ASP-CS*, No. 249, p. 672
Krips, M., Neri, R., García-Burillo, S., *et al.* 2008, *ApJ*, 677, 262
Leroy, A. K., Bolatto, A. D., Ostriker, E. C., *et al.* 2015, *ApJ*, 801, 25
Loenen, A. F., Spaans, M., Baan, W. A., & Meijerink, R. 2008, *A&A*, 488, L5
Martín, S. 2011, in J. Cernicharo & R. Bachiller (eds.), *The Molecular Universe*, Proc. IAU Symposium No. 280, p. 351
Martín, S., Kohno, K., Izumi, T., *et al.* 2015, *A&A*, 573, A116
Martín, S., Krips, M., Martín-Pintado, J., *et al.* 2011, *A&A*, 527, A36
Martín, S., Mauersberger, R., Martín-Pintado, J., Henkel, C., & García-Burillo, S. 2006, *ApJS*, 164, 450
Meier, D. S. & Turner, J. L. 2005, *ApJ*, 618, 259
Meier, D. S. & Turner, J. L. 2012, *ApJ*, 755, 104
Meier, D. S., Walter, F., Bolatto, A. D., *et al.* 2015, *ApJ*, 801, 63
Meijerink, R., Spaans, M., & Israel, F. P. 2007, *A&A*, 461, 793
Nakajima, T., Takano, S., Kohno, K., *et al.* 2015, *PASJ*, 67, 8
Rangwala, N., Maloney, P., Wilson, C., *et al.* 2015, *ApJ*, 806, 17
Sakamoto, K., Aalto, S., Evans, A. S., Wiedner, M. C., & Wilner, D. J. 2010, *ApJ*, 725, L228
Sakamoto, K., Aalto, S., Wilner, D. J., *et al.* 2009, *ApJ*, 700, L104
Salter, C. J., Ghosh, T., Catinella, B., *et al.* 2008, *AJ*, 136, 389
Scoville, N., Sheth, K., Walter, F., *et al.* 2015, *ApJ*, 800, 70
Takano, S., Nakajima, T., Kohno, K., *et al.* 2014, *PASJ*, 66, 75
Tunnard, R., Greve, T. R., Garcia-Burillo, S., *et al.* 2015, *ApJ*, 800, 25
Usero, A., García-Burillo, S., Martín-Pintado, J., Fuente, A., & Neri, R. 2007, *New Astron. Revs*, 51, 75
Wilson, C. D., Rangwala, N., Glenn, J., *et al.* 2014, *ApJ*, 789, L36

From Interstellar Clouds to Star-Forming Galaxies: Universal Processes?
Proceedings IAU Symposium No. 315, 2015
P. Jablonka, Ph. André & F. van der Tak, eds.

doi:10.1017/S1743921316007249

Probing star formation relations of mergers and normal galaxies across the CO ladder

Thomas R. Greve

Department of Physics and Astronomy, University College London,
Gower Street, London WC1E 6BT, United Kingdom
email: t.greve@ucl.ac.uk

Abstract. We examine integrated luminosity relations between the IR continuum and the CO rotational ladder observed for local (ultra) luminous infra-red galaxies ((U)LIRGs, $L_{\rm IR} \geqslant 10^{11}\,M_\odot$) and normal star forming galaxies in the context of radiation pressure regulated star formation proposed by Andrews & Thompson (2011). This can account for the normalization and linear slopes of the luminosity relations ($\log L_{\rm IR} = \alpha \log L'_{\rm CO} + \beta$) of both low- and high-$J$ CO lines observed for normal galaxies. Super-linear slopes occur for galaxy samples with significantly different dense gas fractions. Local (U)LIRGs are observed to have sub-linear high-J ($J_{\rm up} > 6$) slopes or, equivalently, increasing $L_{\rm CO_{high-J}}/L_{\rm IR}$ with $L_{\rm IR}$. In the extreme ISM conditions of local (U)LIRGs, the high-J CO lines no longer trace individual hot spots of star formation (which gave rise to the linear slopes for normal galaxies) but a more widespread warm and dense gas phase mechanically heated by powerful supernovae-driven turbulence and shocks.

Keywords. ISM: lines and bands, ISM: molecules, galaxies: ISM, galaxies: starburst

1. Introduction

Simple luminosity relations between the IR emission ($L_{\rm IR}$) and that of molecular transitions ($L'_{\rm mol}$ – typically low-J rotational lines of CO or of dense gas tracers like HCN, HCO^+, and CS) have seen widespread use as proxies for more fundamental relations between gas density and star formation rate, i.e., so-called Kennicutt-Schmidt 'laws' or star formation relations. However, prior to the launch of the *Herschel* Space Observatory there had been no systematic extragalactic survey of high-J CO lines (i.e., $J_{\rm up} > 7$ and above). For these lines to be strongly excited requires high densities ($n_{\rm crit} \sim 10^4 - 7\times10^5\,{\rm cm^{-3}}$) *and* (in most circumstances) high temperatures ($E_J/k_{\rm B} \sim 55 - 500\,{\rm K}$) – i.e., a hot and dense gas phase that would leave no easily discernible signature on the low-/mid-J lines (and line ratios) of CO and dense gas tracers (e.g., HCN and CS).

The first directly measured $L_{\rm IR} - L'_{\rm CO}$ correlations for $J_{\rm up} > 7$ CO transitions were presented by Greve *et al.* (2014) (hereafter G14), and based on SPIRE-FTS spectra (CO $J = 4-3$ to $J = 13-12$) obtained for 29 local ($z < 0.1$) (U)LIRGs as part of the *Herschel* Comprehensive (U)LIRG Emission Survey (HerCULES, see van der Werf *et al.* (2010) and Rosenberg *et al.* (2015)). G14 also included ground-based $J = 1-0$, $2-1$, and $3-2$ CO line data for a sample of 45 local (U)LIRGs (from Papadopoulos et al. (2012)). Fitting log-linear expression of the form $\log L_{\rm IR} = \alpha \log L'_{\rm CO} + \beta$ to their (U)LIRG sample, G14 found: 1) linear slopes ($\alpha \simeq 1$) for $J = 1-0$ to $6-5$, but increasingly sub-linear ($\alpha < 1$) for higher J-levels; 2) roughly constant normalizations ($\beta \simeq 2$) up to $J = 6-5$, but then increasing with higher J-levels. For $J_{\rm up} \leqslant 6$ the linear slopes are in agreement with the majority of previous studies. Sub-linear CO $J = 7-6$ slopes was also found by Bayet *et al.* (2009), who also predicted increasingly sub-linear slopes at higher J-lines using model extrapolations. More recently Liu *et al.* (2015) and Kamenetzky *et al.*

(2015) (hereafter L15 and K15, respectively) have delineated the $L_{\rm IR} - L'_{\rm CO}$ relations up to $J = 13 - 12$ for much larger samples of local galaxies than that of G14 (spanning a range of $10^7 - 10^{13}\,L_\odot$ in $L_{\rm IR}$). Both studies find increasingly sub-linear α for $J_{\rm up} > 6$, *when fitting only to the (U)LIRG population*, in broad agreement with G14 (Fig. 1a). The three studies also find similar β-values for $J_{\rm up} > 6$ (Fig. 1b). However, L15 and K15 find *linear* slopes for all transitions up to $J = 13 - 12$ (Fig. 1b) when fitting to their full samples, which are dominated by normal star forming galaxies. Finally, most surveys of HCN, HCO^+, and CS towards nearby star forming galaxies ($L_{\rm IR} \sim 10^9 - 10^{12}\,L_\odot$) have established linear slopes for the low-/mid-J transitions (Fig. 1a and b) (e.g., Gao & Solomon 2004; Zhang *et al.* 2014; cf. García-Burillo *et al.* 2012).

2. α and β for low-J CO and HCN/CS/HCO^+ lines

Stars form in dense, highly dust-obscured regions, and the radiation pressure exerted by the strong absorption and scattering of UV light by dust grains is likely to be an important SF-regulating feedback mechanism. Andrews & Thompson (2011) derived the $L_{\rm IR} - L'_{{\rm CO}_{\rm low-J}}$ relations in the case of Eddington-limited SF and found the maximal possible luminosity is given by $L_{\rm Edd} = 4\pi G c \kappa^{-1} X_{\rm CO} L'_{\rm CO}$, where κ is the Rosseland-mean opacity, and $X_{\rm CO}$ is the $L'_{\rm CO}$-to-$M_{\rm H_2}$ conversion factors. This not only accounts for the linear slopes observed for the $L_{\rm IR} - L'_{{\rm CO}_{\rm low-J}}$ relations, but also constrains the overall normalization. Adopting $\kappa = 5 - 30\,{\rm cm^2\,g^{-1}}$ and $X_{\rm CO} \simeq 0.8\,M_\odot\,({\rm K\,km\,s^{-1}\,pc^2})^{-1}$, which are plausible (albeit poorly constrained) values for (U)LIRGs, $\beta_{\rm Edd} = \log(4\pi G c \kappa^{-1} X_{\rm CO}) = 2.5 - 3.3$ (Fig. 1c). For normal star forming galaxies, where $X_{\rm CO} \simeq 4.4\,M_\odot\,({\rm K\,km\,s^{-1}\,pc^2})^{-1}$, $\beta_{\rm Edd} \simeq 3.3 - 4.1$ (Fig. 1d).

As expected, $\beta_{\rm Edd}$ sets an upper limit on the observed β-values, and (U)LIRGs – having a larger fraction of the ISM being dense and actively forming stars – are significantly closer to this maximal limit than normal star forming galaxies.

This 'intermittency' (Andrews & Thompson 2011), i.e., the fraction of the ISM actively forming stars (effectively the dense gas fraction, $f_{\rm dense}$), sets the $L_{\rm IR} - L'_{{\rm CO}_{\rm low-J}}$ normalization for a given galaxy population. By the same token, two galaxy samples with significantly different dense gas fractions ($f_{\rm dense,1}$ and $f_{\rm dense,2}$, say) will have $L_{\rm IR} - L'_{\rm CO}$ relations offset by $\Delta\beta \sim \log(f_{\rm dense,2}/f_{\rm dense,1})$. Thus, an increasing $f_{\rm dense}(L_{\rm IR})$ function (or, equivalently, $\beta(L_{\rm IR})$) can explain the super-linear ($\alpha \simeq 1.1 - 1.3$) $L_{\rm IR} - L'_{{\rm CO}_{\rm low-J}}$ relations derived by some studies of 'mixed' galaxy samples (Fig. 1b).

For HCN and CS, $\beta_{\rm Edd} \sim 3.1 - 5$ and $\sim 3.6 - 5$, respectively, assuming $X_{\rm HCN} = 3 - 35\,M_\odot\,({\rm K\,km\,s^{-1}\,pc^2})^{-1}$ and $X_{\rm CS} = 10 - 40\,M_\odot\,({\rm K\,km\,s^{-1}\,pc^2})^{-1}$. Being superior tracers of the dense, actively star forming gas, the observed β-values for HCN and CS are much closer to (but still enveloped by) the Eddington limit than was the case for (low-/mid-J) CO (Fig. 1d). The issue of 'intermittency' is thereby also all but removed for these tracers, resulting in their approximately linear $L_{\rm IR} - L'_{\rm mol}$ relations across a vast $L_{\rm IR}$ range (e.g., Wu *et al.* 2010).

3. α for high-J CO lines: linear or sub-linear?

Three independent studies (G14; L15; K15) have shown that local (U)LIRGs exhibit sub-linear high-J $L_{\rm IR} - L'_{\rm CO}$ relations, and two of those (L15 and K15) further showed that the slopes are linear for samples dominated by normal star forming galaxies. In the latter case, the high-J CO lines are tracing SF 'hot spots' of warm dense gas being heated either 'calorimetrically' by UV-photons from nearby OB-associations, and/or mechanically by

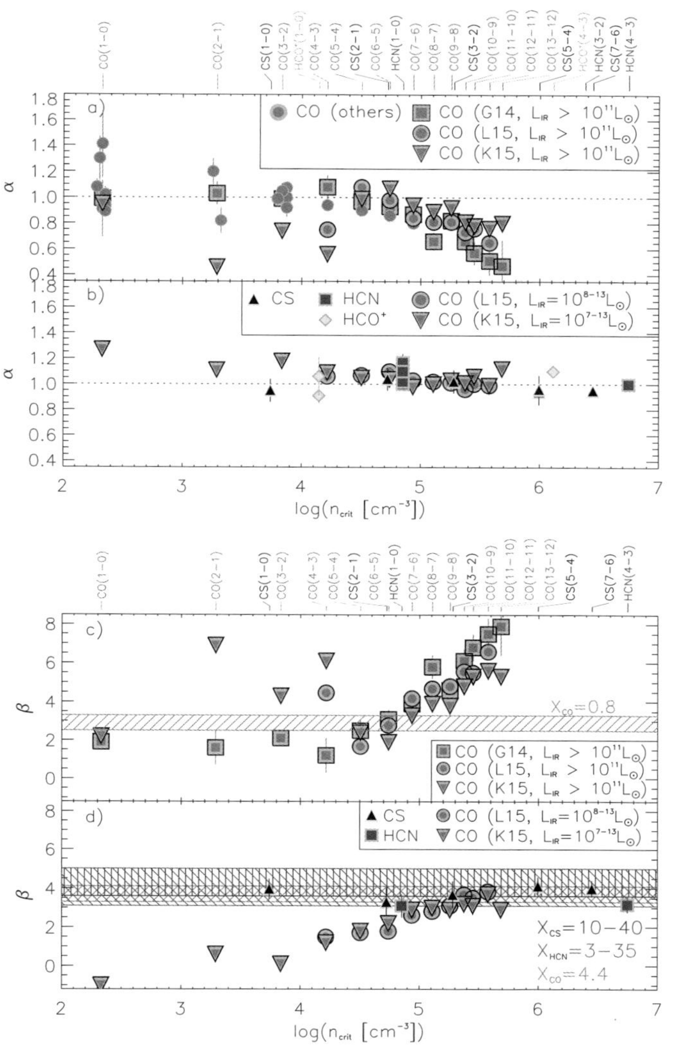

Figure 1. Observed α (top) and β (bottom) vs. critical density for the CO ladder. Panels a)+c) show (α,β)-values inferred from (U)LIRG samples only (Bayet *et al.* 2009; G14; L15; K15), and b)+d) are mainly for normal star forming galaxies (L15; K15). Also shown are (α,β)-values for transitions of the dense gas tracer molecules HCN, CS, and HCO^+. The shaded horizontal regions indicate $\beta_{\rm Edd}$ ($= \log(4\pi G c \kappa X_{\rm mol})$) for CO, HCN, and CS expected for Eddington-limited SF, assuming optically thick FIR opacities in the range $\kappa = 5-30\,{\rm cm^2\,g^{-1}}$ and conversion factors ($X_{\rm mol}$, in units of $M_\odot$ $({\rm K\,km\,s^{-1}\,pc^2})^{-1}$) as indicated in panels c)+d).

SN-driven shocks and outflows from young stellar objects. Just as the HCN/CS lines have β-values close to the Eddington limit, so do the high-J CO lines (Fig. 1d), which may reflect this more direct link with the SF-sites than the lower-J CO lines.

Clearly, local (U)LIRGs (having $\alpha_{\rm CO_{high-J}} < 1$ and $\beta_{\rm CO_{high-J}} > \beta_{\rm Edd}$) do not adhere to this picture, which is not surprising given their extreme ISM conditions. In fact, high-J CO lines are observed to be more strongly excited in local (U)LIRGs than in normal galaxies (e.g., Papadopoulos *et al.* 2014; G14) and, on average, $L_{\rm CO_{high-J}}/L_{\rm IR}$ increases with $L_{\rm IR}$ for LIRG-like luminosities and above (K15). The latter is compatible only with

$\alpha < 1$, since a linear α would imply $L_{\rm CO}/L_{\rm IR} \propto L_{\rm IR}^{1/\alpha-1} \times 10^{-\beta/\alpha} = 10^{-\beta} = constant$. In a similar vein, G14 noted that $\alpha_{{\rm CO}_{J,J-1}}$ can be expressed as:

$$\alpha_{{\rm CO}_{J,J-1}} = \alpha_{{\rm HCN}_{1,0}} \left(1 + \frac{d \log l_{{\rm dense}_{J,J-1}}}{d \log L'_{{\rm CO}_{J,J-1}}}\right) \simeq 1 + \frac{d \log l_{{\rm dense}_{J,J-1}}}{d \log L'_{{\rm CO}_{J,J-1}}}, \quad (3.1)$$

where $l_{{\rm dense}_{J,J-1}} = L'_{{\rm HCN}_{1,0}}/L'_{{\rm CO}_{J,J-1}} \sim f_{\rm dense} r^{-1}_{J,J-1}$, parametrizes deviations in $\alpha_{{\rm CO}_{J,J-1}}$ from unity, and depends on both the dense gas content ($f_{\rm dense} \sim L'_{{\rm HCN}_{1,0}}/L'_{{\rm CO}_{1,0}}$) and the global CO line excitation ($r_{J,J-1} = L'_{{\rm CO}_{J,J-1}}/L'_{{\rm CO}_{1,0}}$). The sub-linear slopes for higher J lines observed in (U)LIRGs is due to an increase in the excitation of these lines – and thereby in the warm and dense gas fraction – with increasing high-J CO luminosity. The presence of a significant warm and dense molecular gas component has been suggested as a general feature of the ISM in extreme merger/starbursts such as local (U)LIRGs by Papadopoulos *et al.* (2012), who argued that high CR energy densities and/or the dissipation of shocks due to strong SN-driven supersonic turbulence can volumetrically heat and maintain significant amounts of high-density gas at temperatures $\gtrsim 100$ K more efficiently than UV radiation, and without being attenuated by dust or readily dissociating CO as UV radiation does.

We end on a few cautionary notes regarding the $L_{\rm IR} - L'_{\rm mol}$ relations. They: 1) are only of use in a statistical sense, and individual sources may show significant departures; 2) do not necessarily apply to high-z dusty star forming galaxies, which are a heterogeneous population and in general *not* scaled-up versions of local ULIRGs; 3) are galaxy-integrated relations. Resolved HCN observations of nearby galaxies have shown significant scatter in the IR/HCN luminosity ratio (Kepley *et al.* 2014); 4) may not have a straightforward interpretation in the case of local (U)LIRGs, where an obscured hot mid-IR core may cause strong self- and continuum-absorption of HCN and CO lines (Aalto *et al.* 2015).

Acknowledgements. I am grateful to the IAU GA commitee for allocating a contributed talk, and to the DeMoGas and HerCULES teams, in particular to Ioanna Leonidaki, Manolis Xilouris, Zhiyu Zhang, Axel Weiß, Paul van der Werf, Padelis Papadopoulos, and Richard Tunnard for their valuable contribution to this work. DeMoGas is implemented under the "ARISTEIA" Action of the "Operational Programme Education and Lifelong Learning", and is co-funded by the European Social Fund and National Resources.

References

Aalto, S., Costagliola, F., González-Alfonso, E., *et al.* 2015 *A&A*, in press
Andrews, B. H. & Thompson, T. A. 2011 *ApJ*, 727, 97
Bayet, E., Gerin, M., Phillips, T. G., *et al.* 2009 *MNRAS*, 399, 264
Gao, Y. & Solomon, P. M. 2004, *ApJ*, 152, 63
García-Burillo, S., Usero, A., Alonso-Herrero, A., *et al.* 2012, *A&A*, 539, A8
Greve, T. R., Leonidaki, I., Xilouris, E. M., *et al.* 2014, *ApJ*, 794, 142
Kamenetzky, J., Rangwala, N., Glenn, J., *et al.* 2015, *ApJ*, submitted
Kepley, A. A., Leroy, A. K., Frayer, D., *et al.* 2014, *ApJ*, 780, 13
Liu D., *et al.* 2015, *ApJ*, 780, 13
Papadopoulos, P. P., van der Werf, P. P., Isaak, K., & Xilouris, E. M. 2012, *ApJ*, 715, 775
Papadopoulos, P. P., Zhang, Z.-Y., Xilouris, E. M., *et al.* 2014 *ApJ*, 788, 153
Rosenberg, M. J. F., van der Werf, P. P., Aalto, S., *et al.* 2015, *ApJ*, 801, 72
van der Werf, P. P., Isaak, K. G., Meijerink, R., *et al.* 2010, *A&A*, 518, L42
Wu, J., Evans, II, N. J., Shirley, Y. L., & Knez, C. 2010 *ApJS*, 188, 313
Zhang, Z.-Y., Gao, Y., Henkel, C., *et al.* 2014 *ApJ*, 784, L31

From Interstellar Clouds to Star-Forming Galaxies: Universal Processes?
Proceedings IAU Symposium No. 315, 2015
P. Jablonka, Ph. André & F. van der Tak, eds.

doi:10.1017/S1743921316007250

Properties of interstellar filaments as derived from *Herschel*, *Planck*, and molecular line observations

Doris Arzoumanian[1], Philippe André[2] and François Boulanger[1]

[1]Institut d'Astrophysique Spatiale (IAS), CNRS (UMR 8617), Université Paris-Sud 11, Bâtiment 121, 91400 Orsay, France

[2]Laboratoire AIM, CEA/DSM–CNRS–Université Paris Diderot, IRFU/Service d'Astrophysique, C.E.A. Saclay, Orme des Merisiers, 91191 Gif-sur-Yvette, France

Abstract. Recent *Herschel* and *Planck* observations of submillimeter dust emission revealed the omnipresence of filamentary structures in the interstellar medium (ISM). The ubiquity of filaments in quiescent clouds as well as in star-forming regions indicates that the formation of filamentary structures is a natural product of the physics at play in the magnetized turbulent cold ISM. An analysis of more than 270 filaments observed with *Herschel* in 8 regions of the Gould Belt, shows that interstellar filaments are characterized by a narrow distribution of central width sharply peaked at $\sim$0.1 pc, while they span a wide column density range. Molecular line observations of a sample of these filaments show evidence of an increase in the velocity dispersion of dense filaments with column density, suggesting an evolution in mass per unit length due to accretion of surrounding material onto these star-forming filaments. The analyses of *Planck* dust polarization observations show that both the mean magnetic field and its fluctuations along the filaments are different from those of their surrounding clouds. This points to a coupling between the matter and the $\vec{B}$-field in the filament formation process. These observational results, derived from dust and gas tracers in total and polarized intensity, set strong constraints on our understanding of the formation and evolution of filaments in the ISM. They provide important clues on the initial conditions of the star formation process along interstellar filaments.

Keywords. ISM, star formation, interstellar filament formation and evolution, magnetic fields

---

1. Omnipresence of filamentary structures in the interstellar medium

While molecular clouds were already known to exhibit filamentary structures (e.g., Schneider & Elmegreen 1979; Abergel *et al.* 1994; Goldsmith *et al.* 2008), the omnipresence of filaments in the interstellar medium (ISM) and molecular clouds has only recently been revealed thanks to the high resolution and the high dynamic range of *Herschel* observations of the submillimeter (submm) dust emission (e.g., André *et al.* 2010; Men'shchikov *et al.* 2010; Molinari *et al.* 2010). Furthermore, the all-sky maps of dust submm emission observed by *Planck* in total intensity, as well as in polarized intensity, emphasize the large-scale, hierarchical filamentary texture of the Galactic ISM (Planck Collaboration Int. XIX 2015; Planck Collaboration Int. XXXII 2014). These *Herschel* and *Planck* data show that interstellar matter is organized in *web-like networks of filaments*, which appear to be formed as a natural result of the physics at play in the ISM. Filamentary structures are also easily produced by any numerical simulation of molecular cloud formation and evolution that includes hydrodynamic or magnetohydrodynamic turbulence (e.g., Federrath *et al.* 2010; Hennebelle *et al.* 2008; Mac Low & Klessen 2004). The presence of interstellar filaments in both star-forming regions as well as in non-star-forming quiescent clouds, alludes to a formation process of filaments preceding any star-forming activity (André *et al.* 2010). The spatial distribution of prestellar cores and protostars extracted

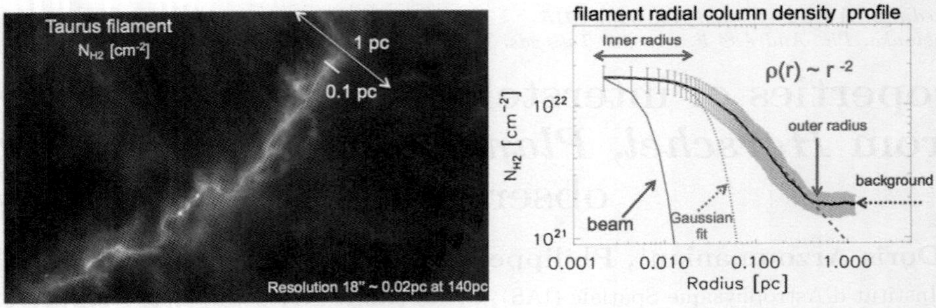

Figure 1. Example of a star-forming filament in the Taurus molecular cloud. Left: Column density map (Palmeirim *et al.* 2013) derived from *Herschel* images observed as part of the *Herschel* Gould Belt survey (André *et al.* 2010). The crest of the filament, over-plotted in white, is traced using the `DisPerSE` algorithm (Sousbie 2011; Arzoumanian *et al.* 2011). Right: Radial column density profile averaged along the filament length. The dispersion of the radial profiles along the length of the filament is shown in yellow. The profile is well fitted with a *Plummer-like* function (red dashed curve) where the density decreases as r^{-2} at large radii (Arzoumanian *et al.* 2011). A Gaussian fit to the inner portion of the average column density profile yields a deconvolved FWHM width of 0.09 ± 0.02 pc (Palmeirim *et al.* 2013).

from *Herschel* images observed mainly on the densest filaments (Könyves *et al.* 2015) indicates that the properties of interstellar filaments may be a key element defining the initial conditions required for the onset of star formation (André *et al.* 2014).

Hence, characterizing the observed filament properties in detail, combining tracers of gas and dust in total and polarized intensities is essential to make progress in our understanding of the physical processes involved in the formation and evolution of interstellar filaments and their role in the star formation process. *Herschel* continuum observations are essential to describe the filament (column) density distribution. These data are complemented with ground based molecular line observations to access to the kinematics of the filamentary structures, while *Planck* dust polarization observations give unprecedented information on the structure of the magnetic field and its connection with interstellar matter.

In the following, I present the main results on the properties of interstellar filaments derived from *Herschel*, *Planck*, and molecular line observations. I discuss these results in the context of a new paradigm of star formation which is closely linked to the formation and fragmentation of self-gravitating filaments.

2. Filament properties as derived from *Herschel* observations of nearby clouds

Statistical analysis of nearby interstellar filaments has been possible thanks to the *Herschel* Gould Belt survey observations (André *et al.* 2010), which are ideal to characterize the filament properties, providing the resolution, the sensitivity, and the statistics (observations of a large number of clouds with different star formation activities and environments) needed for such studies.

Column density and temperature maps are derived from modified blackbody fitting of the *Herschel* five-wavelength images [from 70 to 500 μm] of the optically thin thermal dust emission, covering the peak of cold dust spectral energy distribution (see Fig. 1 and, e.g., Palmeirim *et al.* 2013). These maps are used to derive radial profiles perpendicular to the filament long axes to quantify their column density and temperature structures. The mean radial profile is derived by averaging cuts perpendicular to the crest at each pixel position along the filament (see Fig. 1). The column density profile $N_{H_2}(r)$ is well

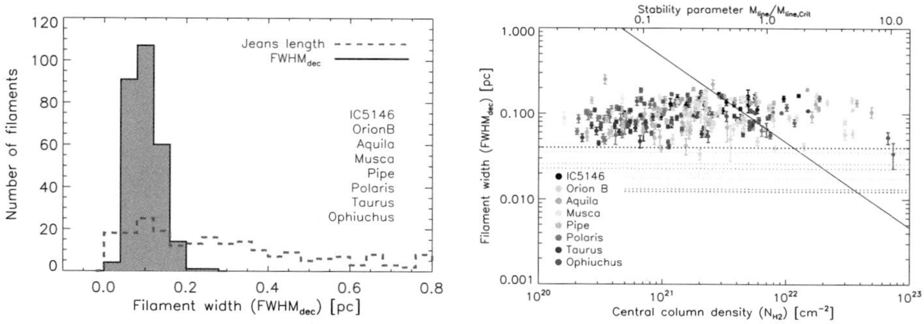

Figure 2. Left: Distribution of deconvolved FWHM widths for the 278 filaments (black solid histogram, filled in orange), with a median value of 0.09 ± 0.04 pc. For comparison, the blue dashed histogram represents the distribution of central (thermal) Jeans lengths of the filaments. Right: Mean deconvolved width versus background subtracted central column density for the same filament sample. The spatial resolutions of the column density maps used in the analyses are marked by the horizontal dotted lines. The solid line running from top left to bottom right shows the central Jeans length as a function of central column density. The upper x-axis scale is an estimate of the filament mass per unit length in units of the thermal critical value $M_{\rm line,crit} = 2c_s^2/G$, where $M_{\rm line} \propto W N^0_{\rm H_2}$ with $W = 0.1$ pc (Arzoumanian *et al.* 2011).

fitted with a *Plummer-like function* of the form: $N_p(r) = N^0_{\rm H_2}/[1 + (r/R_{\rm flat})^2]^{\frac{p-1}{2}}$, where r is the radial distance from the filament crest, $N^0_{\rm H_2}$ is the column density at $r = 0$, $R_{\rm flat}$ is the radius of the flat inner part, and p is the index of the corresponding density profile, which at large radii ($r >> R_{\rm flat}$) approaches a power law: $\rho_p(r) \sim r^{-p}$.

As an example, the observed profile of the Taurus B211/13 filament is shown in Fig. 1. It is characterized by a flat central part with $R_{\rm flat} = 0.03 \pm 0.01$ pc and a density decreasing as r^{-2} at large radii (Palmeirim *et al.* 2013). The filament density profile is thus shallower than the Ostriker (1964) $\rho_p(r) \sim r^{-4}$ model of an isothermal filament in hydrostatic equilibrium. This observed shallower slope is also compatible with the dust temperature profile derived from *Herschel* observations showing a temperature drop towards the central part of the filament well described with a polytropic equation of state (Palmeirim *et al.* 2013). The mass per unit length ($M_{\rm line}$) of the filaments, estimated from their column density profiles, is a very important parameter, which defines "the stability" of a filament: a filament is subcritical, unbound when its mass per unit length is smaller than the critical value $M_{\rm line,crit} = 2\,c_s^2/G \sim 16\,M_\odot$/pc (Ostriker 1964), where $c_s \sim 0.2$ km/s is the isothermal sound speed for $T \sim 10$ K, and G is the gravitational constant. A filament is supercritical, unstable for radial collapse and fragmentation, when $M_{\rm line} > M_{\rm line,crit}$ (Inutsuka & Miyama 1997).

The statistical analysis of a sample of 278 filaments observed by the *Herschel* Gould Belt survey in 8 regions (located at distances from 140 to 460 pc, the different regions are listed in Fig. 2) shows that the filaments span a large range in central column density and mass per unit length (Fig. 2-right), while they have a narrow distribution of FWHM widths centred around a typical value of 0.09 ± 0.04 pc (see Fig. 2–left, Arzoumanian *et al.* 2011; André *et al.* 2014). This characteristic filament width is well resolved by *Herschel* observations of the Gould Belt clouds as can be seen on the radial profile shown in Fig. 1 and Fig. 2-right, where all the measurements of the filament widths lie above the horizontal lines corresponding to the resolution limits. This typical filament width of 0.1 pc is also in contrast with the much broader distribution of central Jeans lengths, $\lambda_J \propto c_s^2/(G N^0_{\rm H_2})$, from 0.02 to 1.3 pc (for $T = 10$ K), implying that these filaments are not in hydrostatic equilibrium.

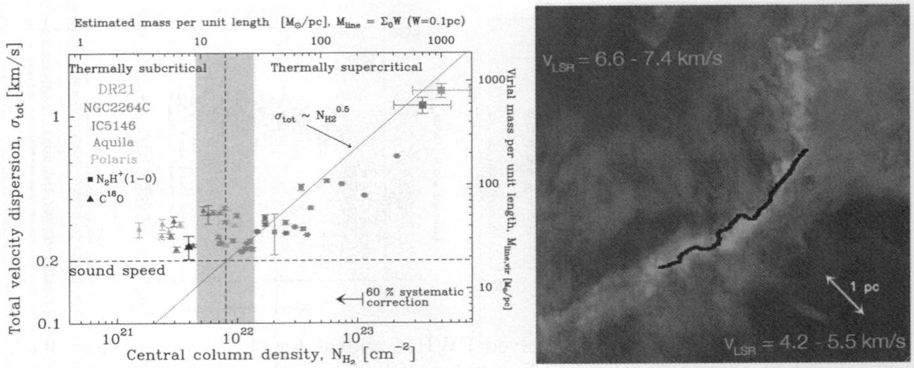

Figure 3. Left: Filament total velocity dispersion versus observed central column density. The horizontal dashed line shows the value of the sound speed (~0.2 km/s for T = 10 K). The vertical dashed line marks the boundary between thermally subcritical and thermally supercritical filaments where the estimated mass per unit length $M_{\rm line}$ is approximately equal to the critical value $M_{\rm line,crit} \sim 16\ {\rm M_\odot/pc}$ for T = 10 K, equivalent to a column density of $8 \times 10^{21}\ {\rm cm^{-2}}$. The grey band shows a dispersion of a factor of 3 around this nominal value. The dotted line running from the bottom left to the top right corresponds to $\sigma_{\rm tot} \propto {\rm N_{H_2}}^{0.5}$ (Arzoumanian *et al.* 2013). Right: CO emission observed towards and around the B211/13 filament (Goldsmith *et al.* 2008). The systemic velocities of the ^{12}CO(1 − 0) emission in the north and south of the filament are redshifted and blueshifted, respectively, with respect to the velocity of the ^{13}CO(1 − 0) emission of the main filament shown in green (taken from Palmeirim *et al.* 2013). The crest of the filament is traced in black.

3. Internal velocity dispersions of interstellar filaments derived from molecular line observations

In the following I present molecular line spectroscopic observations of interstellar filaments and the inferred gas kinematics, which is crucial to understand the physics at play in the formation and evolution of filaments and interstellar structures.

The total velocity dispersion ($\sigma_{\rm tot}$) of selected positions towards a sample of 46 filaments, derived from (^{13}CO, C^{18}O, and N_2H^+) molecular line observations, are presented in Fig. 3–left. Thermally subcritical and nearly critical filaments have transonic velocity dispersions ($c_s \lesssim \sigma_{\rm tot} < 2c_s$) independent of column density and are gravitationally unbound. The velocity dispersion of thermally supercritical filaments increases as a function of their column density (roughly as $\sigma_{\rm tot} \propto {\rm N_{H_2}}^{0.5}$). These measurements confirm that there is a critical threshold in $M_{\rm line}$ above which filaments are self-gravitating and below which they are unbound. The position of this threshold, is consistent within a factor of two with the critical value $M_{\rm line,crit} \sim 16\ {\rm M_\odot/pc}$ for T = 10 K, equivalent to a column density of $8 \times 10^{21}\ {\rm cm^{-2}}$ (Arzoumanian *et al.* 2013). These observations show that the mass per unit length of supercritical filaments is close to their virial mass per unit length $M_{\rm line,vir} = 2\sigma_{\rm tot}^2/G$ (Fiege & Pudritz 2000) where $\sigma_{\rm tot}$ is the observed total velocity dispersion (instead of the thermal sound speed used in the expression of $M_{\rm line,crit}$). We suggest that the large velocity dispersions of supercritical filaments is not a result of the supersonic interstellar turbulence but may be driven by gravitational contraction/accretion (Arzoumanian *et al.* 2013).

Mass accretion is indirectly suggested by transverse velocity gradients observed across the self-gravitating filament in Taurus (the filament shown in Fig. 1). The systemic velocities of the observed emission in the north and south of the filament are redshifted and blueshifted, respectively, with respect to the velocity of the emission observed towards the B211/13 filament (see Fig. 3–right and Palmeirim *et al.* 2013). Such a velocity field

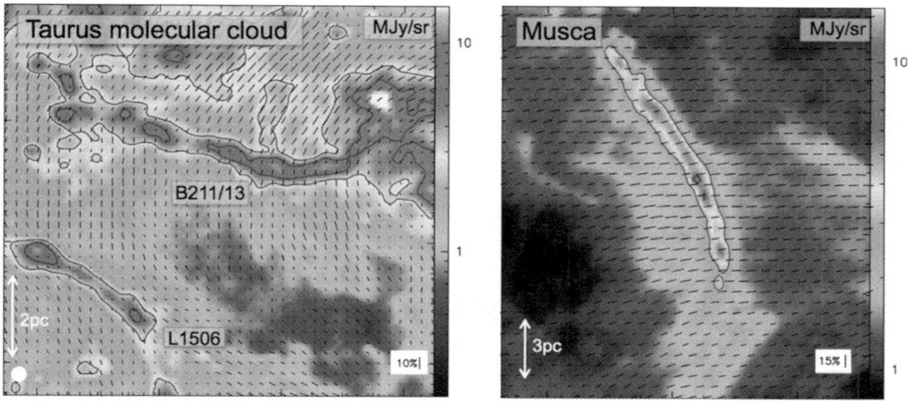

Figure 4. *Planck* 353 GHz (850 μm) total dust intensity (Stokes I) maps at a resolution of 4.′8, towards the Taurus B211/13 and L1506 filaments (Left) and the Musca filament (Right). The maps are in Galactic coordinate system. The blue contours show the levels of 3 and 6 MJy sr^{-1}. The black segments show the $\vec{B}_{\rm POS}$-field orientation (ψ+90°). The length of the pseudo-vectors is proportional to the polarization fraction. The polarization angles and fractions are computed at a resolution of 9.′6 (indicated by the white filled circles on the left hand side map) for increased S/N (Planck Collaboration Int. XXXIII 2014).

pattern (also observed towards other filaments, e.g., Kirk *et al.* 2013; Lee *et al.* 2014; Fernández-López *et al.* 2014) may indicate convergence of matter onto the densest parts of supercritical filaments. This is also compatible with theoretical models for the evolution of supercritical filaments (Hennebelle & André 2013; Heitsch 2013). Such models put forward the role of continuous accretion, which may be a physical reason to explain the observed properties of supercritical filaments.

4. Magnetic field structure as derived from *Planck* dust polarization observations

Dust polarization observations are essential to infer the orientation of the magnetic field ($\vec{B}$) component projected on the plane of the sky (POS). While the observed polarization fraction (p) depends on several parameters (dust polarization properties, grain alignment efficiency, and $\vec{B}$-field structure, Hildebrand 1983; Lee & Draine 1985), the observed polarization angle (ψ) derived from dust polarized emission is perpendicular to orientation of the $\vec{B}$-field component on the POS ($\vec{B}_{\rm POS}$) averaged along the line of sight (LOS). *Planck* observations at 353 GHz provide the first fully sampled maps of the polarized dust emission towards interstellar filaments and their backgrounds, providing unprecedented insight into the $\vec{B}$-field structure.

The first striking result is the impressively ordered structure of $\vec{B}_{\rm POS}$ from the largest Galactic scales down to the smallest scales probed by *Planck* observations, ~0.2 pc in the nearby molecular clouds of the Gould Belt (see Fig. 4, and, e.g., Planck Collaboration Int. XIX 2015; Planck Collaboration Int. XXXIII 2014). Statistical analysis of the *Planck* data indicate that the relative orientation between the angle of $\vec{B}_{\rm POS}$ and that of the intensity filamentary structures changes from preferentially parallel to preferentially perpendicular from low to high column densities (Planck Collaboration Int. XXXII 2014; Planck Collaboration Int. XXXV 2015). These results confirm the rough equipartition between turbulent and magnetic energies in the ISM (e.g., Myers & Goodman 1988) and emphasize the important role that the magnetic field plays in shaping the cold ISM structures.

To quantify the polarized intensity observed towards the filaments we derive radial profiles perpendicular to their crests and averaged along their length (as explained in

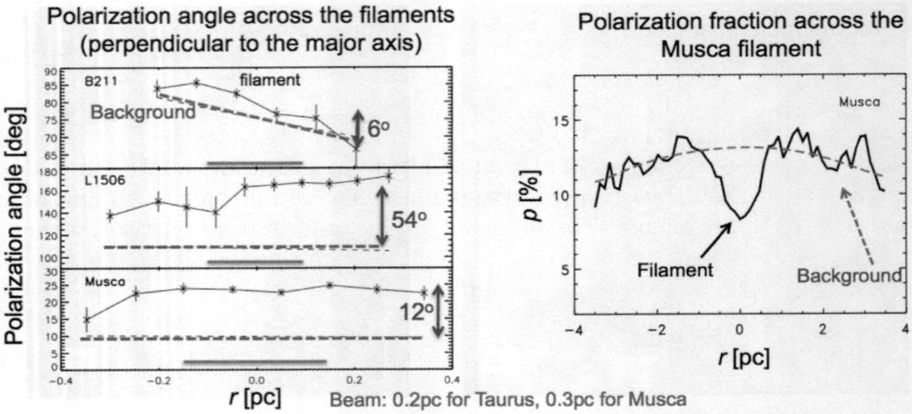

Figure 5. Left: Filament intrinsic polarization angle (tracing the angle of $\vec{B}_{\rm POS}$) across the crests of the B211, L1506, and Musca filaments. The x-axis shows the radial distance from the filament crest. The crosses are data points computed from Q and U background subtracted maps. The dashed line represents the background polarization angle. The difference between the polarization angle of the filament and that of the background is indicated on the plots (taken from Table 2 in Planck Collaboration Int. XXXIII 2014). Right: Observed profile (in black) of the polarization fraction (p) perpendicular to the crest of the Musca filament. The dashed blue curve shows the polarization fraction of the background. It is derived from the Stokes I, Q, and U parameters of the background.

Sect. 2) to increase the signal-to-noise ratio (S/N) in polarization, while keeping the highest resolution (4.′8) of the *Planck* data. To derive the intrinsic polarization properties of the filaments we describe the observations as a two-component model that uses the spatial information in *Planck* images to separate the emission of the filaments from that of their background (i.e., surrounding cloud). This allows us to characterize and compare the polarization properties of each emission component: the filament and its background. This is an essential step in measuring the intrinsic polarization fraction (p) and polarization angle (ψ, which is perpendicular to the plane of the sky $\vec{B}$-field angle) of each emission component. The analyses show that both the polarization angle and fraction measured at the intensity peak of the filaments (before background subtraction) differ from their intrinsic values (after background subtraction) as described in Planck Collaboration Int. XXXIII (2014). The left panel of Fig. 5–left shows the profile of ψ across the Musca, B211, and L1506 filaments. In all three cases, we measure variations in the polarization angle intrinsic to the filaments ($\psi_{\rm fil}$) with respect to that of their backgrounds ($\psi_{\rm bg}$). These variations are found to be coherent along the pc-scale length of the filaments. The differences between $\psi_{\rm fil}$ and $\psi_{\rm bg}$ for two of the three filaments are larger than the dispersion of the polarization angles across and along the filaments (see Table 2 in Planck Collaboration Int. XXXIII 2014). Hence, these differences are not random fluctuations and they indicate a change in the orientation of the POS component of the magnetic field between the filaments and their backgrounds.

The observations show a decrease in the polarization fraction p from the background to the Musca filament (Fig. 5–right), as well as towards the Taurus B211 and L1506 filaments (see Fig. 10 of Planck Collaboration Int. XXXIII 2014). A decrease in p with the total column density $N_{\rm H}$ (i.e., from the backgrounds to the filament crests), has been already shown in previous studies. This decrease has been usually interpreted as due to the turbulent component of the field and/or variations of dust alignment efficiency with increasing column density (e.g., Jones *et al.* 2015; Whittet *et al.* 2008). In our study, the bulk of the drop in p within the filaments cannot be explained by random fluctuations of the orientation of the magnetic field because they are too small ($\sigma_\psi < 10°$). We argue that the observed drop in p towards the filaments may be due to the 3D structure of the magnetic field: both its orientation in the POS and with respect to the POS (Planck Collaboration Int. XXXIII 2014). Indeed, the observed changes of ψ are direct evidence of variations of the orientation of the POS projection of the magnetic field. The systematic variations of ψ suggest changes of the angle of the magnetic field with respect to the POS. This angle must statistically vary as much as ψ, contributing to the observed decrease of p in the filaments. The observed variation of ψ between the filaments and their backgrounds always depolarizes the total emission, due to the integration of the emission along the LOS of two emission components where the angle of $\vec{B}_{\mathrm{POS}}$ varies (see Fig. 5–right and Planck Collaboration Int. XXXIII 2014).

The inner structure of the filaments (as seen with *Herschel*, cf., Palmeirim *et al.* 2013 for B211, Ysard *et al.* 2013 for L1506, and Cox *et al.* 2015 for Musca) is not resolved, but at the smallest scales accessible with *Planck* (~0.2 pc towards the nearby clouds), the observed changes of ψ and p (derived from *Planck* polarization data at the resolution of 4.′8) hold some information on the magnetic field structure within filaments (Planck Collaboration Int. XXXIII 2014). They show that both the mean field and its fluctuations in the filaments are different from those in the background clouds, which points to a coupling between the matter and the $\vec{B}$-field in the filament formation process.

5. Summary and conclusions

The ubiquity of filaments in both quiescent clouds and active star-forming regions, where they are associated with the presence of prestellar cores and protostars (Könyves *et al.* 2015), supports the view that filaments are (first) form in the ISM and the densest of them fragment into star-forming cores (André *et al.* 2014). The observational finding of a filament uniform 0.1 pc width (Arzoumanian *et al.* 2011) sets very strong constraints on the physics at play in the ISM. This result has been recently confirmed by an independent analysis of the same data from another group (Koch & Rosolowsky 2015) and observed (at higher-resolution) in regions farther away than the Gould Belt (Hill *et al.* 2012). Interestingly 0.1 pc corresponds to the sonic scale below which interstellar turbulence becomes subsonic in diffuse, non-star-forming gas (Heyer *et al.* 2009; Pineda *et al.* 2011). This occurrence, along with the observed thermal velocity dispersion of low column density filaments, suggests that large-scale turbulence may be a main player in the formation of the filamentary web observed in molecular clouds (Arzoumanian *et al.* 2011). On the other hand, the increase of the non-thermal velocity dispersion of supercritical, self-gravitating filaments with column density, may indicate the generation of internal turbulence due to gravitational accretion. This may be an explanation for the observed constant width of self-gravitating collapsing filaments (Arzoumanian *et al.* 2013).

While the dissipation of interstellar turbulence provides a plausible mechanism for filament formation, the observed organization between the magnetic field lines and the intensity structures, derived from the analysis of *Planck* data, indicates that the $\vec{B}$-field plays a dynamically important role in shaping the interstellar matter (Planck Collaboration Int. XXXIII 2014; Planck Collaboration Int. XXXV 2015). In particular, they may be a key element to understand the channel of mass flows in the ISM. The fact that most prestellar cores lie in dense, self-gravitating filaments (Könyves *et al.* 2015) suggests that gravity is a major driver in the evolution of supercritical filaments and their fragmentation in star-forming prestellar cores. The detailed analysis of the distribution of matter along the filaments (density fluctuations) may also yield insights into the origin of the shape of the mass distribution of prestellar cores thought to be linked to that of the stars (André *et al.* 2014; Roy *et al.* 2015). The combination of these observational results, derived from dust and gas tracers in total and polarized intensity, give strong constraints on our understanding of the formation and evolution of filaments in the ISM, which provides important clues to the initial conditions of the star formation process along supercritical filaments.

Acknowledgements

DA acknowledges support by the European Research Council grant MISTIC (ERC-267934).

References

Abergel, A., Boulanger, F., Mizuno, A., & Fukui, Y. 1994, *ApJ*, 423, L59
André, P., Di Francesco, J., Ward-Thompson, D., *et al.* 2014, in *Protostars and Planets VI*, p. 27
André, P., Men'shchikov, A., Bontemps, S., *et al.* 2010, *A&A*, 518, L102
Arzoumanian, D., André, P., Didelon, P., *et al.* 2011, *A&A*, 529, L6
Arzoumanian, D., André, P., Peretto, N., & Könyves, V. 2013, *A&A*, 553, A119
Cox, N., Arzoumanian, D., André, P., *et al.* 2015, *A&A* submitted
Federrath, C., Roman-Duval, J., Klessen, R. S., Schmidt, W., & Mac Low, M. 2010, *A&A*, 512, A81+
Fernández-López, M., Arce, H. G., Looney, L., *et al.* 2014, *ApJL*, 790, L19
Fiege, J. D. & Pudritz, R. E. 2000, *MNRAS*, 311, 85
Goldsmith, P. F., Heyer, M., Narayanan, G., *et al.* 2008, *ApJ*, 680, 428
Heitsch, F. 2013, *ApJ*, 769, 115
Hennebelle, P. & André, P. 2013, *A&A*, 560, A68
Hennebelle, P., Banerjee, R., Vázquez-Semadeni, E., Klessen, R. S., & Audit, E. 2008, *A&A*, 486, L43
Heyer, M., Krawczyk, C., Duval, J., & Jackson, J. M. 2009, *ApJ*, 699, 1092
Hildebrand, R. H. 1983, *QJRAS*, 24, 267
Hill, T., Andre, P., Arzoumanian, D., *et al.* 2012, *A&A*, 548, L6
Inutsuka, S. & Miyama, S. M. 1997, *ApJ*, 480, 681
Jones, T. J., Bagley, M., Krejny, M., Andersson, B.-G., & Bastien, P. 2015, *AJ*, 149, 31
Kirk, H., Myers, P. C., Bourke, T. L., *et al.* 2013, *ApJ*, 766, 115
Koch, E. W. & Rosolowsky, E. W. 2015, *MNRAS*, 452, 3435
Könyves, V., André, P., Men'shchikov, A., *et al.* 2015, *A&A* in press [ArXiv:1507.05926]
Lee, H. M. & Draine, B. T. 1985, *ApJ*, 290, 211
Lee, K. I., Fernandez-Lopez, M., Storm, S., *et al.* 2014, ArXiv e-prints
Mac Low, M. & Klessen, R. S. 2004, *Reviews of Modern Physics*, 76, 125
Men'shchikov, A., André, P., Didelon, P., *et al.* 2010, *A&A*, 518, L103+
Molinari, S., Swinyard, B., Bally, J., *et al.* 2010, *A&A*, 518, L100
Myers, P. C. & Goodman, A. A. 1988, *ApJ*, 326, L27
Ostriker, J. 1964, *ApJ*, 140, 1056
Palmeirim, P., André, P., Kirk, J., *et al.* 2013, *A&A*, 550, A38
Pineda, J. E., Goodman, A. A., Arce, H. G., *et al.* 2011, *ApJl*, 739, L2
Planck Collaboration Int. XIX. 2015, *A&A*, in press [ArXiv:1405.0871]
Planck Collaboration Int. XXXII. 2014, *A&A*, in press [ArXiv:1409.6728]
Planck Collaboration Int. XXXIII. 2014, *A&A*, in press [ArXiv:1411.2271]
Planck Collaboration Int. XXXV. 2015, *A&A*, in press [ArXiv:1502.04123]
Roy, A., André, P., Arzoumanian, D., *et al.* 2015, *A&A*, in press [ArXiv:1509.01819]
Schneider, S. & Elmegreen, B. G. 1979, *ApJS*, 41, 87
Sousbie, T. 2011, *MNRAS*, 414, 350
Whittet, D. C. B., Hough, J. H., Lazarian, A., & Hoang, T. 2008, *ApJ*, 674, 304
Ysard, N., Abergel, A., Ristorcelli, I., *et al.* 2013, *A&A*, 559, A133

From Interstellar Clouds to Star-Forming Galaxies: Universal Processes?
Proceedings IAU Symposium No. 315, 2015
P. Jablonka, Ph. André & F. van der Tak, eds.

doi:10.1017/S1743921316007262

The Formation and Destruction of Molecular Clouds and Galactic Star Formation

Shu-ichiro Inutsuka[1], Tsuyoshi Inoue[2], Kazunari Iwasaki[3], Takashi Hosokawa[4] and Masato I. N. Kobayashi[1]

[1]Department of Physics, Graduate School of Science, Nagoya University, Nagoya 464-8602, Japan, email: inutsuka@nagoya-u.jp

[2]Division of Theoretical Astronomy, National Astronomical Observatory of Japan, Osawa, Mitaka, Tokyo 181-8588, Japan

[3]Department of Environmental Systems Science, Doshisha University Tatara Miyakodani 1-3, Kyotanabe City, Kyoto 610-0394, Japan

[4]Department of Physics and Research Center for the Early Universe The University of Tokyo, Tokyo 113-0033, Japan

Abstract. We discuss an overall picture of star formation in the Galaxy. Recent high-resolution magneto-hydrodynamical simulations of two-fluid dynamics with cooling/heating and thermal conduction have shown that the formation of molecular clouds requires multiple episodes of supersonic compression. This finding enables us to create a new scenario of molecular cloud formation through interacting shells or bubbles on galactic scales. We estimate the ensemble-averaged growth rate of individual molecular clouds, and predict the associated cloud mass function. This picture naturally explains the accelerated star formation over many million years that was previously reported by stellar age determination in nearby star forming regions. The recent claim of cloud-cloud collisions as a mechanism for forming massive stars and star clusters can be naturally accommodated in this scenario. This explains why massive stars formed in cloud-cloud collisions follows the power-law slope of the mass function of molecular cloud cores repeatedly found in low-mass star forming regions.

Keywords. stars: formation – ISM: clouds – ISM: bubbles – ISM: magnetic fields – ISM: structure – ISM: kinematics and dynamics

---

1. Introduction

Recent observational studies of the nearby star-forming regions with the *Herschel Space Observatory* have convincingly shown a very remarkable fact that stars are born in self-gravitating filaments (e.g., André *et al.* 2010, 2014) This finding provides a pivot in further theoretical studies of star formation in molecular clouds. The studies of early phases of star formation may set a goal in describing the formation of filamentary clouds where one should explain the apparently universal properties of filamentary clouds such as widths and line-mass of filaments (e.g., Miyama, Narita, & Hayashi 1987a,b; Nagai, Inutsuka, & Miyama 1998; Hennebelle & André 2013). Later phases of star formation correspond to the evolution of filamentary clouds. In contrast to the former, the theoretical study of the latter has been well developed. Indeed, the collapse and fragmentation of filamentary molecular clouds are analyzed in great detail with and without magnetic field (e.g., Nagasawa 1987; Inutsuka & Miyama 1992, 1997). Those studies have essentially provided possible initial conditions for further protostellar collapse calculations, which are now realistically describing the formation of protostars and protoplanetary disks together with the driving of molecular outflows and/or fast collimated jets (e.g., a review by Inutsuka 2012). In this paper we try to explore the former problem, i.e., a theoretical picture for the formation of filamementary molecular clouds.

Before desribing our picture of cloud formation, we point out some implications from the observations of filamentary molecular clouds.

1.1. *Mass Function of Molecular Cloud Cores*

Self-gravitational hydrodynamics is a deterministic process, i.e., the initial conditions and boundary conditions exactly determine the time-evolution of the system. Therefore, the mass function of cores as a result of gravitational fragmentation of a filament should be prescribed essentially by the initial distribution of density fluctuations along the filament. Indeed, Inutsuka (2001) has shown that filamentary clouds with line-mass spectrum $\delta \propto k^n$ with $n \sim -1.5$ will produce a mass function of molecular cloud cores whose power law exponent is close to -2.5, i.e., $dn/dM \propto M^{-2.5}$. Recently Roy *et al.* (2015) showed that the actual power spectrum of fluctuations along filaments in filamentary molecular clouds of the Gould Belt Survey shows $\delta \propto k^{-1.6}$. If this observed exponent is universal in the sense that similar value is expected in any filamentary molecular cloud, we can attribute the origin of the apparently universal core mass function to the origin of the line-mass fluctuations in filamentary clouds. It is interesting to note that the observed Fourier exponent of line-mass fluctuations is close to the exponent, 5/3, of Kolmogorov Law that are generic property of subsonic tubulence driven by a phase transition (Koyama & Inutsuka 2002; Audit & Hennebelle 2005, see the detailed analysis by Hennebelle & Audit 2007).

1.2. *Formation of Massive Stars in the Filament Paradigm?*

The core mass function observed by André *et al.* (2010) covers up to $M_{\rm core} \sim 10 M_{\odot}$. Thus, the success of the filament paradigm explain not only the formation of low mass stars but also the formation of massive stars. This is important in the sense that it means that the formation mechanism of massive stars may be a simple extension of the formation of low mass stars. Since theoretical studies of the formation of massive stars has been controversial for many years (see, e.g., review by Tan *et al.* 2014), it is crucial to identify direct and clear observational evidence for the formation of massive stars in the filament paradigm. This has been shown by ALMA observation of one of the infrared dark clouds, SDC335.579-0.272, where the actual cloud is composed of multiple filaments of coherent line-of-sight velocities. Peretto *et al.* (2013) interpreted their observation in terms of multiple filaments falling along their axes towards one of two potential wells which correspond to progenitors of two massive stars. This interpretation explains why the gravitational motions of filamentary clouds naturally realize large mass accretion rates that are supposed to be required for the formation of massive stars.

2. Formation of Star-Forming Filamentary Clouds

Recent high-resolution magneto-hydrodynamical simulations of two-fluid dynamics with cooling/heating and thermal conduction have shown that the formation of molecular clouds requires multiple episodes of supersonic compression (Inoue & Inutsuka 2008, 2009; Heitsch, Stone & Hartmann 2009). Using three-dimensional MHD simulations including the effects of radiative cooling/heating, chemical reactions, and thermal conductivity, Inoue & Inutsuka (2012) have investigated the formation of molecular clouds in the magnetized ISM. They have considered the formation of a magnetized molecular cloud due to the accretion of HI clouds formed by the thermal instability. Since the mean density of the initial multi-phase HI medium is an order of magnitude larger than the

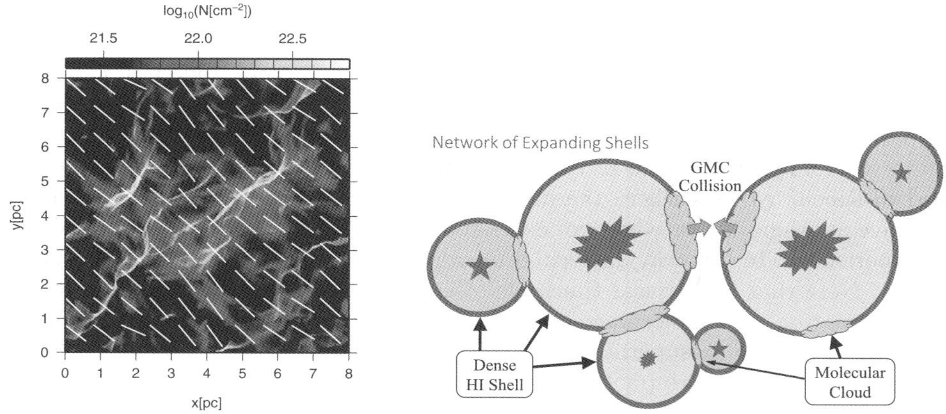

Figure 1. (Left Panel) Face-on column density view of a shock-compressed dense layer in the numerical MHD simulations of Inutsuka *et al.* (2015). The color scale (in cm^{-2}) is shown on the top. The mean magnetic field is in the plane of the layer and its direction is shown by a white bar. Note the formation of dense, magnetized filaments whose axes are almost perpendicular to the mean magnetic field. Fainter "striation"-like filaments can also be seen, which are almost perpendicular to the dense filaments. (Right Panel) A schematic picture of sequential formation of molecular clouds by multiple compressions by overlapping dense shells driven by expanding bubbles. The thick red circles correspond to magnetized dense multi-phase interstellar medium where cold turbulent HI clouds are embedded in warm neutral medium. Molecular clouds can be formed only in the limited regions where the compressional direction is almost parallel to the local mean magnetic field lines or regions experienced excessive number of compressions. An additional compression of a molecular cloud tends to create multiple filamentary molecular clouds that mimics the observation by *Herschel Space Observatory*. Once the line-mass of a filament exceeds the critical value ($\sim 2C_s^2/G$) even in a less massive molecular cloud, star formation starts. In general, star formation in a cloud accelerates with the growth of total mass of the cloud. At limited frequency, giant molecular clouds collide each other. This produces very unstable molecular gas and may trigger very active star formation.

typical diffuse WNM density, the formation timescale is shorter than in the molecular cloud formation through only the accumulation of diffuse WNM. The resulting timescale of molecular cloud formation of ~10 Myr is consistent with the evolutionary timescale of the molecular clouds in the LMC (Kawamura, Mizuno, Minamidani, *et al.* 2009).

We have done numerical simulations of the additional one-sided compression of the already-formed but faint molecular clouds, and found interesting features of more realistic evolution (Inutsuka *et al.* 2015). Figure 1 shows a snapshot of the face-on view of a non-uniform molecular cloud compressed by a shock wave travelling at 10 km/s.

The magnetic field lines are mainly in the dense sheet of the compressed gas. Many dense filaments are created, with axes perpendicular to the mean magnetic field lines. We can also see many faint filamentary structures that mimic observed "striations" and are almost parallel to the mean magnetic field lines. Those faint filaments are feeding gas onto the dense filaments just as envisioned in André *et al.* (2014).

Although further analysis is required to give a conclusive interpretation of the observations, a simple picture based on repeated compression of a molecular cloud may be a main mechanism for the formation of filamentary molecular clouds.

3. Bubble Dominated Scenario of Cloud Formation

HI observation of our galaxy shows many shell-like structures near the Galactic plane (e.g., Hartmann & Burton 1997; Taylor *et al.* 2003). Since we identify repeated interactions of expanding shock waves as a basic mechanism of molecular cloud formation, we depict the overall scenario of cloud formation in our Galaxy as a schematic picture in Figure 1 (right) (Inutsuka *et al.* 2015). In this picture, red circles correspond to the remnant of shock waves due to expanding HII regions or old and slowly expanding supernova remnants. Cold HI clouds embedded in warm neutral medium are almost ubiquitously found in the shell of these remnants. Molecular clouds are expected to be formed in special regions where the mean magnetic field is parallel to the direction of shock wave propagation, or where an excessive number of shock wave sweepings happen to be experienced. In this way, molecular clouds can be found only in limited regions on the shells. Note that the typical timescale of each shock wave is on the order of 1 Myr, but the formation of a molecular cloud requires many Myr. Some bubbles become invisible as HII regions or supernova remnants many Myrs after their births. Therefore, this schematic picture corresponds to a "very long exposure image" of the real structure of the interstellar medium; Each molecular cloud may acquire a random velocity depending on its location in the most recent bubble that interacts the cloud. Interestingly, this multi-generation picture for evolution of the molecular clouds seems to agree with the observational findings by Dawson *et al.* (2011a,b) who investigated the transition of atomic gas to molecular gas in three clouds in the walls of the Galactic supershells.

In limited circumstances, created molecular clouds collide with one another. This produces very unstable molecular gas and may trigger very active star formation (e.g., Fukui *et al.*2013, Inoue & Fukui 2013).

The typical shock velocity of expanding HII regions is 10 km/s. If the shock velocity of a supernova remnant is much higher than 10 km/s, the resultant interaction would result in destruction of molecular clouds. The mean velocity of each molecular cloud should be similar or somewhat smaller than the recently swept shock wave. Therefore the velocity dispersion of molecular clouds should be similar or somewhat smaller than 10 km/s. According to this acquisition process of random velocities, the velocity of a particular cloud is not expected to strongly depend on the mass of the cloud.

4. Evolution of the Molecular Cloud Mass Function

In this section, we try to describe the evolution of the mass function of molecular clouds. For that purpose, we first model the growth of molecular clouds. According to Inoue & Inutsuka (2009, 2012), we need multiple episodes of compression of HI clouds to create molecular clouds. The galactic average timescale of each compression should be on the order of $T_{\rm exp} \sim$ Myr, but can be factor of a few smaller in the spiral arm regions in the Galaxy. Thus, we have to consider the time evolution of cloud mass over a much longer duration than 1 Myr.

After each compression, the cloud may slightly expand because of the reduced pressure of the ambient medium, which may cause the loss of the diffuse component of cloud mass. This process has not yet been theoretically studied in detail as far as the authors are aware. Observationally the average column densities of observed molecular clouds do not seem to change very much and always appear to be that corresponding to a visual extinction of several. This means that mass of the cloud is proportional to the cross section of the cloud. Since the compressional formation of molecular material is expected to be proportional to the cross section of the pre-existing cloud, we can model the rate of increase of molecular cloud mass as

$$\frac{dM}{dt} = \frac{M}{T_{\rm f}}, \tag{4.1}$$

where $T_{\rm f}$ denotes the formation timescale. According to the investigation of molecular cloud formation briefly described in the previous section, Inutsuka *et al.* (2015) estimate the formation timescale $T_{\rm f} \sim 10\,{\rm Myr}$, but this can be factor of a few longer in the inter-arm regions.

Inutsuka *et al.* (2015) also estimated the destruction of molecular clouds, by extending the detailed radiation hydrodynamical calculations of the expanding HII region with photodissociative line radiation (Hosokawa & Inutsuka 2005, 2006a; Hosokawa & Inutsuka 2006b, 2007) to the cases around a magnetized interstellar medium. For example, they show that a $30\,M_\odot$ star can destroy a $10^5 M_\odot$ molecular cloud. The mass of the star most responsible for the cloud destruction depends on the actual shape of stellar initial mass function. However, Inutsuka *et al.* (2015) have shown that the most responsible mass is always $30 \pm 10\,M_\odot$ for the mass function with a high mass slope $-2.3 \sim -2.7$. This is due to the very sharp increase of the disrupted mass as a function of stellar mass M_* for $M_* < 20\,M_\odot$, and its saturation for $M_* > 20\,M_\odot$.

Suppose that a $10^5 M_\odot$ molecular cloud is eventually disrupted by photodissociating radiation from a $30 M_\odot$ star born inside of this cloud, and that star formation is quenched. The condition for this to occur can be written as follows: $10^5 M_\odot \times \epsilon_{\rm SF} > M_{\rm crit}(30\,M_\odot)$, where $\epsilon_{\rm SF}$ is the star formation efficiency (the ratio of the total mass of stars to the mass of the parental cloud) and $M_{\rm crit}(30\,M_\odot)$ is the total mass of stars in a group that include only one star whose mass is larger than $30\,M_\odot$. If the initial mass function is the same as that observed in the solar neiborhood, $M_{\rm crit}(30\,M_\odot)$ should be on the order of $10^3 M_\odot$. If $\epsilon_{\rm SF}$ is smaller than that satisfies the above inequality, photodissociative disruption is not sufficient and star formation continues with the remaining molecular material in the cloud, this in turn raises $\epsilon_{\rm SF}$. Thus, we expect that the actual evolution of a molecular cloud finally satisifies the condition $10^5 M_\odot \times \epsilon_{\rm SF} = M_{\rm crit}(30 M_\odot)$ when the star formation is eventually quenched. This means that the star formation efficiency should be given by

$$\epsilon_{\rm SF} = \frac{M_{\rm crit}(30\,M_\odot)}{10^5 M_\odot} \sim 10^{-2}. \tag{4.2}$$

This may explain the reason for the small star formation efficiency in molecular clouds observationally found many decades ago (e.g., Zuckerman & Evans 1974). For example, according to Hosokawa & Inutsuka (2006b) a $10 M_\odot$ star can photodissociate $10^2 M_\odot$ of the surrounding molecular material, but a $10^2 M_\odot$ molecular cloud is not likely to produce a $10 M_\odot$ star. Therefore the above argument is not possible with $10 M_\odot$ as the mass of the disrupting star.

Now we can estimate the timescale for the destruction of the molecular cloud. The preceding argument has already shown that a $10^5 M_\odot$ molecular cloud can be disrupted within 4 Myr. If we assume the initial mass function does not depend on the parental cloud mass, the destruction timescale is not expected to depend on the mass of the cloud larger than $10^5 M_\odot$. This can be understood as follows. The number of dissociating photons is proportional to the number of massive stars, which increases with the mass of the cloud. But the required number of dissociating photons also increases with the mass of the cloud. Therefore the probability of the cloud destruction does not depend on the mass of the cloud, if the shape of the initial mass function does not depend on the mass of the cloud.

The actual disruption timescale $T_{\rm d}$ should be the sum of the formation timescale of a massive star (T_*) and the expansion timescale of its HII region, i.e., $T_{\rm d} \approx T_* + 4\,{\rm Myr}$. After one cycle of molecular cloud disruption that takes timescale $T_{\rm d}$, only a fraction $\epsilon_{\rm SF}$ of the molecular gas is transformed into stars. Therefore the timescale to completely transform a molecular cloud to stars is $T_{\rm d}/\epsilon_{\rm SF} \sim 1.4\,{\rm Gyr}$. This may explain the so-called "depletion

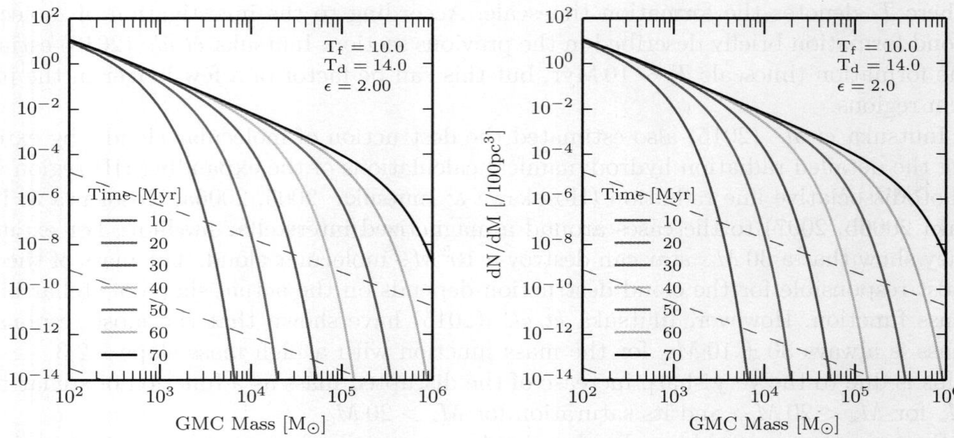

Figure 2. Time evolution of the mass function of molecular clouds. The effect of cloud collisions is not included in the left panel but included in the right panel. The steady state slope of the mass function is essentially determined the ratio of cloud formation and cloud destruction rate as shown in equation (4.6).

timescale" of molecular clouds observationally known as the Schmitt-Kennicutt Law (e.g., Kennicutt 1998).

In order to describe the time evolution of the mass function of molecular clouds over a timescale much larger than Myr (e.g., 10Myr), we adopt coarse graining of short-timescale events of the growth and destruction of clouds, and describe the continuity equation of molecular clouds in mass space.

$$\frac{\partial N}{\partial t} + \frac{\partial}{\partial M}\left(N\frac{dM}{dt}\right) = -\frac{N}{T_\mathrm{d}} + \left(\frac{dN}{dt}\right)_\mathrm{coll}, \tag{4.3}$$

where $N(dM/dt)$ denotes the flux of mass function in mass space, T_d is the cloud disruption timescale, dM/dt describes the growth rate of the molecular cloud as given in Equation (4.1), and the last term on the right-hand side is rate of change of the mass function due to two-body collisions (i.e., cloud-cloud collisions). We assumed the average column density of molecular clouds is always constant ($\sim 10^{22}\mathrm{cm}^{-2}$) and the cross section of a collision can be regarded as a sum of geometrical cross sections of two colliding molecular clouds, and performed time evolution calculations (Kobayashi *et al.* in prep.). Figure (2) shows typical time evolution of the mass function of molecular clouds. In these figures, we confirm that cloud-cloud collisions do not affect the overall mass function of molecular clouds.

If the contribution from cloud-cloud collisions is negligible, a steady state solution of the above equation is

$$N(M) = \frac{N_0}{M_0}\left(\frac{M}{M_0}\right)^{-\alpha}, \tag{4.4}$$

where

$$\alpha = 1 + \frac{T_\mathrm{f}}{T_\mathrm{d}}. \tag{4.5}$$

In typical conditions of spiral arm regions, we expect $T_* \sim T_\mathrm{f}$, and thus, $T_\mathrm{f} \lesssim T_\mathrm{d}$, which corresponds to $1 < \alpha \lesssim 2$. For example, $T_\mathrm{f} = 10\,\mathrm{Myr}$ corresponds to $\alpha \approx 1.7$, which agrees nicely with observations (Solomon *et al.* 1987; Kramer *et al.* 1998; Heyer *et al.* 2001; Roman Duval *et al.* 2010). However, in quiescent regions with very a limited amount of gaseous material away from spiral arms in the Galactic disk, T_f is expected to be large and possibly even larger than $T_d = T_* + 4\,\mathrm{Myr}$. In this case α can be larger than 2. This tendency is actually observed in M33 (Rosolowsky & Blitz 2005; Gratier *et al.* 2012) and in M51 (Colombo *et al.* 2014).

5. Summary

In general, dense molecular clouds cannot be created in shock waves propagating in magnetized warm neutral medium without cold HI clouds. In this paper we identify repeated interactions of shock waves in the dense cold HI medium as a basic machanism for creating filametary molecular clouds, which are ubiquitously observed in the nearby interstellar medium. This suggests an expanding-bubble-dominated picture of the formation of molecular clouds in our Galaxy, which enables us to envision an overall picture of the statistics of molecular clouds and resulting star formation. Our conclusions are summarized as follows:

(*a*) Turbulent cold HI clouds embedded in warm neutral medium can be easily created in the expanding shells of HII regions or the very late phase of supernova remnants. In contrast, the formation of molecular clouds in magnetized ISM need many compression events. One compression corresponds to the order of 1 Myr on average in our Galaxy. The timescale of the cloud formation is a few times 10 Myr.

(*b*) Since the galactic thin disk is occupied by many bubbles, molecular clouds are formed in the overlapping region of (old and new) bubbles. However, since the average lifetime of each bubble is shorter than the timescale of cloud formation, it is difficult to observationally identify the multiple bubbles that create the molecular clouds.

(*c*) The velocity dispersion of molecular clouds should be originated in the expansion velocities of bubbles. This is estimated to be $\lesssim$10 km/s and should not strongly depend on the mass of the molecular cloud.

(*d*) To describe the growth of molecular cloud mass we can temporally smooth out the evolution over the timescale larger than $\sim$ 1 Myr. The resulting mass of each molecular cloud (smoothed over time) is an almost exponentially increasing function of time. Star formation starts even in small molecular clouds, and thus, star formation accelerates on a timescale larger than $\sim$ 1 Myr.

(*e*) Molecular clouds may collide over a timescale of a few times 10 Myr, depending on the relative locations in adjacent bubbles. This molecular cloud collision may result in the active star formation in a giant molecular cloud.

(*f*) The destruction of molecular clouds is mainly due to the dissociative radiation from massive stars of $> 20\,M_\odot$. The probability of cloud destruction is not a sensitive function of the mass of molecular clouds. If the initial mass function does not vary much with the mass of parental molecular clouds, the cloud destruction by $30\,M_\odot$ stars result in a star formation efficiency on the order of 1%. This property explains the observed constancy of gas depletion rate ($\sim [1\,\mathrm{Gyr}]^{-1}$) of giant molecular clouds in our and external galaxies, and hence, corresponds to Kennicutt-Schmidt Law.

(*g*) The steady state of the evolution of cloud mass function corresponds to the power law with exponent n in the range $1 < n \lesssim 2$ in the spiral arm regions of our Galaxy. However, a larger value of the exponent such as $n > 2$ is possible in the inter-arm regions.

In many disk galaxies, star forming regions tend to be located in spiral structures, and thus, many bubbles should be located in spiral structures. It might be interesting to study how this bubble dominated picture of the formation of molecular cloud corresponds to the shaping of spiral structure itself. This requires a more comprehensive study of the dependence of the various timescales relevant in this picture, on the ambient ISM density and pressure, radiation contents, metallicity, etc. We hope that the present picture will provide a useful step for understanding star formation in the Galactic scale.

References

André, P., Men'shchikov, A., Bontemps, S. *et al.* 2010, *A&A*, 518, L102

André, P., Di Francesco, J., Ward-Thompson, D. *et al.* 2014, *Protostars and Planets VI*, (Univ of Arizona Press) eds. H. Beuther, R. Klessen, C. Dullemond, & Th. Henning

Audit, E. & Hennebelle, P. 2005, A&A , 433, 1

Colombo, D., Hughes, A., Schinnerer, E. *et al.* 2014, *ApJ* 784, 3

Dawson, J. R., McClure-Griffiths, N. M., Kawamura, A. *et al.* 2011, *ApJ* 728, 127

Dawson, J. R., McClure-Griffiths, N. M., Dickey, J. M. *et al.* 2011, *ApJ* 741, 85

Gratier, P., Braine, J., Rodriguez-Fernandez, N. J., *et al.* 2012, *A&A*, 542, A108

Hartmann, D. & Burton, W. B. 1997, *Atlas of Galactic Neutral Hydrogen* (New York: Cambridge Univ. Press)

Heyer M. H. *et al.* 2001, *ApJ* 551, 2, 852

Heitsch F., Stone J. M., Hartmann L. W., 2009, *ApJ*, 695, 248

Heitsch F., & Hartmann L. W. (2014) *MNRAS* 443, 230

Hennebelle, P. & André, P. 2013, A&A , 560, 68

Hennebelle, P. & Audit, E., 2007, A&A , 465, 431

Hosokawa, T. & Inutsuka, S. 2005, ApJ , 623, 917

Hosokawa, T. & Inutsuka, S. 2006, ApJ , 646, 240

Hosokawa, T. & Inutsuka, S. 2006, ApJL , 648, L131

Hosokawa, T. & Inutsuka, S. 2007, ApJ , 664, 363

Inoue, T. & Inutsuka, S. 2008, ApJ , 687, 303

Inoue, T. & Inutsuka, S. 2009, ApJ , 704, 161

Inoue, T. & Inutsuka, S. 2012, ApJ , 759, 35

Inutsuka, S. 2001, *ApJ* 559, L149

Inutsuka, S. 2012, *Prog. Theor. Exp. Phys.*, Vol. 2012, 01A307

Inutsuka, S. & Miyama, S. M. 1992, *ApJ*, 388, 392

Inutsuka, S. & Miyama, S. M. 1997 *ApJ*, 480, 681

Inutsuka, S., Inoue, T. Iwasaki, K., & Hosokawa, T. 2015, *A&A* 580, A49

Kawamura, A., Mizuno, Y., Minamidani, T., *et al.* 2009, *ApJS*, 184, 1

Koyama, H. & Inutsuka, S. 2002, *ApJL* 564, Issue 2, L97

Kramer C. *et al.* 1998 *A&A* 329, 249

Kennicutt, R. C. Jr. 1998, *ApJ* 498, 541

Miyama, S. M., Narita, S., & Hayashi, C. 1987, *Prog. Theo. Phys.* 78, 1051

Miyama, S. M., Narita, S., & Hayashi, C. 1987, *Prog. Theo. Phys.* 78, 1273

Nagai, T., Inutsuka, S., & Miyama, S. M. 1998, *ApJ* 506, 306

Nagasawa, M. 1987, *Prog. Theor. Phys.*, 77, 635

Peretto, N., Fuller, G. A. Duarte-Cabral, A. *et al.* 2013, *A&A* 555, A112

Roy, A., André, Ph, Arzoumanian, D. *et al.* 2015, *A&A* in press (arXiv:1509.01819)

Roman Duval J. *et al.* 2010 *ApJ* 723, 492

Rosolowsky, E. & Blitz, L. 2005, *ApJ*, 623, 826

Solomon P. M. *et al.* 1987 *ApJ* 319, 730

Tan, J. C., Beltrán, M. T., Caselli, P. *et al.* (2014) *Protostars and Planets VI*, (Univ of Arizona Press) eds. H. Beuther, R. Klessen, C. Dullemond, & Th. Henning,

Taylor, A. R., Gibson, S. J., Peracaula, M. *et al.* 2003, *AJ* 125, 3145

Zuckerman, B. & Evans, N. J. 1974, *ApJ*, 192, L149

From Interstellar Clouds to Star-Forming Galaxies: Universal Processes?
Proceedings IAU Symposium No. 315, 2015
P. Jablonka, Ph. André & F. van der Tak, eds.

doi:10.1017/S1743921316007274

Connecting the density structure of molecular clouds with star formation

Jouni Kainulainen
Max-Planck-Institute for Astronomy,
Königstuhl 17, 69117 Heidelberg
email: `jtkainul@mpia.de`

Abstract. In the current paradigm of turbulence-regulated interstellar medium (ISM), star formation rates of entire galaxies are intricately linked to the density structure of the individual molecular clouds. This density structure is essentially encapsulated in the probability distribution function of volume densities (ρ-PDF), which directly affects the star formation rates predicted by analytic models. Contrasting its fundamental role, the ρ-PDF function has remained virtually unconstrained by observations. I describe in this contribution the recent progress in attaining observational constraints for the column density PDFs (N-PDFs) of molecular clouds that function as a proxy of the ρ-PDFs. Specifically, observational works point towards a universal correlation between the shape of the N-PDFs and star formation activity in molecular clouds. The correlation is in place from the scales of a parsec up to the scales of entire galaxies, making it a fundamental, global link between the ISM structure and star formation.

Keywords. stars: formation, ISM: clouds, ISM: structure

1. Introduction

How exactly different physical processes give rise to the star formation rates of molecular clouds is a decades old open question. Most of all, we do not understand how the different processes mould the structure and dynamics of the clouds, causing only a small fraction of their gas to have conditions suitable for star formation. Observationally, we know very well that the star formation rates and efficiencies of molecular clouds depend strongly on the *internal structure* of the clouds (e.g., Kainulainen *et al.* 2009, 2014; Lada *et al.* 2012; Evans *et al.* 2014). Therefore, the first step in understanding the regulation of star formation is understanding the regulation of the internal cloud structure.

The major analytic theories that predict star formation rates in the ISM encapsulate the information about the internal cloud structure in the *probability density function* of gas densities (ρ-PDF). The function describes the probability of a unit volume to have a certain density. The function is a basic ingredient in virtually all commonly used analytic theories that start from the basic physical processes acting in the ISM (reviewed in Padoan *et al.* 2014, see also Federrath & Klessen 2012). Recent studies have also established that the distributions of *column* densities (N-PDFs) can carry the same information as the ρ-PDFs (e.g., Brunt *et al.* 2010; Federrath & Klessen 2013; Kainulainen *et al.* 2014), enabling their use as observationally reachable proxies of the ρ-PDFs.

The missing piece of the star formation rate models is that the ρ-PDF function has not been observationally well constrained. And by lack of observational constraints, the models have turned to numerical simulations for information about it. These simulations have heavily concentrated in modelling supersonic turbulence, including also other physics (self-gravity, magnetic fields, chemistry) as computational capabilities have begun to allow it. The basic result of the simulations is that isothermal, supersonic, driven turbulence develops a *lognormal* ρ-PDF (e.g., Vázquez-Semadeni 1994; Federrath *et al.*

2010; Molina *et al.* 2012). Consequently, the star formation rate theories commonly adopt this framework and with it the lognormal ρ-PDF. This means that the key ingredient of the models remains not confronted by systematic observations.

2. Universal link between the N-PDFs and star formation?

Observational works are now in progress of changing this picture; they are beginning to provide systematic constraints for the column density statistics of molecular clouds. This progress has been made possible by the novel dust extinction mapping techniques in near-infrared and *Herschel* dust emission measurements in far-infrared and sub-mm (e.g., Lombardi 2009; André *et al.* 2010). These techniques use dust to trace the column density distributions of the clouds and can provide high-sensitivity maps in resolution that reaches some 0.05 pc in the Solar neighbourhood clouds. Such resolution means that the resulting maps contain hundreds of thousands of independent resolution elements, enabling studies of the column density statistics.

The above techniques have already given us a basic picture of the N-PDFs. Now more than six years ago, we performed the first systematic survey of the N-PDFs in the Solar neighbourhood clouds (Kainulainen *et al.* 2009). In this work, we analysed dust extinction-derived column density maps of practically all clouds closer than 260 pc to the Sun. This work gave rise to two important results. First, we showed that the N-PDFs of most clouds are poorly fitted with log-normal functions when considering their entire column density range. Rather, they showed a possibly log-normal shape only at low-column densities ($N(\mathrm{H}) < 4 - 6 \times 10^{21}$ cm^{-2}), while higher column densities showed a significant, power-law like excess to that shape. This was an important result, as it seemed to at least partially contradict the prediction adopted from the turbulence-regulated ISM framework that molecular clouds might carry log-normal ρ-PDFs (and N-PDFs, e.g., Federrath & Klessen 2013).

The second important result from Kainulainen *et al.* (2009) was that the N-PDFs showed a variety of shapes that correlated with the number of young stars in the clouds. Specifically, clouds with more young stellar objects (YSOs) showed more top-heavy N-PDFs than those with less YSOs. In the extreme of this relation, the couple of clouds in the sample that had no young stars within had very bottom-heavy, close-to log-normal N-PDFs. We quantified this trend better later by showing that the mean star formation rate per unit area in entire molecular clouds correlates with the top-heaviness of their ρ-PDFs (Kainulainen *et al.* 2014).

Studies employing *Herschel* data have found the same trend between the star formation activity and the shape of the N-PDFs (e.g., Schneider *et al.* 2013; Alves de Oliveira *et al.* 2014). They have also raised a question whether the N-PDFs of *any* clouds are log-normal, or are all N-PDFs better described by power-laws (Lombardi *et al.* 2015). The first *Herschel* studies have also reached outside the Solar neighbourhood, towards more massive molecular clouds in the Galaxy (e.g., Schneider *et al.* 2015).

One great advantage of the high resolution of the *Herschel* data is that it allows us to zoom in and study the N-PDFs *within* molecular clouds in the Solar neighbourhood (e.g., Stutz & Kainulainen 2015, see Fig. 1). This allows linking the N-PDFs not only to the *mean* star formation activity of the clouds, but also to their *local* star formation rates. This is possible, because the protostars are identifiable in the Solar neighbourhood clouds, e.g., using *Herschel* data and/or near- and mid-infrared colour selections. The first works taking advantage of this have found that in parsec-scale regions within clouds the number of Class 0 protostars per unit area correlates with the top-heaviness of the

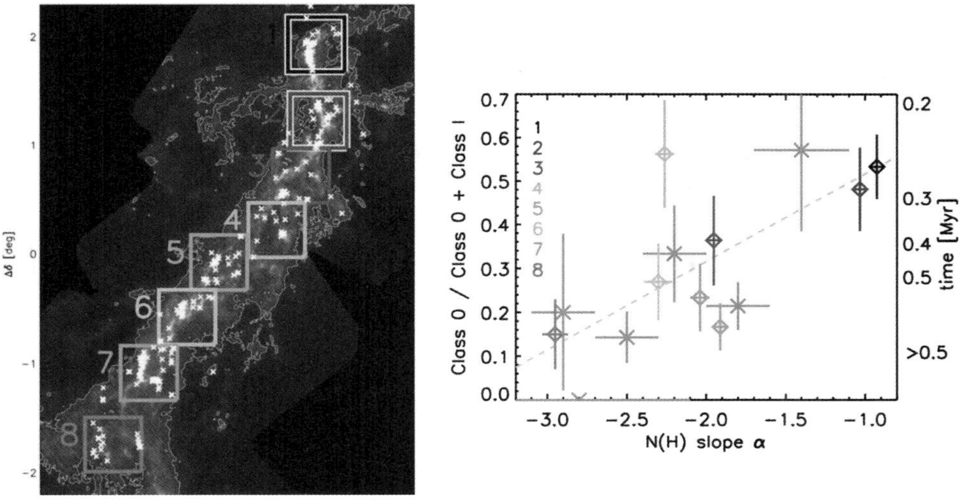

Figure 1. Left: Column density map of Orion A derived using *Herschel* observations (Stutz & Kainulainen 2015). The white crosses show the protostars in the cloud. The boxes show the eight regions in which the relation between protostars and N-PDFs was studied. Right: Class 0 protostar fraction as a function of the N-PDF power-law slope in the eight regions within Orion A (see the left panel). The y-axis on the right gives the evolutionary time-scale assuming a constant star formation rate. The grey data points are for the Perseus cloud (Sadavoy 2013; Sadavoy *et al.* 2014).

N-PDFs (Sadavoy *et al.* 2014; Stutz & Kainulainen 2015). This means that at these scales the *on-going* star formation rate activity is linked to the local gas mass distribution.

Intriguingly, the first small-scale studies have found that the evolutionary time-scale of parsec-scale regions inferred from Class 0 and 1 protostars *anti-correlates* with the top-heaviness of the N-PDFs (e.g., Stutz & Kainulainen 2015). In other words, regions with more top-heavy N-PDFs have *shorter* protostellar time-scales. This is curious, as it suggests that regions with flatter N-PDFs may be younger in terms of their recent star formation. Such a result is not trivial to interpret in the light of current models for N-PDF evolution that predict evolution from a bottom-heavy, turbulence-dominated state to a top-heavy, gravity-dominated state (e.g., Federrath & Klessen 2013; Girichidis *et al.* 2014; Ward *et al.* 2014).

Is the strong relationship between star formation and density distributions also prevalent at Galactic scales? The answer seems to be yes. Abreu-Vicente *et al.* (2015) performed the first systematic Galactic scale study of the N-PDFs, analysing 330 molecular clouds using the ATLASGAL dust emission survey (Fig. 2). The main result of this work was that indeed, objects that showed no signs of on-going star formation had the most bottom-heavy N-PDFs. Star-forming clouds that harbour protostars had more top-heavy N-PDFs, and finally, the most active clouds that contained HII regions had the most top-heavy N-PDFs (Fig. 2). This finding showed that the correlation between the N-PDFs and star formation exists also at Galactic scales; therefore, it must represent a fundamentally important relation describing star formation in the ISM.

Finally, indications of a similar trend have also been detected in external galaxies. Using interferometic observations of CO emission from the entire M51 galaxy, Hughes *et al.* (2013) showed that regions that had higher surface density of star clusters also showed more top-heavy N-PDFs. These observations again evidence a fundamental correlation between the column density structure and star formation activity.

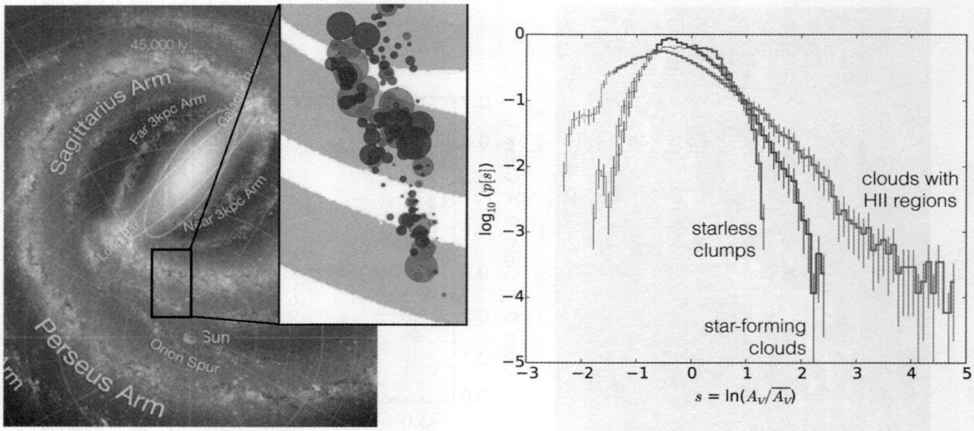

Figure 2. Left: The first Galactic-scale study of the molecular cloud N-PDFs (Abreu-Vicente *et al.* 2015, partial credit: NASA/JPL-Caltech/R. Hurt, SSC-Caltech). The red, blue, and green colours refer to clouds containing HII regions, star-forming clouds, and starless clumps. Right: The total N-PDFs of each object class. The starless clumps have the most bottom-heavy N-PDFs and the clouds containing HII regions the most top-heavy N-PDFs.

While the last half a decade has seen an entirely new set of observational constraints for the column density statistics of the Solar neighbourhood clouds, works currently in progress are expected to extend that set to Galactic volumes. Such an observational view can confront the star formation theories, and the entire turbulence-regulated ISM framework, at the scales relevant for Galactic star formation.

References

Abreu-Vicente, J., Kainulainen, J., Stutz, A., Henning, T., & Beuther, H. 2015, *A&A*, 581, A74
Alves de Oliveira, C., Schneider, N., Merín, B., *et al.* 2014, *A&A*, 568, A98
André, P., Men'shchikov, A., Bontemps, S., *et al.* 2010, *A&A*, 518, L102
Brunt, C. M., Federrath, C., & Price, D. J. 2010, *MNRAS*, 405, L56
Evans, N. J., II, Heiderman, A., & Vutisalchavakul, N. 2014, *ApJ*, 782, 114
Federrath, C. & Klessen, R. S. 2013, *ApJ*, 763, 51
Federrath, C. & Klessen, R. S. 2012, *ApJ*, 761, 156
Federrath, C., Roman-Duval, J., Klessen, R. S., *et al.* 2010, *A&A*, 512, A81
Girichidis, P., Konstandin, L., Whitworth, A. P., & Klessen, R. S. 2014, *ApJ*, 781, 91
Kainulainen, J., Federrath, C., & Henning, T. 2014, *Science*, 344, 183
Kainulainen, J., Beuther, H., Henning, T., & Plume, R. 2009, *A&A*, 508, L35
Lada, C. J., Forbrich, J., Lombardi, M., & Alves, J. F. 2012, *ApJ*, 745, 190
Lombardi, M. 2009, *A&A*, 493, 735
Lombardi, M., Alves, J., & Lada, C. J. 2015, *A&A*, 576, L1
Molina, F. Z., Glover, S. C. O., Federrath, C., & Klessen, R. S. 2012, *MNRAS*, 423, 2680
Padoan, P., Federrath, C., Chabrier, G., *et al.* 2014, in: H. Beuther, R. Klessen, C. Dullemond, T. Henning (eds.), *Protostars and Planets VI*, p. 77
Sadavoy, S. 2013, PhD Thesis, University of Victoria.
Sadavoy, S. I., Di Francesco, J., André, P., *et al.* 2014, *ApJL*, 787, L18
Schneider, N., André, P., Könyves, V., *et al.* 2013, *ApjL*, 766, L17
Schneider, N., Csengeri, T., Klessen, R. S., *et al.* 2015, *A&A*, 578, A29
Stutz, A. M. & Kainulainen, J. 2015, *A&A*, 577, L6
Vázquez-Semadeni, E. 1994, *ApJ*, 423, 681
Ward, R. L., Wadsley, J., & Sills, A. 2014, *MNRAS*, 445, 1575

From Interstellar Clouds to Star-Forming Galaxies: Universal Processes?
Proceedings IAU Symposium No. 315, 2015
P. Jablonka, Ph. André & F. van der Tak, eds.

doi:10.1017/S1743921316007286

An Observational Perspective of the IMF: Progress and Challenges

Stella S. R. Offner
University of Massachusetts,
Amherst, MA, USA
email: soffner@astro.umass.edu

Abstract. The stellar initial mass function (IMF) is a fundamental astrophysical quantity that impacts a wide range of astrophysical problems from heavy element distribution to galactic evolution to planetary system formation. However, the origin and universality of the IMF are hotly debated both observationally and theoretically. I review recent observations of the IMF across a variety of environments. These suggest the IMF is surprisingly invariant between star-forming regions, star clusters, and spiral galaxies but that it may also vary under extreme conditions, including within the Galactic center and early type galaxies.

Keywords. star formation, stars, initial mass function

1. Introduction

The shape of the stellar initial mass function (IMF) has been debated for more than fifty years. Since the underlying physics of star formation is complex and nonlinear, however, our view of the IMF rests on observational studies of individual stars (Bastian *et al.* 2010). Observational biases, dynamical evolution and poor statistics complicate the situation, and the result is an IMF constructed from a jigsaw of old and young stellar populations, dense and open star clusters, and photometric and spectroscopic approaches. Of fundamental importance is whether the IMF is universal (Offner *et al.* 2014). This problem has reached a new urgency with the proliferation of studies of massive elliptical galaxies, with properties very unlike our own, in which individual stars cannot be resolved. Here I briefly review observational studies that form the basis of our knowledge of the IMF and discuss recent progress and current challenges in this field.

2. Functional Forms for the IMF

The IMF is typically expressed as a number of stars per linear or logarithmic bin in mass. The functional form for the IMF is derived empirically from observational studies of Milky Way field stars or star clusters. Observational uncertainties are sufficiently large (see §3) that a variety of analytical forms put forth in the literature are statistically consistent with the data. An early determination of the IMF by Salpeter (1955) found that the number of stars $dN/d\log M \propto M^{-1.35}$. Subsequent studies identified a peak mass around a few tenths of a solar mass. This low-mass turnover can be expressed by broken power-laws (Kroupa 2001), a log-normal (Chabrier 2003, 2005), or an exponential (de Marchi & Paresce 2001). Table 1 summarizes these common functional forms. As shown in Figure 1, sixty years later the Salpeter IMF is widely adopted for stars with masses above 1 $M_\odot$. Meanwhile, the dispersion in the placement of the peak underscores the current level of uncertainty at lower stellar masses. The curves diverge below the brown dwarf mass limit, 0.08 $M_\odot$, which highlights the magnitude of the disagreement regarding

Table 1. Common functional forms for the IMF.

Citation	Peak Mass($M_\odot$)	High-Mass Slope, α	$dN/d\log M$[1]
Salpeter 1955	...	−1.35	$M^{-1.35}$
Kroupa 2001	0.08	−1.3	$0.01 < M < 0.08$: $M^{0.7}$, $0.08 \leqslant M < 0.5$: $M^{-0.3}$, $M \geqslant 0.5$: $M^{-1.3}$
de Marchi & Paresce 2001	0.3	−1.3	$M^{-1.3}(1 - \exp[-(M/0.3)^{2.6}])$
Chabrier 2003	0.08	...[2]	$M \leqslant 1.0 : \exp(-[\log_{10}(M) - \log_{10}(0.08)]^2/(2 \times 0.69^2))$
Chabrier 2005	0.2	−1.35	$M \leqslant 1.0 : \exp(-[\log_{10}(M) - \log_{10}(0.2)]^2/(2 \times 0.55^2))$, $M \leqslant 1.0 : M^{-1.35}$

Notes:
[1] Mass M in units of $M_\odot$. Forms are unnormalized.
[2] Chabrier (2003) does not provide the IMF form above 1 $M_\odot$, but extragalactic studies applying this form commonly assume a Salpeter slope.

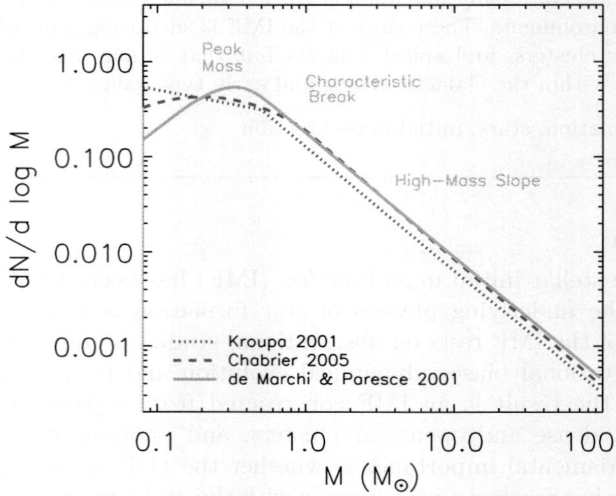

Figure 1. Three proposed functional forms for the IMF, which are calibrated using studies of the population of field stars and star clusters.

very low mass objects. The IMF is sometimes given as a system IMF, which ignores the individual stellar components of binary and higher order star systems in favor of counting the total system mass. However, the single-star and system IMF shapes appear similar (Chabrier 2005).

3. Studies of Resolved Populations

3.1. *Observational Procedures and Uncertainties*

Studies of nearby resolved stars provide the most direct avenue for determining the IMF. However, even when individual stars may be measured, a number of steps are involved, each of which introduces new complications and challenges. The first step of the process entails identifying a volumetrically complete sample of stars and measuring their luminosities. Star clusters are ideal targets since they are spatially concentrated and tend to have similar ages and distances. Next, the luminosity function (LF) must be converted to a present-day mass function using models for the relationship between magnitude and stellar mass. Finally, this "present-day" mass function must be corrected for stars missing due to dynamical or stellar evolution as well as for unresolved binaries.

Despite proximity, studies of resolved population suffer from significant systematic uncertainties. These uncertainties can be divided into those associated with measuring the LF and those associated with theoretical models. Foremost, completeness and contamination can significantly bias the LF. Background galaxies and non-member stars (for a cluster) must be identified and removed. Field star LFs must be corrected for Malmquist bias and distance uncertainties. Star clusters require follow-up proper-motion studies to accurately determine cluster membership. Patchy or differential dust extinction afflicts studies of young regions (<10 Myr). These combined corrections significantly reduce the number of sources, and thus, the remaining sample may have poor number statistics, requiring special care. The centers of dense clusters require corrections for crowding. Finally, since half of all stars are in binary systems (Duchêne & Kraus 2013), statistical corrections must be applied to account for unresolved multiplicity. Stellar multiplicity is observed to vary with age in star forming regions, however, whether the multiplicity distribution varies with environment is poorly constrained.

Theoretical uncertainties are equally critical. Mass-magnitude models are sensitive to underlying assumptions about stellar evolution, and the model choice can result in order unity differences in inferred stellar properties (e.g., Da Rio *et al.* 2012). Pre-main sequence models for stars <10 Myr are especially sensitive to the initial assumptions for the starting stellar properties and how mass is deposited on the star (Hosokawa *et al.* 2011; Baraffe *et al.* 2012). Low-mass stars are magnetically active, which can impact measured temperatures and colors (Stassun *et al.* 2014). Finally, corrections for dynamical interactions, which create mass segregation and can preferentially remove low-mass members, also add uncertainty to older populations.

3.2. *Universality and Outliers*

Whether or not the IMF is universal hinges on the existence of *systematic* changes in the IMF form as a function of some underlying physical variable. If a given observed IMF is a statistical sample drawn from a universal function, then some deviations are expected, especially for small stellar populations. For example, in very rare cases a high-mass star may form in relative isolation (Offner *et al.* 2014, references therein).

As shown in Figure 2, many studies probe only a narrow range of masses and do not fully sample the IMF. Consequently, the size of the observational errors, as noted above, and the difficulty of deriving an IMF from a present-day mass function prohibit a clear determination of universality. However, despite this challenge there is broad consensus that the Milky Way field population, bulge population, and young clusters within the Milky Way all share a common IMF (Bochanski *et al.* 2010; Calamida *et al.* 2015; Alves de Oliveira *et al.* 2013). Likewise, a recent comprehensive study of young intermediate mass star clusters within M31 confirmed an invariant, universal high-mass IMF slope as a function of cluster mass, size and age (Weisz *et al.* 2015). The bottom line is that *the IMF is remarkably invariant across a broad range of environments within observational uncertainties.*

Despite a broad consensus for the IMF form among resolved populations, debate continues for a few individual regions. For example, Taurus, a young low-mass star forming region, appears to have an excess of solar mass stars (Luhman 2012). It is unclear whether this is due to small statistical sampling or has some physical basis. The Taurus region is not physically distinct compared to other young regions, which don't exhibit evidence

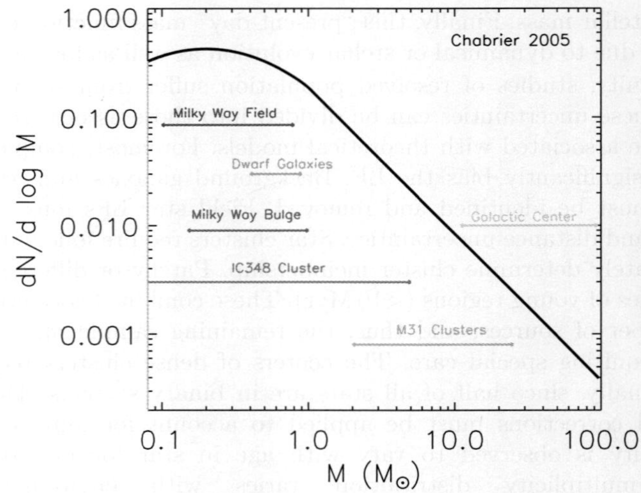

Figure 2. Chabrier IMF over-plotted with horizontal lines indicating stellar mass ranges for specific resolved population studies: the Milky Way field (Bochanski *et al.* 2010), local dwarf galaxies (Geha *et al.* 2013), the Milky Way bulge (Calamida *et al.* 2015), the Galactic center (Lu *et al.* 2013), the young nearby cluster IC348 (Alves de Oliveira *et al.* 2013), and a selection of intermediate mass star clusters in M31 (Weisz *et al.* 2015).

for IMF variations (e.g., Figure 2 in Offner *et al.* 2014), so there's no obvious physical rationale for variation.

Within the Milky Way there is one very conspicuous extreme star-forming environment: the Galactic center. Indeed the inner parsec of the Milky Way appears to have a present-day mass function with an excess of high-mass stars (e.g., Nayakshin & Sunyaev 2005; Bartko *et al.* 2010). Several independent groups have found that the high-mass slope is shallower than the Salpeter slope (e.g., $\alpha = -0.7 \pm 0.2$ versus $\alpha = -1.35$ for $M_* \geqslant 10\,M_\odot$, Lu *et al.* 2013). However, most recently Lu *et al.* (2013) find that after correcting for extinction only one bin in the Kp band LF deviates by more than 2σ from a Milky Way IMF. Considering the extreme temperatures, densities and radiation field in addition to uncertainties in age, dynamics, and statical sampling, the degree of variation is surprisingly small.

Recently, a few groups have constructed mass functions from resolved studies of ultra-faint dwarf galaxies. The smallest dwarfs are dark matter dominated and comprised of mere hundreds to thousands of detectible stars. Given their small size, the observed stars likely formed prior to reionization, at which point their star formation was quenched. Considered together, these environments may show some systematic IMF change and appear to have relatively more high-mass stars than expected (Geha *et al.* 2013). However, as shown in Figure 3, the observations probe a narrow region near the peak of the IMF, where the slope transitions from $\alpha = -1.35$ to $\alpha = 0$. The mass ranges span as little as $0.5\,M_\odot < M_* < 0.8\,M_\odot$. Wyse *et al.* (2002) in fact concluded that the IMF slope of the Ursa Minor dwarf spheroidal was consistent with that of globular clusters, and hence, supported an invariant IMF. Although it's possible that these data in combination with extragalactic unresolved populations signal a broader IMF trend with stellar velocity dispersion (§4), the significance of the effect is very tentative given the magnitude of the observational errors, which place the data within 1σ of the Milky Way IMF.

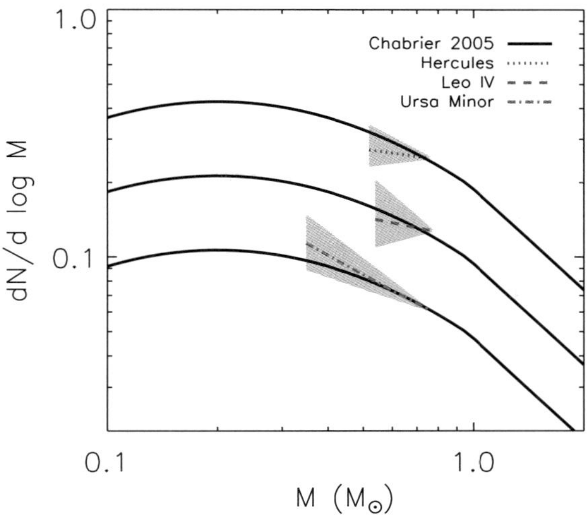

Figure 3. The IMF slopes obtained for three local dwarf galaxies using resolved population studies (Geha *et al.* 2013, Wyse *et al.* 2002): $\alpha = -0.2^{+0.4}_{-0.5}$ (Hercules), $\alpha = -0.3 \pm 0.8$ (Leo IV), and $\alpha = -0.8$ (Ursa Minor). The dwarfs have visual magnitudes $M_v = -6.2$, -5.5 and -9.2, respectively. The 1σ uncertainty in the slope is indicated by the grey shaded area, where 15% uncertainly is adopted for the latter case based on the LF error. The lines are offset for clarity.

4. Studies of Unresolved Populations

4.1. *Observational Procedures and Uncertainties*

IMF studies of unresolved stellar populations probe the large-scale mass distribution and integrated light and then attempt to back out the underlying stellar population. Compared to resolved population studies, these approaches have less sensitivity to particular masses and, consequently, details of the functional form are often lost. Consequently, extragalactic stellar populations are generally characterized as being similar to a Milky Way IMF (e.g., Kroupa 2001 or Chabrier 2003), Salpeter IMF ($\alpha = -1.35$ down to $0.1\,M_\odot$), "top-heavy" IMF ($\alpha < -1.35$ down to $0.1\,M_\odot$), or "bottom-heavy" IMF ($\alpha > -1.35$ down to $0.1\,M_\odot$). The latter are often referred to as "bottom-light" and "top-light" IMFs, respectively. By definition, a Salpeter IMF is also slightly bottom-heavy compared to a Milky Way IMF.

Stellar population synthesis (SPS) plays an important role in most unresolved population studies. Given some functional form for the stellar mass distribution, $\Phi(M)$, SPS calculates the integrated flux as a function of wavelength: $F_\lambda = \int \Phi(M) f_\lambda dM$, where $f_\lambda(M)$ is the spectrum of a star with mass M. This approach requires assumptions for the age and metallicity of the population. Together with mass determinations from gravitational lensing (Shu *et al.* 2015) or stellar dynamics (Cappellari *et al.* 2012), it is possible to construct the mass-to-light ratio. High mass-to-light ratios indicate a larger fraction of low-mass stars and correspond to a steeper, bottom-heavy IMF, while lower mass-to-light ratios indicate a relative excess of high-mass stars and correspond to a top-heavy IMF.

One caveat to integrated approaches is that the total mass is the sum of the stellar and dark matter masses. Extracting the stellar mass requires modeling the underlying distribution of dark matter, which cannot be directly measured. In principle, any variations in the IMF can be removed with an appropriate choice of dark matter profile. However, this is generally discouraged since nontrivial revisions to cosmological models would be necessary in some cases (e.g., Cappellari *et al.* 2012).

An alternative form of SPS involves modeling gravity sensitive spectral features found in stellar atmospheres (Conroy & van Dokkum 2012). This technique is promising since it provides a means of assessing the relative numbers of dwarf and giant stars that is independent of the total mass distribution. In addition, it reproduces the Milky Way IMF found in globular clusters (van Dokkum & Conroy 2011). However, spectral SPS depends on a large number of parameters, including initial elemental abundances and the star formation history. Consequently, it is sensitive to underlying abundance gradients, which are present and not well constrained for most extragalactic targets (McConnell *et al.* 2015). SPS also requires calibration against resolved stellar populations, which are not available for extreme metallicity values.

4.2. *Observational Trends*

The underlying question motivating unresolved studies is whether the IMF varies globally as a function of galaxy type and changes across cosmic time. Due to the challenging nature of the extragalactic observations, studies carried out by various groups using a variety of techniques have historically yielded conflicting, inconsistent and evolving answers to the question of IMF universality.

In recent years, significant attention has been devoted to studies of ancient early type (elliptical) galaxies (ETGs). These galaxies formed most of their stars in a burst in the early universe with little subsequent star formation. Consequently, their present-day stellar populations are old ($\sim$10 Gyr) and contain only relatively low-mass stars ($M_* \lesssim 1\,M_\odot$); they are "red and dead". Their stellar densities and ages are similar to that of globular clusters, but they have higher average metallicities ($Z \gtrsim 0.5\,Z_\odot$). It seems plausible that these extreme conditions hosted significantly different early star formation environments than spiral galaxies like the Milky Way.

Over the last five years a variety of approaches have targeted ETGs and found apparent changes in the IMF for those with the highest stellar velocity dispersions. Spectral feature SPS methods (van Dokkum & Conroy 2010; van Dokkum & Conroy 2012; Spiniello *et al.* 2014), SPS combined with dynamical observations (Callipari *et al.* 2012; Dutton *et al.* 2012; McDermid *et al.* 2014), and SPS combined with gravitational lensing (Treu *et al.* 2010; Spiniello *et al.* 2012; Shu *et al.* 2015) have all concluded that the relative number of low-mass stars increases with increasing velocity dispersion. In concert, these results provide tantalizing evidence for systematic change in the IMF as a function of environment. Confidence is enhanced by the agreement between independent groups using slightly different methods (Conroy *et al.* 2013).

Nonetheless, there are outstanding issues. There is substantial scatter in the inferred IMF slopes with some massive ETGs appearing consistent with a Milky Way IMF. It's unclear whether this is due to poorly characterized uncertainties or arises from actual variation in some additional unidentified physical parameter. There is some evidence that metallicity contributes to the apparent variation (Martin-Navarro *et al.* 2015; McDermid *et al.* 2014), but the extent of the dependence is debated. Upon closer inspection, the methods frequently disagree on the IMF slope of individual galaxies (e.g., Smith & Lucey 2013; Smith *et al.* 2015). This suggests that the IMF of individual objects remains very uncertain.

The confluence of uncertainties allows leeway for alternative explanations for the apparent IMF trends. Maccarone (2014) suggests that variations in the Wing-Ford spectral feature used by Conroy & van Dokkum can be explained by extrinsic S stars (a class of star with enhanced s-process element abundances). These types of stars are neglected by SPS models but could be enhanced in the cores of giant ellipticals. However, it is noted that they might also enhance the calcium triplet line, which is observed to be weak. If ETGs have bottom-heavy IMFs, Peacock *et al.* (2014) note that they should also have smaller populations of neutron stars and black holes, which affect the mass-to-light ratio. They use K-band light to study the low-mass X-ray binary populations of a number of ETGs and find that it is uniform across the sample. This, too, undermines the argument for IMF variation. Many observations of integrated light have poor resolution within a galaxy and thus average over large areas. Consequently, variation as a function of age or metallicity gradients within the galaxy may masquerade as a varying IMF (McConnell *et al.* 2015). If gradients in spectroscopy features do correlate with IMF variations, then a bottom-heavy IMF may be limited to a relatively small volume within the galaxy (Martín-Navarro *et al.* 2015). At minimum, this emphasizes that the lingering uncertainties are formidable.

Consequently, there is significant room for variation. Given the age of the stellar populations in ETGs, the integrated light probes a relatively narrow range of stellar masses (e.g., $0.1\,M_\odot \leqslant M < 1\,M_\odot$; Conroy & van Dokkum 2012). As a result, current evidence for slope variation cannot distinguish between a changing characteristic mass, a changing slope or some combination of the two. This presents a significant challenge for connecting observations with the prior physical conditions of star formation. Making this connection is essential for building a firm theoretical framework for the IMF and understanding the basis of IMF features.

5. Progress and Challenges

In summary, *the IMF is surprisingly invariant, but it may vary under extreme conditions.* The apparent robustness of the IMF form both within the Milky Way and across a broad range of extragalactic environments is surprising given variations in metallicity, gas density, stellar density, and radiative background over several orders of magnitude. There are a few intriguing locations where the IMF may differ. These include the inner parsec of the Galactic center, giant elliptical galaxies (although possibly only their centers), and ultra faint dwarfs. However, it is worth noting that all these environments combined comprise a small fraction of the universe, thus reinforcing that the IMF in the majority of environments and across cosmic time is remarkably uniform.

One major piece of progress is an emerging consensus that the fraction of low-mass stars increases with increasing stellar velocity dispersion. While disagreement continues around the IMF of specific galaxies, this trend is found by several independent methods, including stellar population synthesis, gravitational lensing and dynamical arguments.

Despite increasing observational agreement, the physical basis for IMF variation remains unconstrained. The underlying physics responsible for star formation and the origin of the IMF is debated. One particular challenge entails linking present-day stellar observables to the natal conditions of star formation. Power-law fits to the observations are currently incapable of discriminating between a changing characteristic stellar mass and evolution in the relative production of high- and low-mass stars. However, with continuing improvements in stellar models this may be within reach of future work.

A second challenge involves reconciling inconsistencies in the observations. Trends with velocity dispersion show a large amount of scatter; some ellipticals with very large dispersions appear consistent with the Milky Way IMF. This may be due to uncertainties in the techniques or to some other unidentified variable. Globular clusters, which resemble massive elliptical galaxies in terms of their stellar densities, display apparently normal IMFs. Determining why globular clusters do not exhibit variation will shed light on their origin and provide revealing clues about the process of star formation, generally.

Finally, a comprehensive, predictive theory for the IMF remains elusive. A successful theory must explain both the apparent universality of the IMF across a broad range of environments and deviation in select extreme environments. The theory must take into account gas metallicity, cloud properties, and stellar feedback. Until current models succeed in fully explaining star formation within the Milky Way, progress towards understanding extreme environments will be evasive.

References

Alves de Oliveira, C., Moraux, E., Bouvier, J., *et al.* 2013, *A&A*, 549, A123

Baraffe, I., Vorobyov, E., & Chabrier, G. 2012, *ApJ*, 756, 118

Bartko, H., Martins, F., Trippe, S., *et al.* 2010, *ApJ*, 708, 834

Bastian, N., Covey, K. R., & Meyer, M. R. 2010, *ARA&A*, 48, 339

Bochanski, J. J., Hawley, S. L., Covey, K. R., *et al.* 2010, *AJ*, 139, 2679

Calamida, A., Sahu, K. C., Casertano, S., *et al.* 2015, *ApJ*, 810, 8

Cappellari, M., McDermid, R. M., Alatalo, K., *et al.* 2012, *Nature*, 484, 485

Chabrier, G. 2003, *ApJL*, 586, L133

Chabrier, G. 2005, The Initial Mass Function 50 Years Later, 327, 41

Conroy, C., Dutton, A. A., Graves, G. J., Mendel, J. T., & van Dokkum, P. G. 2013, *ApJL*, 776, L26

Da Rio, N., Robberto, M., Hillenbrand, L. A., Henning, T., & Stassun, K. G. 2012, *ApJ*, 748, 14

de Marchi, G. & Paresce, F. 2001, *Astronomische Gesellschaft Meeting Abstracts*, 18, 551

Duchêne, G. & Kraus, A. 2013, *ARA&A*, 51, 269

Dutton, A. A., Mendel, J. T., & Simard, L. 2012, *MNRAS*, 422, L33

Geha, M., Brown, T. M., Tumlinson, J., *et al.* 2013, *ApJ*, 771, 29

Hosokawa, T., Offner, S. S. R., & Krumholz, M. R. 2011, *ApJ*, 738, 140

Kroupa, P. 2001, *MNRAS*, 322, 231

Lu, J. R., Do, T., Ghez, A. M., *et al.* 2013, *ApJ*, 764, 155

Luhman, K. L. 2012, *ARA&A*, 50, 65

Maccarone, T. J. 2014, *MNRAS*, 442, L5

Martín-Navarro, I., Barbera, F. L., Vazdekis, A., Falcón-Barroso, J., & Ferreras, I. 2015, *MNRAS*, 447, 1033

McConnell, N. J., Lu, J. R., & Mann, A. W. 2015, arXiv:1506.07880

McDermid, R. M., Cappellari, M., Alatalo, K., *et al.* 2014, *ApJL*, 792, L37

Nayakshin, S. & Sunyaev, R. 2005, *MNRAS*, 364, L23

Offner, S. S. R., Clark, P. C., Hennebelle, P., *et al.* 2014, Protostars and Planets VI, 53

Peacock, M. B., Zepf, S. E., Maccarone, T. J., *et al.* 2014, *ApJ*, 784, 162

Salpeter, E. E. 1955, *ApJ*, 121, 161

Shu, Y., Bolton, A. S., Brownstein, J. R., *et al.* 2015, *ApJ*, 803, 71

Smith, R. J., Lucey, J. R., & Conroy, C. 2015, *MNRAS*, 449, 3441

Smith, R. J. & Lucey, J. R. 2013, *MNRAS*, 434, 1964

Spiniello, C., Trager, S. C., Koopmans, L. V. E., & Chen, Y. P. 2012, *ApJL*, 753, L32

Spiniello, C., Trager, S., Koopmans, L. V. E., & Conroy, C. 2014, *MNRAS*, 438, 1483

Stassun, K. G., Scholz, A., Dupuy, T. J., & Kratter, K. M. 2014, *ApJ*, 796, 119

Treu, T., Auger, M. W., Koopmans, L. V. E., *et al.* 2010, *ApJ*, 709, 1195

van Dokkum, P. G. & Conroy, C. 2011, *ApJL*, 735, L13

Weisz, D. R., Johnson, L. C., Foreman-Mackey, D., *et al.* 2015, *ApJ*, 806, 198

Wyse, R. F. G., Gilmore, G., Houdashelt, M. L., *et al.* 2002, *NA*, 7, 395

From Interstellar Clouds to Star-Forming Galaxies: Universal Processes?
Proceedings IAU Symposium No. 315, 2015
P. Jablonka, Ph. André & F. van der Tak, eds.

doi:10.1017/S1743921316007298

The role of interstellar filaments in regulating the star formation efficiency and shaping the initial mass function

Vera Könyves and Philippe André

Laboratoire AIM, CEA/DSM-CNRS-Université Paris Diderot, IRFU/Service d'Astrophysique, CEA Saclay, 91191 Gif-sur-Yvette, France
emails: `vera.konyves@cea.fr`, `philippe.andre@cea.fr`

Abstract. Recent surveys at infrared and submillimeter wavelengths with the *Spitzer* and *Herschel* space observatories suggest that star formation in dense molecular gas is governed by essentially the same "laws" in nearby Galactic clouds and distant external galaxies. This raises the possibility of a unified picture for star formation in the Universe from individual-cloud scales to galaxy–wide scales. We summarize the star formation scenario favored by *Herschel* studies of the nearest molecular clouds of the Galaxy which point to the key role of the quasi-universal filamentary structure pervading the cold interstellar medium.

Keywords. stars: formation, ISM: clouds, ISM: structure, ISM: individual objects (Aquila Rift complex), submillimeter

---

1. The filamentary interstellar medium

Interstellar filaments were already seen earlier in molecular clouds (e.g., Schneider & Elmegreen 1979; Hartmann 2002; Myers 2009), after which *Herschel* observations not only confirmed their omnipresence, but the ubiquity of filaments also revolutionised our view of star formation in the Galaxy.

The *Herschel* Gould Belt survey (HGBS, André *et al.* 2010) observations confirm the omnipresence of filaments in nearby molecular clouds and suggested a close connection between the filamentary structure of the cold ISM and the formation process of prestellar cores (André *et al.* 2010, Men'shchikov *et al.* 2010). Besides, other *Herschel* imaging programs such as HOBYS (Motte *et al.* 2010), HiGAL (Molinari *et al.* 2010), and EPoS (Henning *et al.* 2010) demonstrated the presence and importance of filaments in the interstellar medium. These filaments present a high degree of universality, including their typical inner radial width of $\sim$0.1 pc (e.g., Arzoumanian *et al.* 2011, Palmeirim *et al.* 2013, Koch & Rosolowsky 2015), and they likely play a central role in the star formation process (see for a review André *et al.* 2014).

2. Spatial distribution of prestellar cores

One of the main goals of the HGBS is to elucidate the physical mechanisms responsible for the formation and evolution of prestellar cores in molecular clouds. In the Aquila cloud complex based on *Herschel* data† we have identified a complete sample of 651 starless cores extracted with the multi-scale, multi-wavelength source finding algorithm *getsources* (Men'shchikov *et al.* 2012). Approximately 60% ± 10% of this sample are

† The *Herschel* products of the Aquila molecular cloud, including maps and catalogs, can be found at the *Herschel* Gould Belt Survey Archive, http://gouldbelt-herschel.cea.fr/archives.

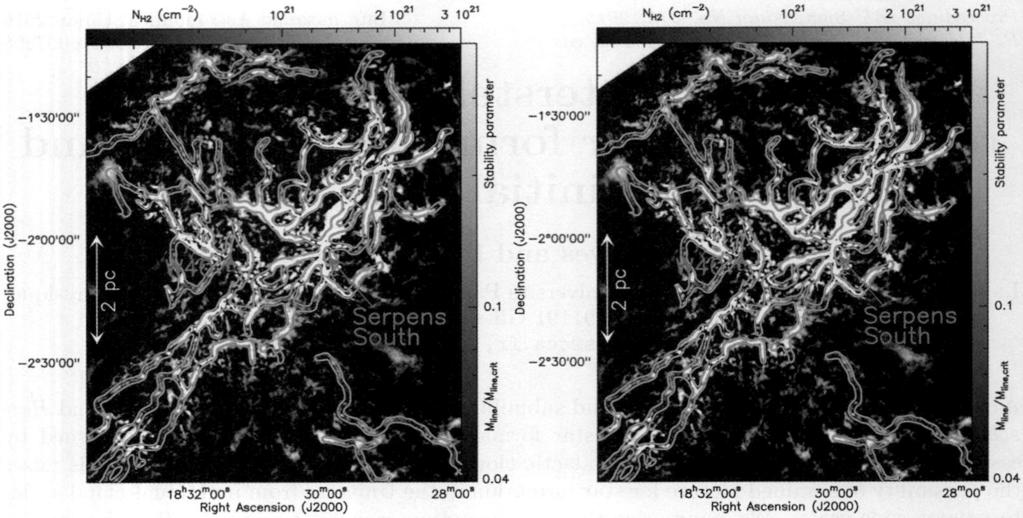

Figure 1. Left: Curvelet component of the *Herschel* column density map of Aquila, showing the main region in equivalent mass per unit length along the filaments (cf. André *et al.* 2010), which are likely gravitationally unstable (white region) above half the critical value $M_{\rm line,crit} = 2c_s^2/G$ (Inutsuka & Miyama 1997). The blue contours outline the 0.1 pc–wide footprints of the filaments (see details in Könyves *et al.* 2015). Right: Same map as in the left panel with the locations of candidate prestellar cores (blue triangles) and protostellar cores (green up-side-down triangles).

gravitationally bound prestellar cores, which may form (proto)stars in the future. We also detected 58 protostellar cores (Könyves *et al.* 2010, 2015; Maury *et al.* 2011).

The comparison of the census of dense prestellar cores and extracted filaments in Aquila suggest their close connection via the core/star formation process. We find that a high fraction of the dense gas mass is in the form of filaments above $A_{\rm V} \sim 7$. Furthermore, ~90% of our prestellar cores are located above an equivalent background column density of $\sim 7 \times 10^{21}$ cm^{-2}, and a similarly high portion of them lie within the densest filaments with supercritical masses per unit length $>16\,M_\odot$/pc (Könyves *et al.* 2015).

Fig. 1 shows the filamentary structure of the Aquila main field, traced by the curvelet transform component (Starck *et al.* 2003) of the column density map in the form of an equivalent stability parameter (mass per unit length) along the filaments. In the left panel the filaments are represented by their 0.1 pc–wide footprints traced with DisPerSE (Sousbie 2011), while in the right panel the candidate prestellar cores and protostellar cores are overplotted. The spatial correlation between the densest filaments and the prestellar cores is strong (see details in Könyves *et al.* 2015).

3. Prestellar CMF and star formation efficiency

In Aquila the core mass function of the large population of prestellar cores confirms earlier findings that the shape of the CMF is very similar to the stellar initial mass function (IMF), and there is a close physical link between the two (Fig. 2/left, Könyves *et al.* 2015).

The local core formation efficiency at the prestellar core level is high in Aquila, $\epsilon_{\rm core} \sim 30 - 40\%$, and only a small fraction (~15%) of the gas mass is in the form

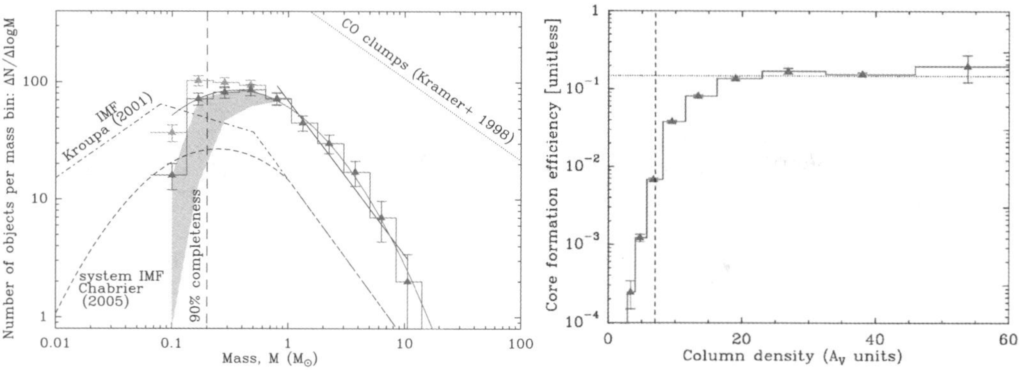

Figure 2. Left: Differential core mass function ($dN/d\log M$) of 651 starless cores (upper green histogram) and 446 candidate prestellar cores (lower blue histogram) identified with *Herschel* in the Aquila field. The shaded area reflects a fraction of $60\% \pm 10\%$ of the starless cores classified as gravitationally bound (see details in Könyves *et al.* 2015). The lognormal fit to the prestellar cores (red curve) peaks at $\sim 0.5\,M_\odot$, which is very similar in shape to the IMF of multiple systems (e.g., Chabrier 2005). The power-law fit (black line) has a slope of -1.33 ± 0.06. Right: Histogram of observed differential core formation efficiency (CFE) as a function of background column density in A_V units; $\mathrm{CFE_{obs}}(A_V) = \Delta M_{\mathrm{cores}}(A_V)/\Delta M_{\mathrm{cloud}}(A_V)$. The vertical dashed line marks the threshold at $A_V^{\mathrm{bg}} \sim 7$. The horizontal dotted line marks the rough asymptotic value of the CFE $\sim 15\%$, corresponding to a star formation rate per unit gas mass $\sim 5 \times 10^{-8}\ \mathrm{yr}^{-1}$.

of prestellar cores above $A_V \sim 7$ (André *et al.* 2014, Könyves *et al.* 2015). Invoking the typical filament width of $\sim$0.1 pc, this observed column density core formation "threshold" corresponds –within a factor of 2– to the threshold above which interstellar filaments become gravitationally unstable (Fig. 1, André *et al.* 2010, 2014).

In order to support further the threshold for core formation we found from the distribution of background cloud column densities of the prestellar cores (e.g., André *et al.* 2014, Könyves *et al.* 2015), we can inspect the observed core formation efficiency (CFE). The right panel of Fig. 2 shows the CFE in Aquila, obtained by dividing the mass in the form of prestellar cores in a given column density bin by the cloud mass observed in the same bin; $\mathrm{CFE_{obs}}(A_V) = \Delta M_{\mathrm{cores}}(A_V)/\Delta M_{\mathrm{cloud}}(A_V)$. The observations indicate a very steep rise in the range of $A_V \sim 5 - 15$, below which the prestellar CFE is negligible, and approximately constant, above. Despite of the fact that such realizations of the threshold are not infinitely sharp, we argue for the presence of a physical threshold for prestellar core formation around a fiducial value of $A_V^{bg} \sim 7$ (André *et al.* 2014, Könyves *et al.* 2015). With *Spitzer*, similar threshold has been observed in the spatial distribution of young stellar objects (YSOs) in nearby clouds (Heiderman *et al.* 2010, Lada *et al.* 2010, 2012, Evans *et al.* 2014).

Altogether, our *Herschel* findings support a picture according to which interstellar filaments and prestellar cores represent two fundamental steps in the formation process of solar-type stars (André *et al.* 2010, 2014): first, the dissipation of kinetic energy in large-scale MHD flows generates $\sim$0.1 pc–wide filaments in the cold ISM; then the densest filaments –above a critical threshold of $\sim$16 $M_\odot$/pc in mass per unit length– form prestellar cores (and ultimately protostars) by gravitational instability. The peak of the CMF (at $\sim 0.5\,M_\odot$ in Aquila) may also result primarily from the gravitational fragmentation of marginally critical filaments.

4. Filaments-driven universal star formation law in dense gas?

Our *Herschel* data in the Aquila Rift cloud complex allows us to directly infer the "efficiency" of the star formation process from the physics of prestellar core formation within filaments (see André *et al.* 2014, Könyves *et al.* 2015). The star formation rate per unit mass of dense gas above the threshold is estimated to be $\mathrm{SFR}/M_{\mathrm{dense}} \sim 5^{+2}_{-2} \times 10^{-8}\,\mathrm{yr}^{-1}$ (Fig. 2/right), similar to which was also found independently in nearby clouds by near-/mid infrared studies (Heiderman *et al.* 2010, Lada *et al.* 2010, 2012, Evans *et al.* 2014), and in other galaxies (Gao & Solomon 2004).

These works use different tracers of the dense gas which can make the comparison somewhat uncertain. We also have to bear in mind that the presented star formation processes may be different in some extreme environments, for instance, in the central molecular zone close to the Galactic center, where star formation is known to be more inefficient above the same density threshold (Longmore *et al.* 2013).

In spite of the uncertainties, the similar core formation thresholds may be interpreted as the consequence of the ubiquitous filamentary structure of the nearby molecular clouds, manifesting in a quasi-universal "star formation law" within dense gas above the (column) density threshold (see André *et al.* 2014).

Acknowledgements

This work has received support from the ERC 'ORISTARS' project (ERC Advanced Grant Agreements no. 291294).

References

André, P., Men'shchikov, A., Bontemps, S., *et al.* 2010, *A&A*, 518, L102+
André, P., Di Francesco, J., Ward-Thompson, D., *et al.* 2014, *in PPVI, ed. H. Beuther et al.*, 27
Arzoumanian, D., André, P., Didelon, P., *et al.* 2011, *A&A*, 529, L6
Chabrier, G. 2005, *ASSL*, 327, 41
Evans, II, N. J., Heiderman, A., & Vutisalchavakul, N. 2014, *ApJ*, 782, 114
Gao, Y. & Solomon, P. M. 2004, *ApJ*, 606, 271
Hartmann, L. 2002, *ApJ*, 578, 914
Heiderman, A., Evans, II, N. J., Allen, L. E., *et al.* 2010, *ApJ*, 723, 1019
Henning, T., Linz, H., Krause, O., *et al.* 2010, *A&A*, 518, L95
Inutsuka, S.-I. & Miyama, S. M. 1997, *ApJ*, 480, 681
Koch, E. W. & Rosolowsky, E. W. 2015, *MNRAS*, 452, 3435
Könyves, V., André, P., Men'shchikov, A., *et al.* 2010, *A&A*, 518, L106+
Könyves, V., André, P., Men'shchikov, A., *et al.* 2015, *A&A*, in press, arXiv:1507.05926
Kramer, C., Stutzki, J., Rohrig, R., & Corneliussen, U. 1998, *A&A*, 329, 249
Kroupa, P. 2001, *MNRAS*, 322, 231
Lada, C. J., Lombardi, M., & Alves, J. F. 2010, *ApJ*, 724, 687
Lada, C. J., Forbrich, J., Lombardi, M., & Alves, J. F. 2012, *ApJ*, 745, 190
Longmore, S. N., Bally, J., Testi, L., *et al.* 2013, *MNRAS*, 429, 987
Maury, A. J., André, P., Men'shchikov, *et al.* 2011, *A&A*, 535, A77
Men'shchikov, A., André, P., Didelon, P., *et al.* 2010, *A&A*, 518, L103
Men'shchikov, A., André, P., Didelon, P., *et al.* 2012, *A&A*, 542, A81
Molinari, S., Swinyard, B., Bally, J., *et al.* 2010, *A&A*, 518, L100
Motte, F., Zavagno, A., Bontemps, S., *et al.* 2010, *A&A*, 518, L77
Myers, P. C. 2009, *ApJ*, 700, 1609
Palmeirim, P., André, P., Kirk, J., *et al.* 2013, *A&A*, 550, A38
Schneider, S. & Elmegreen, B. G. 1979, *ApJs*, 41, 87
Starck, J. L., Donoho, D. L., & Candès, E. J. 2003, *A&A*, 398, 785
Sousbie, T. 2011, *MNRAS*, 414, 350

From Interstellar Clouds to Star-Forming Galaxies: Universal Processes?
Proceedings IAU Symposium No. 315, 2015
P. Jablonka, Ph. André & F. van der Tak, eds.

doi:10.1017/S1743921316007316

Molecular cloud fragmentation and core collapse

Shantanu Basu

Department of Physics and Astronomy, The University of Western Ontario,
London, Ontario N6A 3K7, Canada
email: basu@uwo.ca

Abstract. I review some steps in the conversion of molecular cloud gas into stars and planets, with an emphasis in this presentation on the early stage molecular cloud fragmentation that leads to elongated filaments/ribbons. Magnetic fields can play a crucial role in all stages and need to be invoked particularly for early stage fragmentation as well as in late core collapse where it may control disk formation. I also review some elements of hydrodynamic modeling of disk evolution.

Keywords. ISM: clouds, ISM: magnetic fields, ISM: molecules, MHD, stars: formation

1. Introduction

The conversion of molecular gas into stars and planets can for simplicity be conceptually divided into four stages: (1) the fragmentation of clouds into large scale structures (filaments/ribbons); (2) the formation of dense cores within these large structures; (3) core collapse to form a hydrostatic protostar; (4) disk formation and evolution leading to outflows, multiplicity, planets, etc. In this review I emphasize step (1), given recent developments in the literature, and comment briefly on the others.

2. Cloud fragmentation and filaments

Molecular clouds are assembled from H I clouds in the interstellar medium that have a mass-to-flux ratio M/Φ that is significantly less than the critical value required for gravitational fragmentation and collapse (Heiles & Troland 2005). If molecular clouds are assembled from these subcritical components, then the observed star formation in the clouds leads to two possibilities: (1) the molecular clouds are entirely supercritical, having been assembled from a sufficient amount of subcritical material, by motions along Galactic magnetic field lines; (2) the molecular clouds are largely subcritical but that the breakdown of flux freezing due to ambipolar diffusion allows subregions to become subcritical and produce dense cores. The first scenario places challenging demands on the accumulation length of matter in molecular clouds, and also on the speed of streaming motions along the magnetic field (Mestel 1999). The second scenario has the drawback that the ambipolar diffusion timescale is at least 10^7 yr, or even much longer, in diffuse molecular gas. However, the addition of turbulent initial conditions leads to the formation of dense structures on a much shorter turbulent crossing time. Within the compressed gas, rapid ambipolar diffusion can take place due to the elevated density and magnetic field gradient. This scenario has been explored in many studies over the last decade (Nakamura & Li 2005; Elmegreen 2007; Kudoh & Basu 2008; Nakamura & Li 2008; Basu *et al.* 2009; Kudoh & Basu 2011) and also provides a natural way to understand filamentary structures in clouds. By filaments we do not mean objects with cylindrical symmetry,

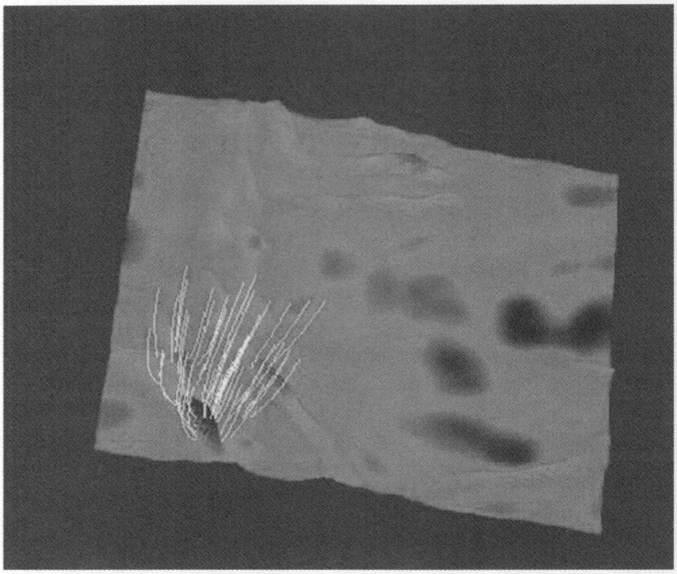

Figure 1. An image of column density in a thin sheet perturbed with turbulent initial conditions. There is a large scale magnetic field initially perpendicular to the sheet, and is illustrated with white lines in the maximally compressed region only. Turbulence and magnetic fields lead to elongated structures (ribbons) with magnetic field largely perpendicular to the long axis of the ribbon. From a simulation by Basu *et al.* (2009).

but rather more ribbon-like structures, with flattening along the ambient magnetic field direction, and compression by turbulence in primarily one direction perpendicular to the magnetic field. The overall scenario is illustrated in Figure 1, taken from a simulation of Basu *et al.* (2009). The filament formation is thus quite distinct from the cosmological filament formation simulations (e.g., Springel *et al.* 2005). in which the gravitational field of cold dark matter is able to collapse into filamentary structures on a local dynamical time. The magnetic turbulent picture allows the formation of filaments but not necessarily in a volume filling network as in cosmological simulations, since large parts of clouds may remain in a subcritical common envelope with little to no star formation. Models of magnetohydrodynamic (MHD) wave propagation in inhomogeneous clouds (Kudoh & Basu 2003; Kudoh & Basu 2006) also show that the velocity dispersion is essentially Alfvénic and that low density regions with high local Alfvén speed support large amplitude velocity fluctuations, in agreement with the observed linewidth-size relations of molecular clouds (Solomon *et al.* 1987). Magnetic fields can also lower the dissipation rate of turbulence in some cases, although not necessarily dramatically (Kim & Basu 2013), and allow more efficient distribution of outflow power across the cloud (Wang *et al.* 2010).

Recent observations by the *Herschel Space Observatory* have established that the dense cores and young stellar objects are often threaded along networks of elongated structures seen in projection (André *et al.* 2010; Men'schikov *et al.* 2010; Arzoumanian *et al.* 2011). When magnetic field morphology can be detected through polarized dust emission, the magnetic field is generally aligned perpendicular to the long axis of the filaments (Palmeirim *et al.* 2013). These features are at least qualitatively consistent with the magnetic turbulent scenario described above. Another interesting result is that the measurement of average FWHM widths across filaments in three different clouds shows a cluster of values around ~ 0.1 pc, even though the Jeans length calculated from the peak column densities are varying by about two orders of magnitude (Arzoumanian *et al.* 2011).

It seems that the filament widths require an explanation that goes beyond the simple Jeans length prescription. They are likely to be quasi-equilibrium objects based on the amount of fragmentation along their lengths to form dense cores. Recall that for a non-magnetized isothermal cylinder, there is a critical line mass $2c_s^2/G$ (where c_s is the isothermal sound speed) below which radial equilibrium is possible but fragmentation occurs along the axis of the cylinder. However if the line mass exceeds this critical value, then indefinite collapse to a spindle occurs before any fragmentation can take place (Inutsuka & Miyama 1997). Modified values of critical line mass, actually critical line mass to line flux ratio, for magnetized cylinders with a lateral magnetic field have recently been obtained by Tomisaka (2014) and Hanawa & Tomisaka (2015). Given the fragmentation along their length, it is likely that the line mass to line flux ratio of observed ribbons is subcritical.

Hennebelle & André (2013) developed an analytic model of filament width in which the energy input from accretion onto a filament is balanced by energy dissipation by ion-neutral friction, and there is mechanical equilibrium between gravity and internal turbulence. They obtained a mean filament width of ~ 0.1 pc for a range of column densities. An alternate model of filament formation was explored by Kudoh & Basu (2014) in which a filament (really ribbon) is created by turbulent compression of subcritical gas and halts contraction when magnetic pressure balances ram pressure. However, their semi-analytic model was used to estimate time scales for core formation and not to estimate filament widths.

3. Cores to star-disk systems

Once a dense core is formed by fragmentation of a larger clump or filament, it will begin runaway collapse as long as its mass-to-flux ratio is supercritical. The hydrodynamic collapse leads to a well known set of outcomes: development of an r^{-2} density profile and a shrinking central flat density peak that eventually forms the first hydrostatic core, later collapsing to the second core of stellar dimensions (e.g., Larson 1969). The addition of rotation and/or magnetic fields leads to a flattened collapsing core, especially at high densities. Once a central protostar is formed at the pivotal moment $t = 0$, several interesting things happen. An expansion wave moves outward from the center (Shu 1977), changing the character of flow variables within it. When rotation is added, a centrifugal disk is able to form, however an additional flux-frozen magnetic field leads to catastrophic magnetic braking that shuts off disk formation (Allen *et al.* 2003; Galli *et al.* 2006).

A straightforward solution to catastrophic magnetic braking is the presence of non-ideal MHD effects, notably ambipolar diffusion and Ohmic dissipation. These will weaken the magnetic field strength and reduce the effectiveness of magnetic braking. In Dapp & Basu (2010) and Dapp *et al.* (2012) we used the thin-disk approximation to follow collapse to the formation of a second stellar core while including non-ideal MHD effects. Both studies show that non-ideal MHD effects allow the formation of a disk immediately after the pivotal moment $t = 0$. In Dapp *et al.* (2012) we used a chemical network model to calculate the ionization fraction and coefficients of ambipolar diffusion and Ohmic dissipation. Figure 2 shows a series of column density profiles at various times, with the last one at the moment immediately following the formation of a second core of stellar dimensions. Each distinct region of the column density profile is captured, including the MHD accretion shock created by outward moving magnetic flux (Li & McKee 1996; Contopoulos *et al.* 1998). Figure 3 shows the magnetic field lines in cases with ideal

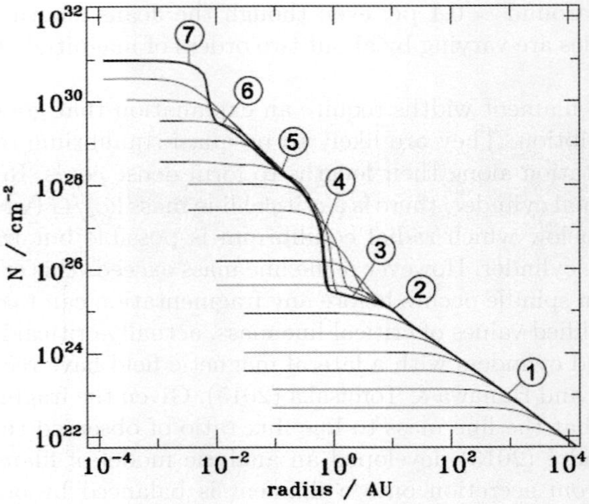

Figure 2. Column number density (N) profile versus radius from Dapp *et al.* (2012). The thin lines are profiles in ascending order of time. Several features are identifiable via their associated breaks in the profile (1) Prestellar infall profile with $N \propto r^{-1}$. (2) Magnetic wall at ≈10 AU, where the bunched-up field lines decelerate material before it continues the infall. (3) Expansion wave profile with $N \propto r^{-1/2}$ outside the first core. (4) First core at 1 AU. (5) Infall profile onto the second core with $N \propto r^{-1}$. After the first core has reached ≈1,000 K, it starts to collapse, as H_2 is dissociated. (6) Expansion wave profile with $N \propto r^{-1/2}$ outside the second core. (7) Second core at $\approx 1\ R_\odot$.

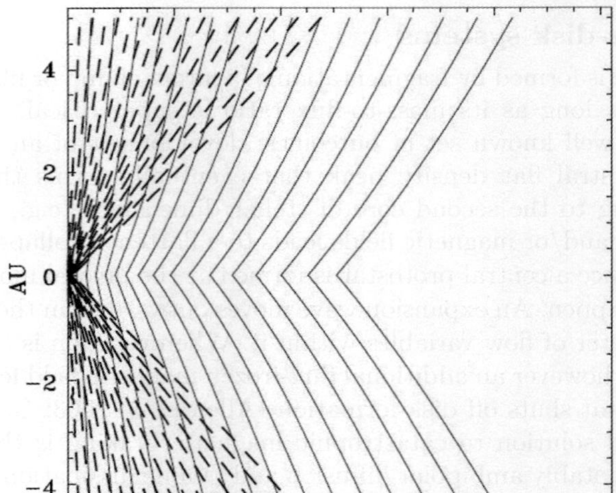

Figure 3. Magnetic field lines on 10 AU scales from Dapp *et al.* (2012). The dashed lines represent the flux-freezing model, while the solid lines show *the same* field lines for the model including non-ideal MHD effects for a grain size $a_{\rm gr} = 0.038\ \mu$m. In both cases, the second core has just formed and is on the left axis midplane. The field lines straighten out significantly on small scales in the non-ideal MHD model compared to the flux-frozen model.

and non-ideal MHD. The former case has extreme flared field lines and no disk is able to form due to catastrophic magnetic braking. The non-ideal MHD model yields the

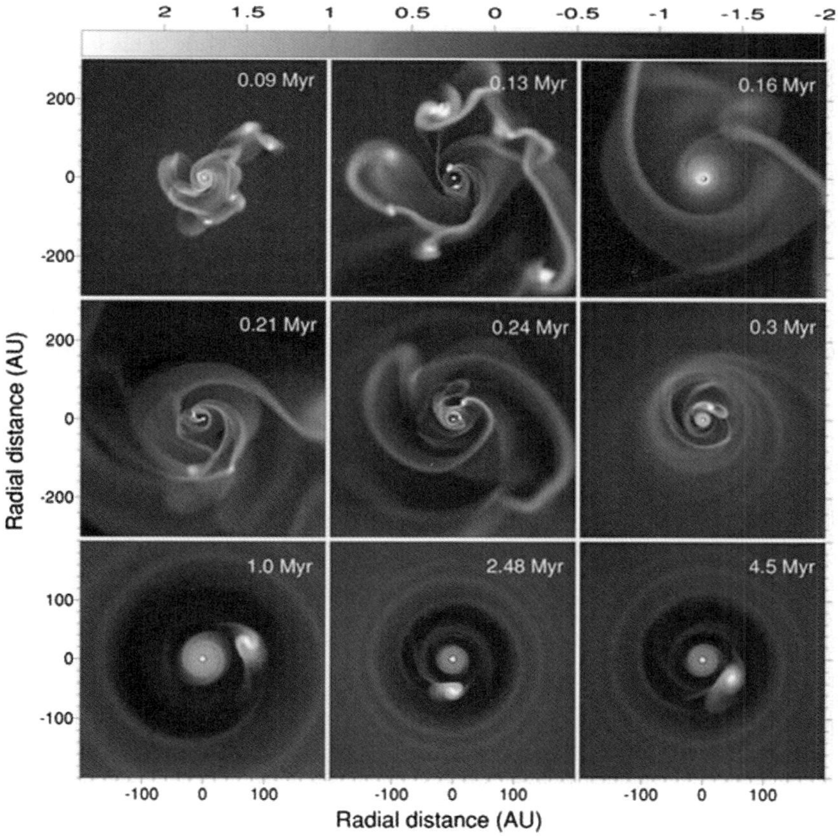

Figure 4. Gas surface density maps (g cm^{-2}, log units) at six times after the formation of the central star (bright circle in the coordinate center), from Vorobyov & Basu (2010a). Note the zoom-in as the time increases. The top two rows contain images of size 600 AU on each side, while the bottom row contains images of size 400 AU on each side.

development of a centrifugal disk immediately after the formation of the second core, however the loss of some angular momentum due to magnetic braking is expected to keep disks at relatively small sizes $\lesssim$10 AU in the early Class 0 phase ($\lesssim 4 \times 10^4$ yr).

4. Disks and multiplicity

As shown in the previous section, the collapse of a prestellar core leads to a very focused collapse with a power-law density profile. As a result, the probability of fragmentation during the runaway collapse phase is low. However, a centrifugally supported disk can form after a central protostar has formed. The existence of a central point mass potential allows a centrifugal barrier to be encountered by an infalling rotating fluid element that conserved its angular momentum (see discussion in Basu & Mouschovias 1995). A protostellar disk is a quasi-equilibrium structure and evolves on a time scale much longer than its dynamical time. Therefore, it has time for fragmentation if conditions are suitable, e.g., the criteria for gravitational instability and fragmentation are satisfied (Toomre 1981). Disk fragmentation is the most likely mechanism for the formation of multiple stellar systems and the formation of brown dwarfs and giant planets at wide orbits of several tens of AU. We show here the results of hydrodynamic numerical simulations of

Vorobyov & Basu (2010a), who modeled the formation of the disk from the collapse of a prestellar core, as well as its subsequent evolution on Myr time scales. Figure 4 shows images of the disk column density at a series of times after formation of the central protostar. During the first 10^5 yr, there is vigorous episodic gravitational instability in the disk, driven by mass accretion from the parent core. Clumps form in spiral arms through nonlinear gravitational instability, and are eventually driven in to the center through gravitational interaction with trailing spiral arms. In this model, one clump is able to survive to late times and eventually carves out a gap in the disk and settles into a stable orbit by 4.5 Myr. The clump is essentially a first hydrostatic core and is the progenitor of a low mass companion star with an orbit radius ≈ 52 AU. This mechanism can also produce wide orbit proto-brown-dwarfs and giant protoplanets. Details of the numerical model and the overall scenario of the migrating embryo model can be found in Vorobyov & Basu (2006), Vorobyov & Basu (2010b), and Vorobyov & Basu (2015).

References

Allen, A., Li, Z.-Y., & Shu, F. H. 2003, *ApJ*, 599, 363

André, P., *et al.* 2010 *A&A*, 518, L102

Arzoumanian, D., *et al.* 2011, *A&A*, 529, L6

Basu, S., Ciolek, G. E., Dapp, W. B., & Wurster, J. 2009, *New Astron.*, 14, 483

Basu, S. & Mouschovias, T. Ch. 1995, *ApJ*, 452, 386

Contopoulos, I., Ciolek, G. E., & Königl, A. 1998, *ApJ*, 504, 247

Dapp, W. B. & Basu, S. 2010, *A&A*, 521, L56

Dapp, W. B., Basu, S., & Kunz, M. W. 2012, *A&A*, 541, A35

Elmegreen, B. G. 2007, *ApJ*, 668, 1064

Galli, D., Lizano, S., Shu, F. H., & Allen, A. 2006, *ApJ*, 647, 374

Hanawa, T. & Tomisaka, K. 2015, *ApJ*, 801, 11

Heiles, C. & Troland, T. H. 2005, *ApJ*, 624, 773

Hennebelle, P. & André, P. 2013, *A&A*, 560, A68

Inutsuka, S.-i. & Miyama, S. M. 1997, *ApJ*, 480, 681

Kim, C.-G. & Basu, S. 2013, *ApJ*, 778, 88

Kudoh, T. & Basu, S. 2003, *ApJ*, 595, 842

Kudoh, T. & Basu, S. 2006, *ApJ*, 642, 270

Kudoh, T. & Basu, S. 2008, *ApJ*, 679, L97

Kudoh, T. & Basu, S. 2011, *ApJ*, 728, 123

Kudoh, T. & Basu, S. 2014, *ApJ*, 794, 127

Larson, R. B. 1969, *MNRAS*, 145, 271

Men'schikov, A., *et al.* 2010, *A&A*, 518, L103

Mestel, L. 1999, *Stellar Magnetism* (Oxford: Oxford Univ. Press)

Nakamura, F. & Li, Z.-Y. 2005, *ApJ*, 631, 411

Nakamura, F. & Li, Z.-Y. 2008, *ApJ*, 687, 354

Palmeirim, P., *et al.* 2013, *A&A*, 550, A38

Shu, F. H. 1977, *ApJ*, 214, 488

Solomon, P. M., Rivolo, A. R., Barrett, J., & Yahil, A. 1987, *ApJ*, 319, 730

Springel, V. *et al.* 2005, *Nature*, 435, 629

Tomisaka, K. 2014, *ApJ*, 785, 24

Toomre, A. 1981, in: S. M. Fall, & D. Lynden-Bell (eds.), *The Structure and Evolution of Normal Galaxies* (Cambridge Univ. Press, Cambridge), p. 111

Vorobyov, E. I. & Basu, S. 2006, *ApJ*, 650, 956

Vorobyov, E. I. & Basu, S. 2010a, *ApJ*, 714, L133

Vorobyov, E. I. & Basu, S. 2010b, *ApJ*, 719, 1896

Vorobyov, E. I. & Basu, S. 2015, *ApJ*, 805, 115

Wang, P., Li, Z.-Y., Abel, T., & Nakamura, F. 2010, *ApJ*, 709, 27

From Interstellar Clouds to Star-Forming Galaxies: Universal Processes?
Proceedings IAU Symposium No. 315, 2015
P. Jablonka, Ph. André & F. van der Tak, eds.

doi:10.1017/S1743921316007328

SCUBA2 observations of prestellar cores

Derek Ward-Thompson and Kate Pattle,
on behalf of the JCMT and Herschel-SPIRE Gould Belt Consortia

Jeremiah Horrocks Institute, University of Central Lancashire,
Preston PR1 2HE, United Kingdom
email: dward-thompson@uclan.ac.uk

Abstract. We show data from the SCUBA2 camera on JCMT, of molecular clouds. We focus on starless cores within the clouds. We present data of the Taurus region and show how the environment is affecting some cores' appearance in this region. We compare the SCUBA2 data with Herschel data and discuss the sensitivity of SCUBA2 to surface brightness in the submillimetre. We show how this leads to its ability to pick out the densest cores at a given temperature. Hence SCUBA2 preferentially picks out gravitationally bound pre-stellar cores. We discuss the effects of the magnetic field, and how this lends support to a model for the formation and evolution of cores in filamentary molecular clouds.

Keywords. stars: formation – ISM: clouds – ISM: individual objects: L1495 – submillimeter: ISM

1. Introduction

Stars form in dense cores in molecular clouds. Exactly how these cores form is still a matter of debate (e.g. André *et al.*, 2014). Cores which do not contain any stars or newly-formed protostars are called starless cores. Those starless cores which are gravitationally bound are known as pre-stellar cores (Ward-Thompson *et al.*, 2007a; Di Francesco *et al.*, 2007). These pre-stellar cores then collapse to form Class 0 protostars (André *et al.*, 1993).

The James Clerk Maxwell Telescope (JCMT) Gould Belt Legacy Survey (GBLS) set out to map all nearby star-forming regions in various spectral lines, and also in the continuum at 850 μm with the Sub-mm Common-User Bolometer Array 2 (SCUBA2) camera (Ward-Thompson *et al.*, 2007b). This data-set can be used in comparison with the Herschel Gould Belt Survey (HGBS) at wavelengths from 70 to 500 μm (André *et al.*, 2010; Pattle *et al.*, 2015).

Cores are a significant stage in star formation. Recent work by the HGBS and the JCMT GBLS has led to new insight into core formation (André *et al.*, 2014). One of the goals of the GBLS was to catalogue and map all of the pre-stellar cores within about 0.5 kpc (Ward-Thompson *et al.*, 2007b). The mass function of cores can then be modelled onto the IMF of stars (Goodwin *et al.*, 2008).

2. Taurus

Figure 1 shows the L1495 region in Taurus (Ward-Thompson *et al.*, 2016). Five Herschel wavebands and one waveband from SCUBA2 are shown. A number of cores and filamentary structures can be seen. In particular a triangle-shaped arrangement of filaments can be seen just below centre of each image. The ubiquity of filaments is a feature seen in almost all of the Herschel data of star-forming regions (André *et al.*, 2010).

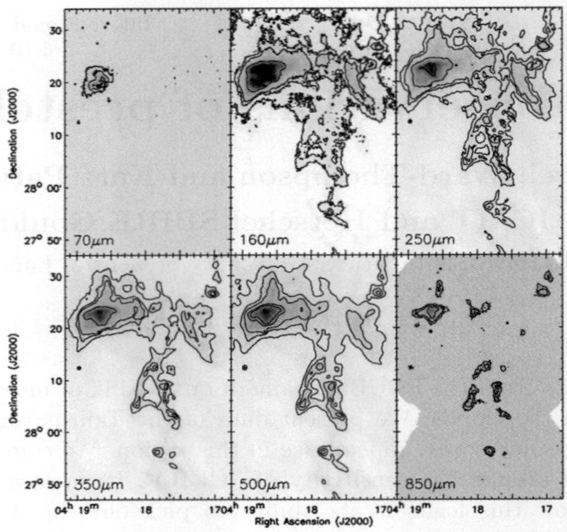

Figure 1. Herschel and SCUBA2 images of the L1495 molecular cloud in Taurus. Top row, left to right: 60, 170 and 250 μm respectively. Lower row, left to right: 350 and 500 μm (all from Herschel) and 850 μm (from SCUBA2) respectively. Note how SCUBA2 only detects some of the cores seen at the other wavelengths by Herschel (Ward-Thompson *et al.*, 2016).

The bright region towards the upper part of the image that is seen at most wavebands is the L1495A core, and is exactly coincident with L1495A-S. This lies at the head of a large, filamentary structure that is seen clearly in the Herschel data of the Taurus region (Marsh *et al.*, 2016). We measured a temperature gradient across this core, with the hotter material lying to the south (Ward-Thompson *et al.*, 2016).

There is a bright star, slightly to the south of L1495A-S, which is known as V892 Tau (IRAS04155+2812). This is a Herbig Ae/Be star, and it is clearly heating L1495A-S, which is otherwise starless, and causing the temperature gradient across the core. So we see that environment is affecting the appearance of some cores in the far-infrared and sub-mm. Similar effects were seen in cores in Cepheus (Nutter *et al.*, 2009; Pattle *et al.*, 2016). Note that only some of the cores visible in the Herschel images are seen in the SCUBA2 image. For example, the filament to the upper right in the Herschel images appears to Herschel as no different from the other filaments, but it is almost invisible in the SCUBA2 image.

Figure 2 shows a plot of density versus temperature for the starless cores in the L1495 region. The cores detected by SCUBA2 are indicated by black crosses. Those that are detected by Herschel are indicated by green crosses. The cores detected by SCUBA2 all lie in the upper part of this plot (Ward-Thompson *et al.*, 2016). However, note that both sets of cores occupy a similar range of temperatures.

Figure 3 shows a plot of mass versus deconvolved full-width at half-maximum (FWHM) size for the starless cores in the L1495 region. The cores detected by SCUBA2 are indicated by black points. Those that are detected by Herschel are indicated by green points. The cores detected by SCUBA2 all lie in the upper part of this plot (Ward-Thompson *et al.*, 2016). However, both set of cores occupy a similar range of sizes.

Figures 2 and 3 taken together would appear to indicate that SCUBA2 is only sensitive to the densest cores, or to the most massive cores for a given size, whereas Herschel sees more of the cores (Ward-Thompson *et al.*, 2016). This can be understood in terms of a surface brightness effect in the SCUBA2 data. For a given, relatively narrow, range of temperatures, such as is observed in starless cores, SCUBA2 surface brightness traces

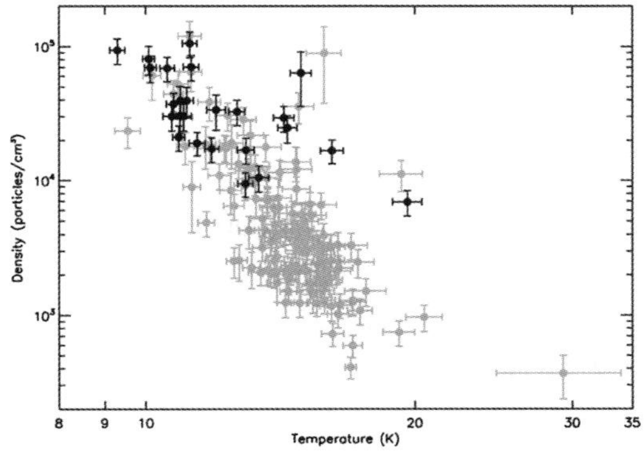

Figure 2. Density plotted against temperature for Herschel (grey crosses) and SCUBA2 (black crosses) data of starless cores. Note how both data-sets span a similar range of temperatures, but that the SCUBA2 cores only occupy the higher density parts of the plot (Ward-Thompson *et al.*, 2016).

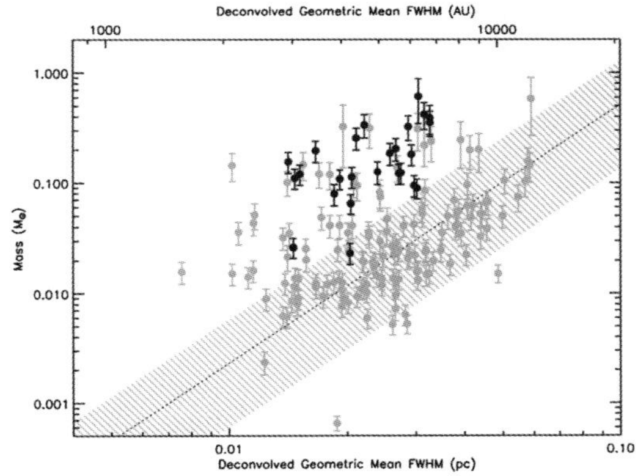

Figure 3. Mass plotted against deconvolved FWHM for Herschel (grey crosses) and SCUBA2 (black crosses) data of starless cores. Note how both data-sets span a similar range of FWHM size, but that the SCUBA2 cores only occupy the higher mass parts of the plot (Ward-Thompson *et al.*, 2016).

column density. Furthermore, for normally non-elongated structures, such as starless cores, column density effectively traces volume density. Hence the surface brightness of a starless core seen in a SCUBA2 image traces the density of that core. All of this implies that SCUBA2 will preferentially detect the densest cores – i.e. those closest to gravitational collapse – pre-stellar cores (Ward-Thompson *et al.*, 2016).

3. Magnetic fields

The magnetic field direction (in the plane of the sky) in a molecular cloud can be inferred from sub-mm dust continuum polarization measurements. Figure 4 shows results from the Ballon-borne Large Aperture Sub-mm Telescope Polarimeter (BLASTPol) of a filamentary structure in the Lupus molecular cloud (Matthews *et al.*, 2014). The curved line traces the axis of the filament. The vectors trace the direction of the magnetic field inferred from the polarization measurements. Note how the B-field lies largely perpendicular to the direction of the filament axis (Matthews *et al.*, 2014).

Observations such as these have led to a new model being proposed for core formation (André *et al.*, 2014). In this new scenario, material is funnelled onto filaments by the

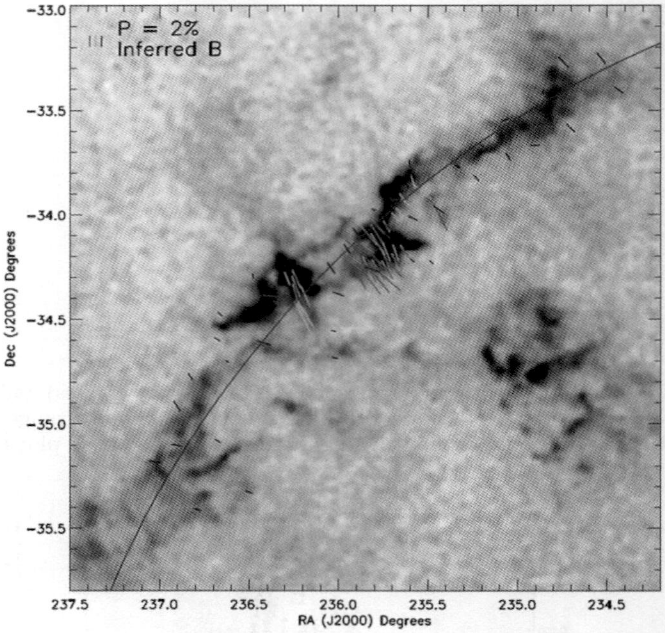

Figure 4. BLASTPol B-field vectors superposed on a sub-mm image of the Lupus molecular cloud. The vectors trace the direction of the magnetic field. The curved line traces the axis of the filament. Note how the B-field lies largely perpendicular to the direction of the filament axis (Matthews *et al.*, 2014). *(A color version of this figure is available online.)*

magnetic field (Palmeirim *et al.*, 2013). Dense parts of the filaments form cores by accreting further matter along the filaments. The filaments accrete mass until they exceed the critical mass per unit length against gravitational instability (Kawachi & Hanawa 1998). At this point they collapse under self-gravity to form proto-stellar cores. Herschel observed that roughly two-thirds of cores form in this manner (Könyves *et al.*, 2015).

References

André, P., Ward-Thompson, D., & Barsony, M. 1993, *ApJ* 406 122

André, P., Men'shchikov, A., Bontemps, S., *et al.* 2010, *A&A* 518 102

André, P., Di Francesco, J., Ward-Thompson, D., *et al.* 2014, *Protostars & Planets VI* p. 27

Di Francesco, J., Evans, N. J. II, Caselli, P., *et al.* 2007, *Protostars & Planets V* p. 17

Goodwin, S. P., Nutter, D., Kroupa, P., *et al.* 2008, *A&A* 477 823

Kawachi, T. & Hanawa, T., 1998, *PASJ* 50 577

Könyves, V., André, P., Mensh'chikov A., *et al.* 2015, *A&A*, 584, A91

Marsh, K. M., Kirk, J. M., André, P., *et al.* 2016, *MNRAS* submitted

Matthews, T. G., Ade, P. A. R., Angilè, F. E., *et al.* 2014, *ApJ* 784 116

Nutter, D., Stamatellos, D., Ward-Thompson, D. 2009 *MNRAS* 396 1851

Palmeirim, P., André, P. Kirk, J. M., Ward-Thompson, D., *et al.* 2013, *A&A* 550 A38

Pattle, K., Ward-Thompson, D., Kirk, J. M., White, G. J. *et al.* 2015, *MNRAS* 450 1094

Pattle, K., Ward-Thompson, D., Keown, J., *et al.* 2016 *MNRAS* in prep.

Ward-Thompson, D., André, P., Crutcher, R., *et al.* 2007, *Protostars & Planets V* p.33

Ward-Thompson, D., Di Francesco, J., Hatchell, J., *et al.* 2007, *PASP* 119 855

Ward-Thompson, D., Pattle, K., Kirk, J. M. *et al.* 2016, *MNRAS* in prep.

From Interstellar Clouds to Star-Forming Galaxies: Universal Processes?
Proceedings IAU Symposium No. 315, 2015
P. Jablonka, Ph. André & F. van der Tak, eds.

doi:10.1017/S174392131600733X

Observational studies of the formation and evolution of dense cores

Mario Tafalla

Observatorio Astronómico Nacional (IGN)
Alfonso XII 3, E-28014 Madrid, Spain
email: m.tafalla@oan.es

Abstract. Dense cores are the simplest star-forming sites. They represent the end stage of the fragmentation hierarchy that characterizes molecular clouds, and they likely control the efficiency of star formation via their relatively low numbers. Recent dust continuum observations of entire molecular clouds show that dense cores often lie along large-scale filamentary structures, suggesting that the cores form by some type of fragmentation process in an approximately cylindrical geometry. To understand the formation mechanism of cores, additional kinematic information is needed, and this requires observations in molecular-line tracers of both the dense cores and their surrounding cloud material. Here I present some recent efforts to clarify the kinematic structure of core-forming regions in the nearby Taurus molecular cloud. These new observations show that the filamentary structures seen in clouds are often more complex than suggested by the maps of continuum emission, and that they consist of multiple fiber-like components that have different velocities and sonic internal motions. These components likely arise from turbulent fragmentation of the large-scale flows that generate the filamentary structures. While not all these fiber-like components further fragment to form dense cores, a small group of them does so, likely by gravitational instability. This fragmentation produces characteristic chain-like groups of dense cores that further evolve to form stars.

Keywords. stars: formation, ISM: clouds, ISM: kinematics and dynamics, ISM: molecules

1. Introduction

Star formation is largely a fragmentation process. A typical molecular cloud contains tens of thousands of solar masses of material, and to form stars, it needs to fragment into parcels with typically the mass of our Sun. Understanding star formation therefore requires following the process of cloud fragmentation over multiple spatial scales, and identifying the different structures that appear at each of these scales.

In the hierarchy of cloud fragmentation, dense cores occupy a special place. These objects have typical sizes of ≈ 0.1 pc, and in nearby clouds like Taurus, Perseus, or Ophiuchus, they can be easily identified by their obscuration in optical images, by the emission of specific molecular tracers like NH_3 and N_2H^+, or by their concentrated emission in the millimeter-submillimeter dust continuum (Benson & Myers 1989; di Francesco *et al.* 2007; Bergin & Tafalla 2007)

With typical masses of a few $M_\odot$, dense cores have internal motions that are subsonic, and this indicates that thermal pressure dominates over non-thermal motions as a source of support against gravity. Cores, in addition, often harbor stellar objects at the earliest evolutionary stages (Beichman *et al.* 1986), a sign that the formation of a core represents the final step in the process of cloud fragmentation just prior to gravitational collapse. This close connection between core formation and star formation is further strengthened by the strong similarity between the distribution of core masses and the stellar initial

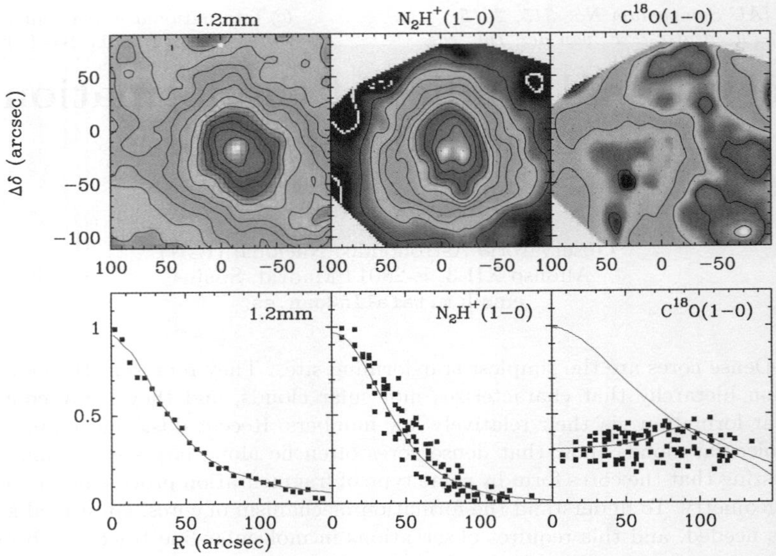

Figure 1. Molecular differentiation in the L1517B core. The top panels show the spatial distribution of three tracers (dust, N_2H^+, and $C^{18}O$), while the bottom panels show their radial profiles assuming spherical symmetry. The dust emission (left) traces the column density of the core, and its radial profile has been used to infer a core density profile. The N_2H^+ emission (middle) shows a similar distribution as the dust emission, and its radial profile is well fitted using a constant abundance law (red line). Finally, the $C^{18}O$ emission (right) seems to avoid the core peak, and the radial profile can only be fitted with a model having a large drop of abundance toward the central part of the core (blue line). Adapted from Tafalla *et al.* (2004). *(A color version of this figure is available online.)*

mass function, which suggests that the mass of the future star is already set at the time of core formation (Motte *et al.* 1998; Alves *et al.* 2007).

2. The need for molecular-line data

Large-scale optical images of clouds have often revealed their strikingly complex and filamentary geometry (Barnard 1907; Schneider & Elmegreen 1979), but only with the arrival of the new dust continuum images from the Herschel Space Observatory has it become clear that a filamentary geometry is a common feature of all clouds, both low and high mass, and both star-forming and quiescent (André *et al.* 2010, Molinari *et al.* 2010). While clouds appear as complex webs of filaments having different sizes and orientations, the dense cores embedded in them tend to be aligned with the filaments like beads along a string, providing a strong impression that the cores must form by some type of fragmentation process that occurs inside the filaments (Arzoumanian *et al.* 2011; André *et al.* 2014; Könyves *et al.* 2015).

Understanding how filaments form and how they later fragment into dense cores and stars requires observations that go beyond the Herschel continuum data. These data, while truly spectacular, are insensitive to the velocity of the gas, so they only provide a limited view of the physical state of the clouds. To complete this view, the Herschel data needs to be combined with observations of molecular line tracers that are sensitive to the internal kinematics of the emitting gas.

Molecular line tracers provide additional information on the state of the gas. For example, the gas chemical composition is sensitive to its evolutionary state, since it

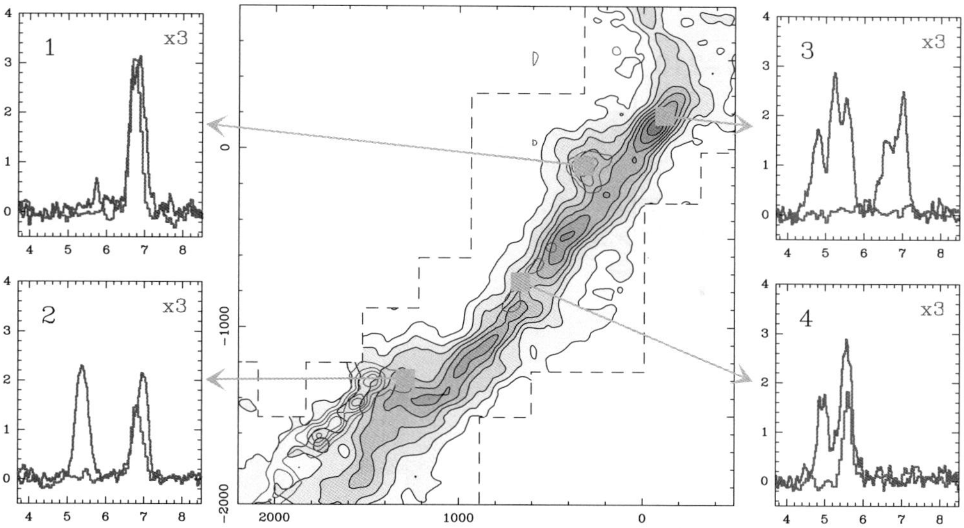

Figure 2. Evidence for multiple velocity components in L1495/B213. The central panel shows a map of $C^{18}O(1–0)$ integrated intensity toward the central part of the cloud. Although the emission appears to arise from a single filamentary region, the spectra at individual positions (left and right panels) reveal the presence of multiple velocity components (blue histograms). These multiple peaks do not arise from self-absorption or similar optical-depth effects, since the isolated $N_2H^+(1–0)$ feature, when detected, coincides with one of the $C^{18}O$ peaks (red histograms). From Hacar *et al.* (2013). *(A color version of this figure is available online.)*

changes significantly during the phase of contraction due to a number of processes, most notably the freeze out molecules onto the cold dust grains (Bergin & Tafalla 2007). This chemical inhomogeneity of the cloud gas can be used to separate its different density regimes by applying a tomographic analysis that combines selected molecular tracers. The densest parts of the cloud (the cores), for example, are chemically evolved and heavily depleted in carbon-bearing molecules like CO, so they are invisible in even the thinnest CO isotopologues, but they are very bright in N-bearing molecules like N_2H^+. The surrounding outer gas, on the other hand, is chemically younger and therefore very weak in N_2H^+, but can easily be traced with CO isotopologues, that in addition of being abundant, have a low-dipole moment that makes them easily thermalized. An example of this technique can be seen in Fig. 1 for L1517B in Taurus, which is a typical representative of a starless core at an intermediate state of evolution.

3. From filaments to fibers

One of the first core-forming regions studied with a combination of N_2H^+ and $C^{18}O$ observations has been L1495/B213 in Taurus (Hacar *et al.* 2013). This region was previously studied in detail using IR extinction and dust continuum observations by Schmalzl *et al.* (2010) and Palmeirim *et al.* (2013), respectively, and appears in the dust images as a single 10 pc-long filament with a number of dense cores and young stellar objects (YSOs) embedded in its interior.

Observations of the L1495/B213 cloud in molecular lines show that the single-filament picture inferred from the dust absorption/emission data is overly simplistic. As Fig. 2 illustrates, even lines of sight where the integrated emission appears as a single filament have $C^{18}O$ spectra with multiple velocity components separated by up to 2 km s^{-1}, which

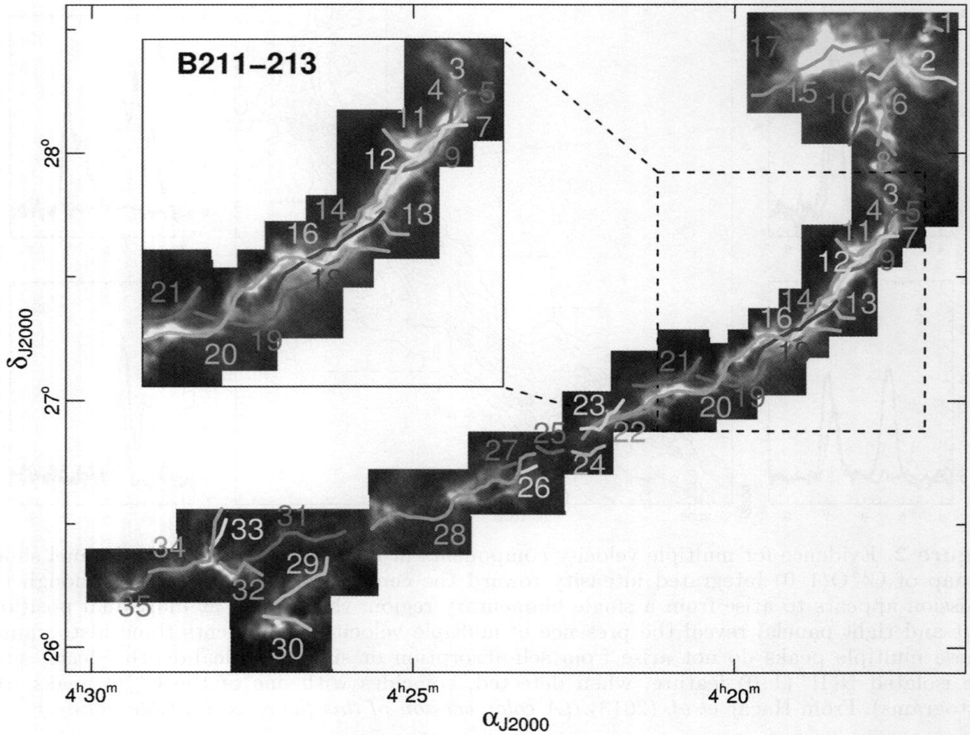

Figure 3. Location of the fibers in L1495/B213. Each colored line represents the central axis of one of the fibers identified from the $C^{18}O(1–0)$ emission by Hacar *et al.* (2013). The background grey scale image is the SPIRE 250 μm continuum map from Palmeirim *et al.* (2013). The good agreement between the two components indicates that the dust emission arises from a complex web of velocity-coherent fibers. Figure from Hacar *et al.* (2013). *(A color version of this figure is available online.)*

correspond to a Mach number of 10 for molecular gas at a temperature of 10 K (Hacar *et al.* 2013). This multiplicity of velocity components separated by supersonic speeds already indicates that the fragmentation process by which the L1495/B213 filament has formed its cores must have been far more complex than expected from the dust emission analysis.

To understand the underlying structure of the L1495/B213 region, Hacar *et al.* (2013) disentangled the overlapping velocity components. They first fitted multiple Gaussian profiles to each $C^{18}O$ spectrum, estimating the number of velocity components present at each position, and for each component, deriving its velocity centroid and velocity dispersion. Then, they connected the velocity components at different positions using a friends-of-friends algorithm similar to that used in redshift surveys to identify clusters of galaxies. With this procedure, Hacar *et al.* (2013) found that the $C^{18}O$ emission of the L1495/B213 region can be understood as resulting from 35 velocity components that lie more or less evenly spread along the full length of the "single" filament seen in the dust maps. Fig. 3 shows a representation of these components overlayed on the dust continuum map from Palmeirim *et al.* (2013).

While the velocity components in L1495/B213 have different sizes and masses, they tend to share a number of properties. First of all, they tend to be elongated, and their aspect ratios are typically larger than 4. For this reason, they are often referred to as

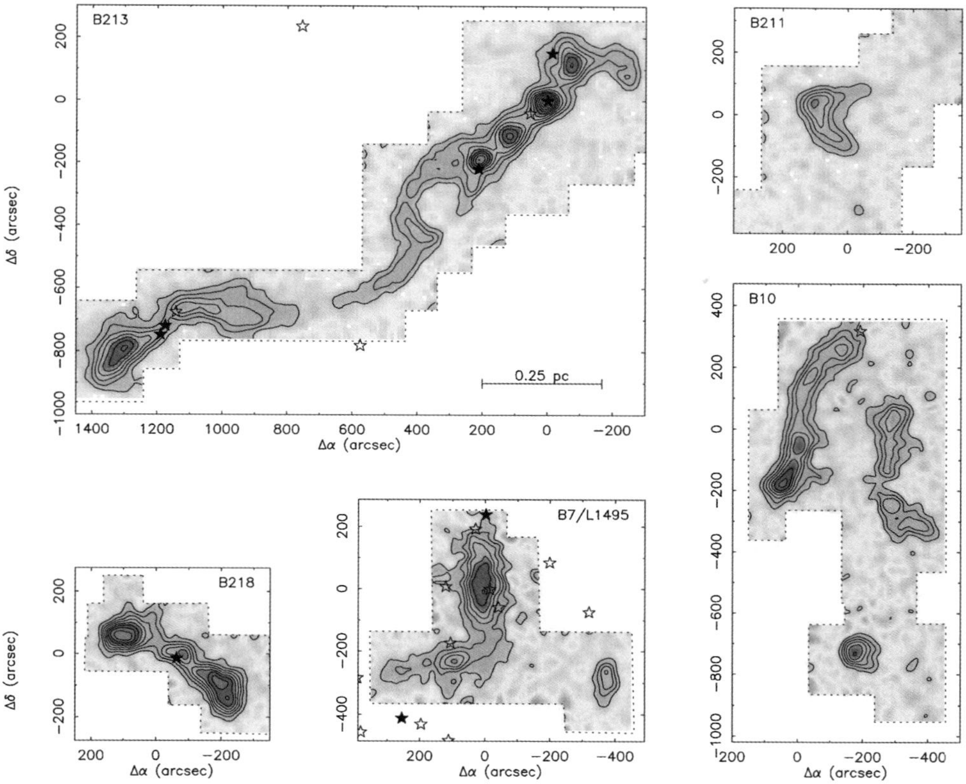

Figure 4. Maps of $N_2H^+(1\text{–}0)$ emission toward the regions with dense cores in L1495/B213. Note how the cores mostly form small linear groups of several compact cores connected by extended emission. These "chains" of cores seem to form from the fragmentation of the $C^{18}O$ fibers. From Tafalla & Hacar (2015).

"fibers," to distinguish them from the large-scale L1495/B213 "filament." The internal velocity field of these fibers presents large-scale oscillations, and the non-thermal component of the linewidth lies near the sonic-transonic boundary. In addition, a number of fibers have mass per unit length ratios of about 16 $M_\odot$ pc^{-1}, which approximately corresponds to the expected value for a cylinder in hydrostatic equilibrium at 10 K (e.g., Stodółkiewicz 1963; Ostriker 1964).

The discovery of multiple fibers in what initially appeared to be a single filament may seem surprising, but a number of recent simulations have shown that fiber-like structures form naturally. Moeckel & Burkert (2015), for example, have presented simulations of a self-gravitating turbulent cloud, and have found multiple fiber-like structures with oscillatory velocity fields similar to those seen in L1495/B213. Other simulations that show evidence for fiber-like structures are those by Kritsuk *et al.* (2013), Smith *et al.* (2016), and Li *et al.* (2015), who have simulated the formation and evolution of turbulent molecular clouds with different masses and sizes. While not all these fiber-like structures may have the same properties as the L1495/B213 fibers, their common sight in simulations suggests that the formation of multiple velocity components in the gas does not require a special tuning in the initial conditions. Understanding in detail how these structures form remains however an important issue.

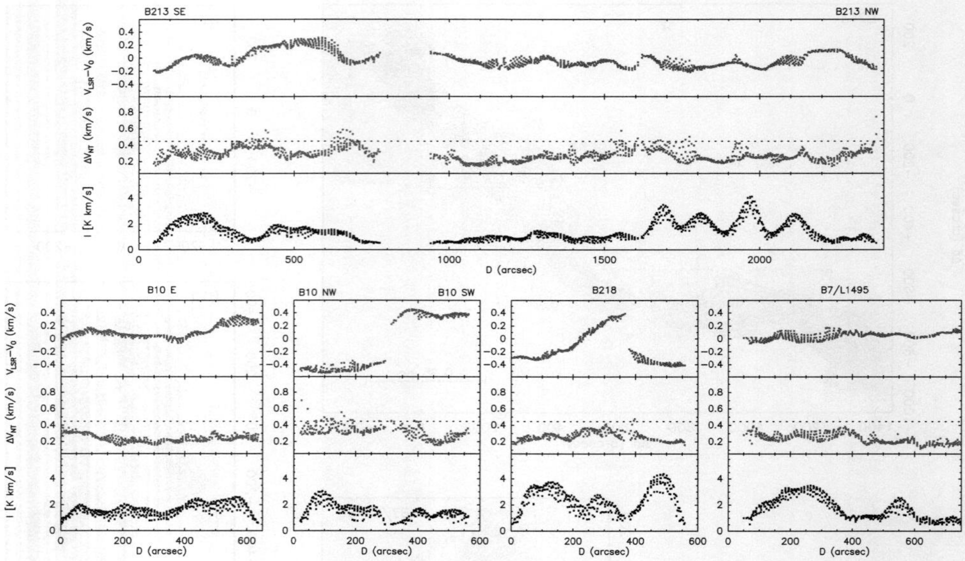

Figure 5. Internal kinematics of the core chains. Central velocity (top, blue symbols), non-thermal linewidth (middle, red symbols), and integrated intensity (bottom, black symbols) of N_2H^+ (1–0) along the axis of each chain shown in Fig. 4. Note the large-scale coherence of the velocity and the mostly subsonic internal motions. A few positions with enhanced kinematics seem to be associated with stellar feedback. See Tafalla & Hacar (2015) for further details. *(A color version of this figure is available online.)*

4. From fibers to cores

So far, we have seen the behavior of the $C^{18}O$ emission which, for chemical reasons, traces the low density gas in L1495/B213. To study the population of dense cores in the cloud, we now use N_2H^+, which is immune to depletion at typical core densities and therefore traces the dense gas. Fig. 4 shows N_2H^+ (1–0) maps of the dense cores in L1495/B213 obtained with the IRAM 30m telescope by Tafalla & Hacar (2015). As can be seen, most dense cores belong to small linear groups of a few condensations each connected by more extended emission. These core groups, which we will refer to as "chains," seem to be associated with some of the fibers identified in $C^{18}O$, since they coincide with them in both position, orientation, and velocity. Quantitatively, of a total of 19 cores in the L1495/B213 region, 14 ($\approx$70%) are associated with just 4 fibers, while the remaining 5 cores are more of less isolated.

This association of the majority of cores with a small number of fibers suggests that fibers can be classified in two groups. On the one hand, about 28 of the 35 fibers in the cloud (i.e., 80%) are not associated with any dense core, and therefore seem to be sterile. The remaining 20% of the fibers, on the other hand, is responsible for the whole core population, which means that each of these fibers has produced on average almost 3 dense cores. Fibers, therefore, seem to either be sterile and form no cores, or produce a small group of 2 or 3 cores, and in this case we classify them as fertile.

The fertile/sterile dichotomy suggests that some fibers are more prone than others to form dense cores, and therefore, that core formation results from an intrinsic property of the fiber as a whole. The property most likely to separate fertile and sterile fibers is the mass per unit length, although our line data are not ideal to determine masses due to the depletion of CO and the uncertain abundance of N_2H^+. The $C^{18}O$ data of Hacar *et al.* (2013), however, suggest that that core-forming fibers have a slightly higher mass

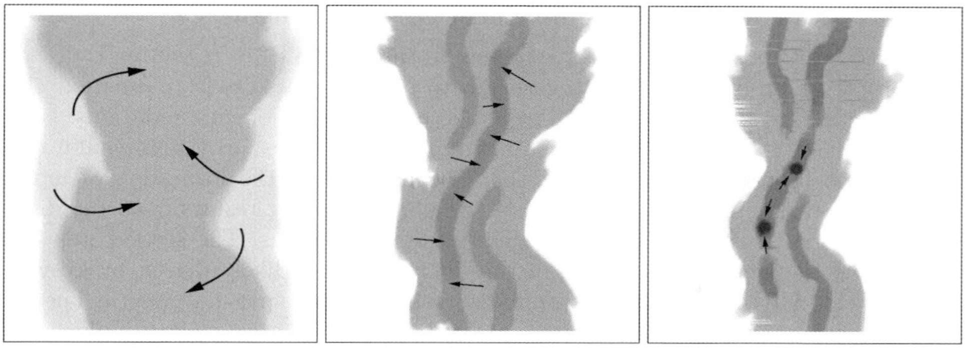

Figure 6. *Fray and fragment* scenario of core formation in L1495/B213 (Tafalla & Hacar 2015).

per unit length than the fibers that do not form cores, and that the fertile fibers lie close to the 16 $M_\odot$ pc^{-1} threshold of gravitational instability predicted by a simple model of a pressure-supported isothermal cylinder at 10 K (Stodółkiewicz 1963; Ostriker 1964).

To further explore the mechanism of core formation in fertile fibers, Fig. 5 presents a series of velocity profiles along each of the core chains. As can be seen, most chains have oscillatory velocity profiles (top panels, blue symbols), with the only exception of B218, where the dense gas seems to be accelerated by the outflow from the central source. The coherent velocity profiles of the chains match those of the larger-scale $C^{18}O$ fibers, indicating that core formation has occurred without a significant change in the gas kinematics.

Further evidence for a quiescent process of core formation comes from the analysis of the non-thermal linewidth profiles (middle panels, red symbols). These profiles show that the overwhelming majority of the points have subsonic internal motions, and that the few points with supersonic linewidths are mostly associated to feedback from the embedded YSOs (see Tafalla & Hacar 2015 for further details). Such quiescent state of the gas over chain scales strongly suggests that core formation has occurred without appreciable changes in the velocity field, or in other words, without the presence of internal shocks. These characteristics are most consistent with a process of gravitational fragmentation that starts in a rather quiescent state.

While the gas in the fibers and cores is extremely quiescent, the process of core formation deviates from that predicted by the theory of exponential growth of infinitesimal perturbations. In this simplest instability mode, all cores are expected to form simultaneously by the non-linear growth of a sinusoidal initial perturbation. As the maps in Fig. 4 show, however, cores with stars alternate with cores without stars in the same chain without displaying a systematic pattern. Chains B213SE and B218, for example, have a YSO at the center and two starless cores at the sides, which is the opposite to what is expected in the case of fragmentation of a truncated cylinder, where star formation should occur at the edges because they are more unstable due to the asymmetry in the gravitational field (Burkert & Hartmann 2004).

If cores with and without stars alternate in the same chain (e.g., B213NW), the formation and evolution of the cores must not occur simultaneously across the chain. Most likely, the fibers form with non-negligible density perturbations imprinted on them, and the growth of these perturbations leads to core formation. To test these ideas, a more systematic effort to study core formation in irregular cylindrical systems is needed.

5. A "fray and fragment" scenario of core formation

Our analysis of core formation has focussed on the L1495/B213 region in Taurus, but preliminary work on other regions, together with published results from different clouds (Henshaw *et al.* 2014) and the results from the already-mentioned simulations, suggest that the L1495/B213 results are likely to common. If so, core formation would appear to involve two different physical processes. The first one is the formation of multiple velocity-coherent fibers, which as shown by simulations, likely results from the dissipation of turbulence in shocks between large-scale streams of gas. The second step in the process is the fragmentation of some of the fibers, which is likely driven by gravity in those fibers that have reached the instability threshold of mass per unit length. These two different processes are illustrated schematically in Fig. 6 for a geometry similar to that of L1495/B213. We call this two-step scenario "fray and fragment," with "fray" referring to the production of multiple fibers and "fragment" referring to their gravitational fragmentation. If correct, it suggests that dense cores are, by formation, gravitationally unstable, and that their evolution can only further lead to the formation of stars.

Acknowledgements

It is a pleasure to thank Álvaro Hacar for a fruitful and ongoing collaboration from which many of the results presented here have originated. I also thank the Spanish MINECO for funding support from grants AYA2012-32032 and FIS2012-32096, and the organizers of the meeting for their invitation to present this work.

References

Alves, J., Lombardi, M., & Lada, C. J. 2007, *A&A*, 462, L17
André, P., Men'shchikov, A., Bontemps, S., *et al.* 2010, *A&A*, 518, L102
André, P., Di Francesco, J., Ward-Thompson, D., *et al.* 2014, *Protostars and Planets VI*, 27
Arzoumanian, D., André, P., Didelon, P., *et al.* 2011, *A&A*, 529, L6
Barnard, E. E. 1907, *ApJ*, 25, 218
Beichman, C. A., Myers, P. C., Emerson, J. P., *et al.* 1986, *ApJ*, 307, 337
Benson, P. J. & Myers, P. C. 1989, *ApJS*, 71, 89
Bergin, E. A. & Tafalla, M. 2007, *ARAA*, 45, 339
Burkert, A. & Hartmann, L. 2004, *ApJ*, 616, 288
di Francesco, J., Evans, N. J., II, Caselli, P., *et al.* 2007, *Protostars and Planets V*, 17
Hacar, A., Tafalla, M., Kauffmann, J., & Kovács, A 2013, *A&A*, 554, A55
Henshaw, J. D., Caselli, P., Fontani, F., Jiménez-Serra, I., & Tan, J. C. 2014, *MNRAS*, 440, 2860
Könyves, V., André, P., Men'shchikov, A., *et al.* 2015, *A&A*, 584, A91
Kritsuk, A. G., Lee, C. T., & Norman, M. L. 2013, *MNRAS*, 436, 3247
Li, P. S., McKee, C. F., & Klein, R. I. 2015, *MNRAS*, 452, 2500
Moeckel, N. & Burkert, A. 2015, *ApJ*, 807, 67
Molinari, S., Swinyard, B., Bally, J., *et al.* 2010, *A&A*, 518, L100
Motte, F., André, P., & Neri, R. 1998, *A&A*, 336, 150
Ostriker, J. 1964, *ApJ*, 140, 1056
Palmeirim, P., André, P., Kirk, J., *et al.* 2013, *A&A*, 550, A38
Schmalzl, M., Kainulainen, J., Quanz, S. P., *et al.* 2010, *ApJ*, 725, 1327
Schneider, S. & Elmegreen, B. G. 1979, *ApJS*, 41, 87
Smith, R. J., Glover, S. C. O., Klessen, R. S., & Fuller, G. A. 2016, *MNRAS*, 455, 3640
Stodólkiewicz, J. S. 1963, *AcA*, 13, 30
Tafalla, M., Myers, P. C., Caselli, P., & Walmsley, C. M. 2004, *A&A*, 416, 191
Tafalla, M. & Hacar, A. 2015, *A&A*, 574, A104

From Interstellar Clouds to Star-Forming Galaxies: Universal Processes?
Proceedings IAU Symposium No. 315, 2015
P. Jablonka, Ph. André & F. van der Tak, eds.

doi:10.1017/S1743921316007341

Numerical simulation of star formation in filamentary dark molecular clouds

Pak Shing Li[1], **Richard I. Klein**[1,2] **and Christopher F. McKee**[1,3]

[1]Astronomy Department, University of California,
501 Campbell Hall #3411, Berkeley, CA 94720-3411, USA
email: psli@berkeley.edu

[2]Lawrence Livermore National Laboratory,
P.O. Box 808, L-23, Livermore, CA 94550, USA,
email: klein@astron.berkeley.edu

[3]Physics Department, University of California,
Berkeley, CA 94720, USA,
email: cmckee@astro.berkeley.edu

Abstract. Numerical simulations of star formation faces challenges including the huge spatial dynamic range and the presence of multiply coupled highly non-linear physics, such as magnetic field, supersonic turbulence, gravitation, radiation and protostellar outflow feedback. We present in this symposium our latest high resolution adaptive mesh refinement numerical simulations of the formation of filamentary dark molecular clouds from 4.55 pc size scale down to the formation of a protostellar cluster with maximum resolution at 28 AU. The physical properties of the long braided filamentary dark cloud formed in the simulation, the magnetic field properties of the cloud clumps, and the protostellar mass function in the simulations match well with the latest observations.

Keywords. Magnetic fields-MHD-ISM: magnetic fields-ISM: kinematics and dynamics-stars: formation

1. Introduction

Observations show that the star formation process starts from the giant molecular cloud (GMC) complex of several tens of parsecs (pcs) all the way down to the formation protostars at the size scale of solar radius (9 orders of magnitude in spatial scale). In between, we observe long filamentary dark clouds of a few to more than 10 pc in length, large molecular cloud clumps about 0.1 to 1 pc in size, dense cloud cores less than 0.1 pc which house protostellar formation, and protostellar accretion disks $\leqslant$1,000 AU. Multiply coupled highly non-linear physics are involved in the process, including magnetic field, supersonic turbulence, gravity, radiation and protostellar outflow feedback. It is a challenge to simulate the entire process in a single numerical simulation. To achieve such a long term challenge, we need efficient adaptive codes, such as adaptive mesh refinement (AMR) code, that can manage memory usage efficiently and are highly scalable on large-scale high performance supercomputers. Here we present some of the highlights of our simulation results using our radiative-magnetohydrodynamics AMR code ORION2 (Li *et al.* 2012) on formation of filamentary dark clouds, the magnetic properties of the dense cloud clumps, and protostellar cluster formation.

2. Simulations

Our simulation starts with a large scale turbulence region with 3D thermal Mach number of 10. Using line width-size relation, the length of the periodic cubical region is 4.55 pc. We use the mean density of the GMC at such size scale and the total mass of the whole simulation region is 3110 $M_{\odot}$. Motivated by observations, the other two dimensionless parameters we chose are the Alfvénic Mach number ($\mathcal{M}_{\rm A}$) and the virial parameter, both of which are unity. We assume isothermal conditions and ideal MHD. The corresponding plasma $\beta = 0.02$ and the mass-to flux ratio is 1.62. This corresponds to a relatively strong initial magnetic field that is found to be more appropriate in describing the magnetic field properties of molecular clouds than our other model with exactly the same initial conditions (Li *et al.* 2015a) but 10 times weaker initial field strength. The base grid is 512^3 with two levels of grid refinement. We drive the system using a solenoidal field for two crossing times, defined as the box length divided by the rms velocity. At this time the system has reached a turbulent equilibrium state. We turn on gravity and continue driving the system with constant energy injection to maintain a global 3D value of Mach 10. The simulation continues for 2/3 of a free-fall time ($1t_{\rm ff} = 1.4 \times 10^6$ yrs). We use this simulation to study the formation of filamentary IRDCs.

To study the formation of a protostellar cluster, we restart the IRDCs simulation at about $0.25t_{\rm ff}$ when no cell has violated the Jeans condition yet. We allow maximum refinement to 6 levels in a specified 4.2 pc^3 zoom-in region that encompasses the main filament to make the zoom-in simulation computationally feasible. We attain a maximum 28 AU resolution in this zoom-in region. Anywhere outside this region will be simulated at the base level resolution. The protostellar star particles used in the simulation, evolve based on subgrid model of protostars. They are created at the highest level of grid refinement. As a result, no protostar is created outside the zoom-in region. Nevertheless, more than 75% of the most massive clumps exist in the zoom-in region. In the zoom-in simulation, we include radiation transfer and protostellar outflow feedback to study protostellar cluster formation.

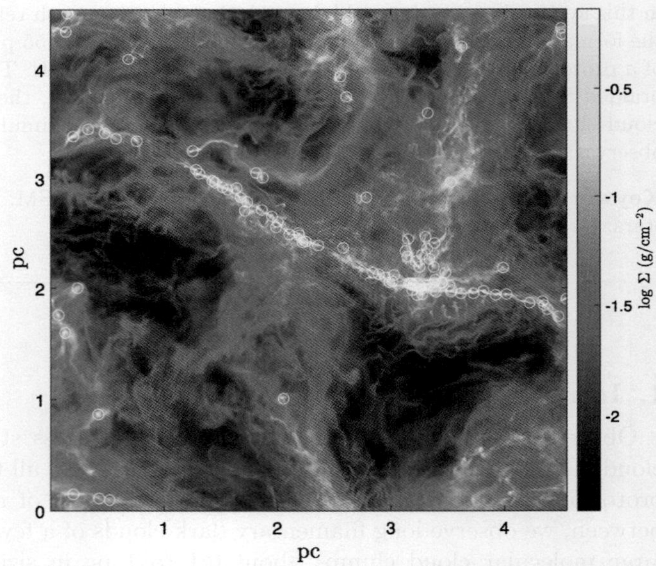

Figure 1. Column density map of the simulated region showing the main long massive filament cloud near the center and the 100 most massive clumps (white circles).

3. Highlights of the simulation results

Here we discuss three highlights from our simulations. In Fig. 1, we show the column density map of the entire region along the mean field direction from the IRDCs simulation at 0.5 $t_{\rm ff}$. A dense and long filamentary cloud is located near the center. The filament is ~ 4.5 pc long and has a mass of $\sim 660\,M_{\odot}$. The volume and column densities are about $2 \times 10^4 \mathrm{cm}^{-3}$ and $4 \times 10^{22} \mathrm{cm}^{-2}$, similar to some observed IRDCs. The mass per unit length is $\sim 77.6\,M_{\odot}\,\mathrm{pc}^{-1}$ within 0.2 pc radius from the filament axis, about five times of the critical density $2c_s^2/G$ and gravitationally bounded. In the map, we plot the 100 most massive clumps marked by the white circles. Most of them are along the main filament. Note the appearance of another filament near the upper right connected to the main filament. The two filaments are in fact colliding. This configuration is strikingly similar in appearance to the dark cloud SDC13 (Peretto *et al.* 2014). The comparison with SDC13 will be reported in Li *et al.* (2016).

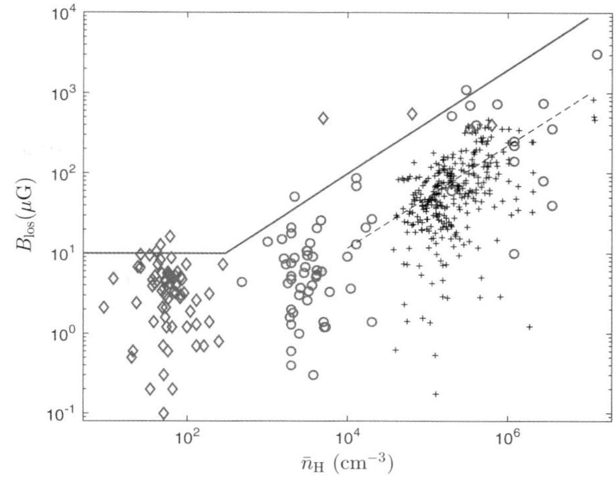

Figure 2. $B_{\rm los}$-n relation of Zeeman observations from Crutcher *et al.* (2010) (HI green diamond, OH+CN green circles) and the 100 most massive clumps (black plus) from the initially strong magnetic field simulation. The dashed line is the best power law fit to the simulation data.

In Li *et al.* (2015a), we have performed a detail statistical comparison of the magnetic field properties of the 100 most massive clumps with the Zeeman observations of molecular clouds summarized in Crutcher *et al.* (2010). In Fig. 2, we plot the B-n relation of the observed line-of-sight (los) magnetic field and volume density of the HI, OH, and CN sources with the 100 most massive clumps from the IRDCs simulation. The distribution of clumps from simulation resembles the power-law-like B-n relation of the molecular OH and CN sources well. In fact, the B-n relation of the *total* magnetic field from the simulation is consistent with $B_{\rm tot} \propto n^{0.65}$ concluded in Crutcher *et al.* (2010). The large scattering in the observations is the result of dispersion in total field strength and the line of sight projection effect. In another simulation with initially 10 times weaker field, the B-n relation of the clumps is very similar to the initially strong field simulation and visually indistinguishable in the $B_{\rm los}$ vs n plot. It is because of the field amplification in the weak field model due to turbulent driving. Despite the similarity in the B-n relation between the two simulations, there is a clear difference in the distribution of field orientation of the clumps with respect to the mean field direction, as shown in Fig. 3. The distribution from the initially strong field simulation is consistent with the recent observational study of molecular clouds magnetic field orientation by Li *et al.* (2009). For the initially weak field simulation, the

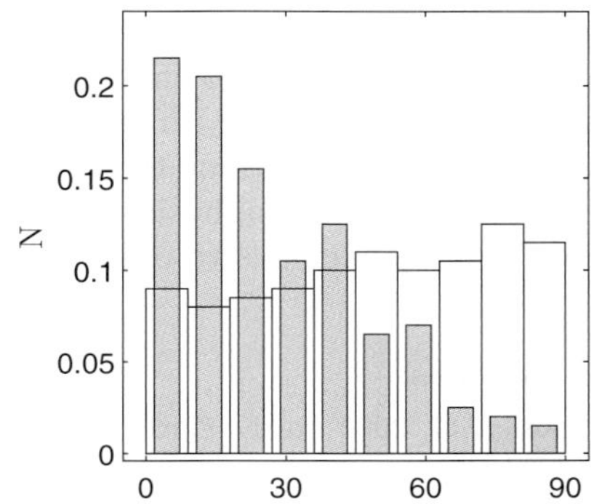

Figure 3. Distribution of magnetic field orientation of the 100 most massive clumps, w.r.t. the mean field direction, of the initially strong (open) and weak (shaded) field simulations.

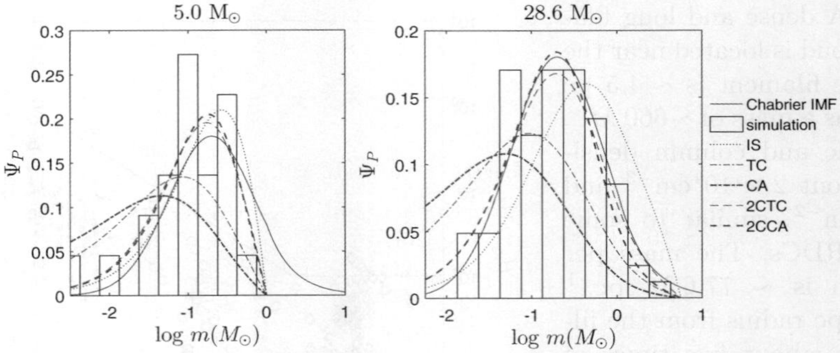

Figure 4. The PMFs at the time when the total masses of the protostars are 5 and 28.6 $M_\odot$ with 4 theoretical PMF models and the Chabrier IMF. The 2CTC model best fits the PMF from the simulation.

distribution is basically random which is not consistent with the strong alignment of the global B field of the surrounding intercloud medium.

In the 6-level zoom-in simulation, at the time of 0.5 t_{ff}, there are total 82 protostellar particles formed. Fig. 4 shows the protostellar mass functions (PMFs) of the protostars when the total masses of protostars are 5 $M_\odot$ and 28.6 $M_\odot$ (at the end of the simulation), compared with several theoretical PMFs (McKee & Offner 2010). Using a reduced χ^2 comparison, the two-component turbulent core PMF best fits the simulation results for all time. At the end of the simulation, the PMF also matches the Chabrier stellar initial mass function (IMF) (Chabrier 2005) very well. The median of the PMF is 0.196 $M_\odot$, very close to the expected 0.2 $M_\odot$ from Chabrier IMF. The results of the zoom-in simulation will be reported in Li *et al.* (2015b).

4. Conclusions

Using AMR combined with a zoom-in method, we achieve more than 4 orders of magnitude in spatial resolution dynamic range in these star formation simulations. We successfully recreate many observational results of the star formation process including the large scale structure and dynamics of filamentary IRDCs, the magnetic field properties of cloud clumps, and the formation of the protostellar cluster. New simulations with an even larger dynamic range of spatial resolution of 6 orders of magnitude are undertaken in order to continue the simulation down to scales of the formation of protostellar accretion disks. Multiphysics AMR simulations are crucial in following star formation from large scale clouds to protostellar formation.

References

Chabrier, G. 2005, in: E. Corbelli & F. Palla (eds.), *The Initial Mass Function 50 Years Later* Ap&SSL Conf. Ser., 327, 41
Crutcher R. M., Wandelt B., Heiles C., Falgarone E., & Troland T. H. 2010, *ApJ*, 725, 466
Li H.-b., Dowell C. D., Goodman A., Hildebrand R., & Novak, G. 2009, *ApJ*, 704, 891
Li, P. S., Martin, D. F., Klein, R. I., & McKee, C. F. 2012, *ApJ*, 745, 139
Li, P. S., McKee, C. F., & Klein, R. I. 2015a, *MNRAS*, 30, 490
Li, P. S., Klein, R. I., & McKee, C. F. 2015b, in preparation
Li, P. S., Klein, R. I., & McKee, C. F. 2016, in preparation
McKee, C. F. & Offner, S. S. R.. 2010, *ApJ*, 716, 167
Peretto, N., Fuller, G. A., André, Ph., *et al.* 2014, *A&A*, 561, A83

From Interstellar Clouds to Star-Forming Galaxies: Universal Processes?
Proceedings IAU Symposium No. 315, 2015
P. Jablonka, Ph. André & F. van der Tak, eds.

doi:10.1017/S1743921316007353

From disks to planets: observational insights

Andrea Isella

Physics & Astronomy Department, Rice University,
MS-108, 6100 Main St., Houston, TX, USA, 77005
email: isella@rice.edu

Abstract. The unprecedented sensitivity and imaging capabilities offered by the *Atacama Large Millimeter Array* (ALMA) are transforming our understanding of protoplanetary disks and, hence, of planet formation. In this brief chapter, I first discuss the main results and caveats related to the measurement of the mass of solids in protoplanetary disks based on millimeter-wave observations. I then present a recent analysis of the ALMA observations of the HL Tau disk, which suggests that the observed circular rings might be due to the tidal interaction between Saturn mass planets and the circumstellar material. In the conclusion, I argue that the existing observations of protoplanetary disks suggest that planets might form very early on, perhaps at the same time as the formation of the disk itself.

Keywords. Protoplanetary disks, Planets and satellites: formation, Submillimeter: planetary systems, stars: formation

1. Introduction

In the last few years, the study of the formation of planets has been revolutionized by the unprecedented imaging capabilities of new infrared and microwave observing facilities. In particular, the Atacama Large Millimeter Array (ALMA) is now capable of delivering high fidelity images at an angular resolution of a few tens of milliarcseconds, which resolve the submillimeter emission from nearby protoplanetary disks on scales as small as 1 AU.

ALMA observations of protoplanetary disks are revealing surprising structures which are thought to be caused by the tidal interaction between yet unseen giant planets and the circumstellar material. These observations probe the earliest stages of the formation of new planetary systems and enable us to study the physical processes responsible for the assembly of planets.

At the same time, the superior sensitivity of ALMA observations allows to quickly measure millimeter fluxes for hundreds of protoplanetary disks in the nearest star forming regions. These measurements can be used to estimate the amount of gas and dust in protoplanetary disks and study how the density of the circumstellar material depends on the properties of the host star.

In this brief paper, I first review the main results concerning the measurement of the amount of solids in protoplanetary disks and then discuss a recent analysis of the ALMA observations of HL Tau. A discussion about how these observations inform about the formation of planets is presented in the Conclusions.

2. Amount of solids in protoplanetary disks: caveats and results

The study of the structure and evolution of protoplanetary disks mostly relies on photometric measurements of their continuum emission, which, at wavelengths spanning from the near-infrared to the microwave regimes, originates from the thermal emission of solid particles. If the column density of such particles is low, then the continuum emission is optically thin, and it can be used to measure the column density of particles themselves. In the opposite regime, i.e., when the column density of solids is high, the continuum emission is optically thick and probes instead the temperature of the $\tau \simeq 1$ layer. For the typical densities of protoplanetary disks, the transition between the optically thick and thin regimes happens between the wavelength of 1 mm and 1 cm.

The flux density of the optically thin continuum emission at a wavelength λ depends on the mass of solids M_s through the relation

$$F(\lambda) = M_s \times \kappa_s(\lambda) \times B(\lambda, T_s)/d^2, \tag{2.1}$$

where $\kappa_s(\lambda)$ is the opacity of the solids, $B(\lambda, T_s)$ is the emission of a black body at the temperature T_s of the solids, and d is the distance of the source. Deriving M_s from measurements of $I(\lambda)$ therefore requires assumptions on the distance, opacity, and temperature of the solid materials.

The temperature of protoplanetary disks is measured by comparing the disk spectral energy distribution to radiative transfer models. Typical values range from a few hundred degrees Kelvin within 1 AU from solar type stars, to a few degrees Kelvin in the outermost disk regions. The measurement of the disk temperature is affected by relatively small errors (a factor of two at most) and will not be discussed further in this paper. Similarly, the distance of nearby protoplanetary disks relies on parallax measurements that are accurate at a 10% level, and will not be discussed here.

The opacity of solids is much more uncertain and dominates the overall uncertainty on the measurement of the mass of solids. It can be demonstrated that the solid particles that most efficiently emit thermal radiation at a wavelengths λ are those with a radius a such that $2\pi a = \lambda$. Therefore, the mm- and cm-wave thermal emission mostly arises from particles with radii between about 0.1 mm and 1 mm, i.e., dust grains. The opacity of these grains can be calculated once the composition and structure is known using, for example, the Mie theory.

However, the population of solids in a protoplanetary disk likely comprises a broader range of sizes. An estimate of the total mass of solids therefore requires assuming a grain size distribution. A common, though not well justified choice, is to a adopt a MRN distribution, $n(a) \propto a^{-3.5}$ (Mathis *et al.* 1977). In this case, most of the mass is locked into the largest grains ($a = a_{max}$), while most of opacity, and therefore emission, at a wavelength λ is provided by the grains with sizes comparable to λ. It is easy to demonstrate that, in the case of a MRN size distribution, the mass of solids derived from a single wavelength measurement of the flux density at millimeter-wavelengths is proportional to $\sqrt{a_{max}}$.

Multi-wavelengths observations at wavelengths between about 1mm-1cm help in constraining a_{max}. They measure the spectral slope α of the continuum emission $[I(\lambda) \propto \lambda^{\alpha}]$, which, if the emission is optically thin, is related to the spectral slope of the dust opacity β $[\kappa(\lambda) \propto \lambda^{-\beta}]$, which, in turn, depends on a_{max}. However, the relation between a_{max} and β is uncertain because it depends on the dust composition and size distribution, and, more importantly, it is degenerate for values of a_{max} larger than a few centimeters (see, e.g., Figure 4 in the PPVI review chapter by Testi *et al.* 2014). In practice, observations of the thermal dust continuum emission cannot measure the amount of solids locked into particles larger than a few centimeters.

Given all these caveats and assumptions, we should be doubtful about disk mass estimates and, in particular, we should avoid over interpreting mass estimates for single objects. However, mass measurements for homogeneous samples of disks can still provide important clues about the dependence between the mass of solids and the properties of

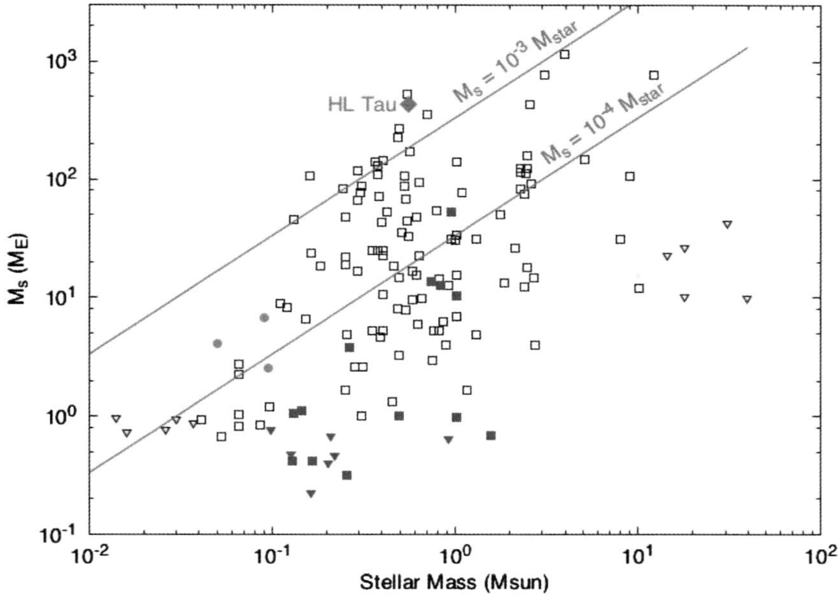

Figure 1. Mass of solids particles in units of the Earth mass measured form millimeter-wave observations of nearly protoplanetary disks as function of the stellar mass. Empty black triangles and squares indicate upper limits and detections, respectively, for disks in Taurus and Ophiuchus (Williams & Cieza 2011 and references therein). Filled triangles and squares indicate upper limits and detections, respectively, for disks in the Upper Scorpius association (Carpenter *et al.* 2014). Purple circles indicate disks around brown dwarfs (Ricci *et al.* 2014). *(A color version of this figure is available online.)*

the central star (such as age, mass, and metallicity), and help understanding the demographics of planetary systems.

For example, millimeter observations of nearby star forming regions have delivered estimates of the mass of solids for tens of protoplanetary disks spanning ages between 1-10 Myr and masses of the host star between 0.01-30 $M_\odot$ covering stellar spectral types from brawn dwarfs to O-type stars (Figure 1). In a recent review of this topic, Williams & Cieza (2011) argue that there is a linear correlation between the mass of solids in disks (M_s) and the mass of the central star ($M_\star$), such that, on average, $M_s = 10^{-4} \times M_\star$. Protoplanetary disks around solar mass stars contain on average about 30 Earth masses (M_E) of solids. As a reference, this value is equal to the minimum mass of solids required to form the planets in our Solar system.

Taken at face value, the relatively low mass of solids measured toward solar type stars suggest that these systems might be able to form only low mass planets. However, one has to keep in mind that this mass estimate is affected by all the uncertainties discussed above, and in particular, by the lack of knowledge about the fraction of mass of solid locked into large (>10 cm) bodies.

Furthermore, the correlation between the mass of solids and the stellar mass is characterized by a large scatter. For example, the mass of solids around solar mass stars varies by at least two orders of magnitude between 1 and 100 M_E. The lower limit of this range is set by the sensitivity of current observations and it is poised to decrease in the near future as the results of ALMA observations of nearby star forming regions are released (see, e.g., Carpenter *et al.* 2014, Ricci *et al.* 2014). Vice versa, the upper limit of the mass range is significant and implies that very few disks contain solids in excess of 10^{-3} $M_\star$, or 300 M_E for a Solar type star. Although this value is about 10× the minimum mass required to form the planets in the Solar system, it nevertheless suggests that either the planet formation process is very efficient in converting dust into planets, or that planetesimals, and perhaps planets, are already formed at the age of 1–5 Myr.

This result comes with two important caveats. First, Figure 1 shows mass measurements only for Class II disks. Mass estimates for protostellar systems in earlier phases of formation, e.g., Class 0 and I systems, would populate the uppermost part of the diagram which is now mostly empty. Second, as the mass of solids increases, the optical depth of the millimeter-wave emission also increases, and the mass estimates become more uncertain. In particular, if the emission is not completely optically thin, the mass of solids derived from Equation 2.1 provides only a lower limit of the true mass. If protoplanetary disks have the same size and temperature, this optical depth problem becomes more severe as the disk mass increases.

3. What have we learned from HL Tau?

A major milestone in the study of protoplanetary disks was achieved in December 2014 with the successful commissioning of the ALMA 10 km baselines and subsequent release of the dust continuum images of the HL Tau system (ALMA partnership, 2015). After one year from the release of the HL Tau images, the scientific community is debating whether the observed concentric rings are caused by sudden jumps in the radial profile of the dust opacity at radii where the gas temperature drops below the freeze out temperature of the water and CO (Banzatti *et al.* 2015; Okuzumi *et al.* 2015; Zhang *et al.* 2015), or whether the rings are caused by the tidal interaction with unseen planets (Akiyama *et al.* 2015; Di Pierro *et al.* 2015; Gonzalez *et al.* 2015; Picogna *et al.* 2015). At the time of writing, the planet-disk interaction model is favored because it better reproduces the main observable of the HL Tau system.

In a recent analysis of the ALMA data, Jin *et al.* (2016) find that the radial profile of the 1 mm flux density and of the spectral index measured between 0.87 and 1.3 mm are quantitatively consistent with the tidal interaction between a rather massive disk ($M_d/M_\star = 0.13$, $M_s/M_\star = 0.0013$) and three Saturn mass planets orbiting at about 13 AU, 33 AU, and 69 AU, respectively (Figure 2).

The planets' perturbations on the distribution of gas and dust is shown in the upper panel of Figure 2: the planets open shallow gaps in the gas density and deeper gaps in the distribution of the grains that mostly emits at the wavelength of 1 mm. This difference is due to the fact that, as the planets start opening gaps in the gas density, the sub-millimeter dust grains located within the gaps radially migrate toward the gap edges. As a result of the planet-disk interaction, the dust-to-gas ratio changes from an initial constant value of 0.01 to about 0.02 in the intra-planets regions, and 0.005 within the gas depleted gaps (top right panel).

Jin *et al.* (2016) find that most of the 1 mm continuum emission arising form the HL Tau disk is optically thick (bottom right panel of Figure 2). Theoretical models suggest optical depths between 1 and 3 in the innermost and middle bright rings, respectively, and values larger than a few in the innermost 10 AU from the central star. The estimate of the optical depth results from modeling the dust emission in the regions where the optical depth is less than 1 and needs to be confirmed by observing the HL Tau disk at longer wavelengths. For example, if the analysis of Jin *et al.* (2006) is correct, the dust emission at 7 mm from the innermost and middle rings should have optical depth of about 0.4 and 0.2, respectively.

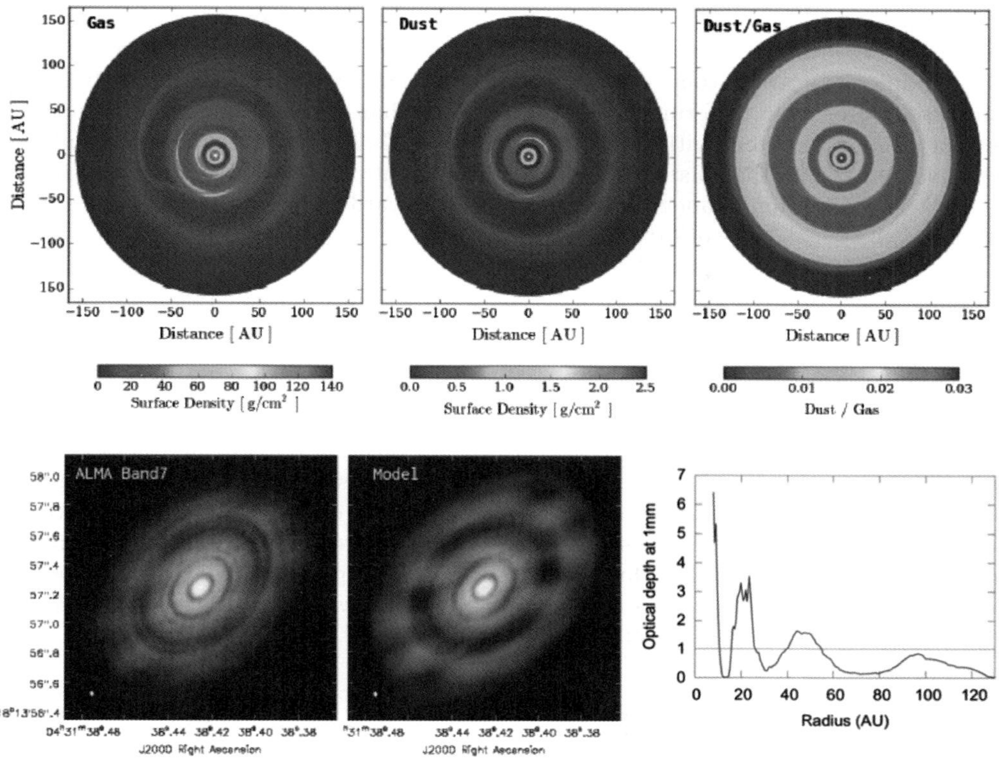

Figure 2. From Jin *et al.* (2006). Top: model for the gas and dust surface density, as well as dust-to-gas ratio for the HL Tau disk. Concentric rings of dust and gas are caused by the tidal perturbation between the circumstellar material and three Saturn mass planets orbiting at 13 AU, 33 AU, and 69 AU, respectively. Bottom left: comparison between model and observations. The color scale in the two panels is the same and it indicates the flux density. Bottom right: radial profile of the dust opacity at the wavelength of 1 mm inferred form the observations. *(A color version of this figure is available online.)*

Since a large fraction of the 1 mm continuum emission from the HL Tau disk is optically thick, the estimate of the disk mass is quite uncertain. A similarly large uncertainty likely affects the masses of the other millimeter-bright disks that occupy the top portion of the diagram shown in Figure 1. However, Jin *et al.* (2016) find that if the HL Tau disk mass were greater than about 0.2 $M_\star$, the disk would be gravitationally unstable. The measurement of the flux density on one side and dynamical arguments on the other, bracket the total mass of the HL Tau disk between 0.13-0.2 $M_\star$.

4. Conclusions

The relatively small mass of solids measured in the majority of protoplanetary disks with ages between 1–10 Myr, on one side, and the morphology of the HL Tau disk, on the other, suggest that planet formation might start very early on and proceed very quickly, such that the embryos of giant planets already exists in disk younger than 1 Myr. Planet formation might therefore be contemporaneous with the formation of disks themselves, or at least start when the disks are very young, massive, and perhaps gravitationally unstable.

Gravitational instabilities might explain the existence of Saturn mass planets orbiting at tens of AU from HL Tau. Indeed, gravitational instabilities arise naturally in the cold disk midplane far away from the central star. Furthermore, gravitational instabilities are supposed to be able to form planets much faster than the core accretion model. This would also be consistent with the young age of the HL Tau system. Whereas it might be reasonable to think that planets form when the parent disks are young and massive, it is unclear how the same planets might survive the inward radial migration caused by the interaction with the gaseous disk.

The observations discussed above demonstrate the transforming capabilities of ALMA. In a few years from now, many more systems like HL Tau will be mapped in great details and millimeter fluxes for most of the protoplanetary disks in nearby star forming regions will be measured. At that time, we will be able to derive firmer conclusion on the processes responsible for the formation of planets.

Finally, current ALMA observations stress the need to map protoplanetary disks at centimeter wavelengths in order to peer into the regions that are optically thick at ALMA wavelengths and achieve better estimates of the mass of solid particles. Centimeter-wave observations for the few brightest disks can already be obtained with the *Jansky Very Large Array* (JLVA), but more sensitive telescopes are required to map the majority of nearby protoplanetary disks. One of such future telescopes is the *next generation Vary Large Array* (ngVLA), which is discussed in a series of white papers available on arXiv (see, e.g., Carilli *et al.* 2015, Isella *et al.* 2015).

Acknowledgement

A.I. acknowledge support from NSF (grant no. 1535809) and NASA (grant no. NNX15AB06G).

References

ALMA Partnership, Brogan, C. L., Prez, L. M., Hunter, T. R., Dent, W. R. F., Hales, A. S., Hills, R. E., Corder, S., Fomalont, E. B., Vlahakis, C., Asaki, Y., Barkats, D., Hirota, A., Hodge, J. A., Impellizzeri, C. M. V., Kneissl, R., Liuzzo, E., Lucas, R., Marcelino, N., Matsushita, S., Nakanishi, K., Phillips, N., Richards, A. M. S., Toledo, I., Aladro, R., Broguiere, D., Cortes, J. R., Cortes, P. C., Espada, D., Galarza, F., Garcia-Appadoo, D., Guzman-Ramirez, L., Humphreys, E. M., Jung, T., Kameno, S., Laing, R. A., Leon, S., Marconi, G., Mignano, A., Nikolic, B., Nyman, L.-A., Radiszcz, M., Remijan, A., Rodn, J. A., Sawada, T., Takahashi, S., Tilanus, R. P. J., Vila Vilaro, B., Watson, L. C., Wiklind, T., Akiyama, E., Chapillon, E., de Gregorio-Monsalvo, I., Di Francesco, J., Gueth, F., Kawamura, A., Lee, C.-F., Nguyen Luong, Q., Mangum, J., Pietu, V., Sanhueza, P., Saigo, K., Takakuwa, S., Ubach, C., van Kempen, T., Wootten, A., Castro-Carrizo, A., Francke, H., Gallardo, J., Garcia, J., Gonzalez, S., Hill, T., Kaminski, T., Kurono, Y., Liu, H.-Y., Lopez, C., Morales, F., Plarre, K., Schieven, G., Testi, L., Videla, L., Villard, E., Andreani, P., Hibbard, J. E., & Tatematsu, K. 2015, *ApJ*, 808, 3

Banzatti, A., Pinilla, P., Ricci, L., Pontoppidan, K. M., Birnstiel, T., & Ciesla, F. 2015, *ApJ*, 815, 15

Carilli, C. L., McKinnon, M., Ott, J., Beasley, A., Isella, A., Murphy, E., Leroy, A., Casey, C., Moullet, A., Lacy, M., Hodge, J., Bower, G., Demorest, P., Hull, C., Hughes, M., di Francesco, J., Narayanan, D., Kent, B., Clark, B., & Butler, B. 2015, *eprint arXiv:1510.06438*

Carpenter, J. M., Ricci, L., & Isella, A. 2014, *ApJ*, 787, 42

Dipierro, G., Price, D., Guillaume, L., Hirsh, K., Cerioli, A., & Lodato, G. 2015, *MNRAS*, 453, 73

Gonzalez, J.-F., Laibe, G., Maddison, S. T., Pinte, C., & Mnard, F. 2015, *MNRAS*, 454, 36

Isella, A., Hull, L. H., Moullet, A., Galván-Madrid, R., Johnstone, D., Ricci, L., Tobin, J., Testi, L., Beltran, M., Lazio, J., Siemion, A., Liu, Hauyu B., Du, F., berg, K. I., Bergin, T., Caselli, P., Bourke, T., Carilli, C., Perez, L., Butler, B., de Pater, I., Qi, C., Hofstadter, M., Moreno, R., Alexander, D., Williams, J., Goldsmith, P., Wyatt, M., Loinard, L., Di Francesco, J., Wilner, D., Schilke, P., Ginsburg, A., Snchez-Monge, Zhang, Q., & Beuther, H. 2015, *eprint arXiv:1510.06444*

Jin, S., Li, S., Isella, A., Li, H., & Ji, J. 2016, *ApJ in press*

Mathis, J. S., Rumpl, W. & Nordsieck, K. H. 1997, *ApJ*, 217, 425

Okuzumi, S., Momose, M., Sirono, S., Kobayashi, H., & Tanaka, H. 2015, *ApJ submitted*, arxiv:1510.03556

Picogna, G., & Wilhelm, K. 2015, *A&A*, 584, 110

Ricci, L., Testi, L., Natta, A., Scholz, A., de Gregorio-Monsalvo & I., Isella, A. 2014, *ApJ*, 791, 20

Testi, L., Birnstiel, T., Ricci, L., Andrews, S., Blum, J., Carpenter, J., Dominik, C., Isella, A., Natta, A., Williams, J. P., & Wilner, D. J. 2014, *Protostars and Planets VI*, 339

Williams, J. P. & Cieza, L. A. 2011, *ARA&A*, 49, 67

Zhang, Ke, Blake, Geoffrey A., Bergin, & Edwin A. 2015, *ApJL*, 806, 7

From Interstellar Clouds to Star-Forming Galaxies: Universal Processes?
Proceedings IAU Symposium No. 315, 2015
P. Jablonka, Ph. André & F. van der Tak, eds.

doi:10.1017/S1743921316007365

Small-scale properties of Class 0 protostars from the CALYPSO IRAM-PdBI survey

Anaëlle Maury[1], Philippe André[1], Sébastien Maret[2], Arnaud Belloche[3], Claudio Codella[4], Sylvie Cabrit[2], Frédéric Gueth[5] and CALYPSO collaboration[1]

[1]Laboratoire AIM, CEA/DSM-CNRS-Université Paris Diderot, France
email: anaelle.maury@cea.fr
[2]IPAG, Grenoble, France
[3]MPIfR Bonn, Germany
[4]INAF, Osservatorio Astrofisico di Arcetri, Firenze, Italy
[5]IRAM, Grenoble, France

Abstract. Because the formation of protostars is believed to be closely tied to the angular momentum problem of star formation, characterizing the properties of the youngest disks around Class 0 objects is crucial. However, not much is known on the structure of the youngest protostellar envelopes, on the small scales at which disks and multiple systems are observed around more evolved YSOs, due to a lack of comprehensive high angular resolution observations (probing <100 AU). In order to tackle this issue, we conducted a large observing program with the IRAM Plateau de Bure interferometer (PdBI): the CALYPSO survey, providing us with detailed maps of molecular lines and millimeter continuum emission, probing scales down to ~30–50 au towards a sample of 17 Class 0 protostars. Here we present our analysis of the CALYPSO dust continuum emission maps, constraining disk properties of the Class 0 protostars in our sample. We show that large, $r > 50$ au, disk structures are not observed in most Class 0 protostars from our sample, which can be described by various envelope models reproducing satisfactorily the intensity distribution of the dust emission at all scales from 50 au to 5000 au.

Keywords. Class 0 Protostars, Circumstellar Disks, IRAM-PdBI, CALYPSO

1. Introduction

Young Class 0 protostars (André *et al.* 2000), observed only $t \lesssim 5 \times 10^4$ yr after their formation, still have most of their mass in the form of a dense core/envelope collapsing onto the central protostellar embryo: they are representative of the main accretion phase, during which most of the final stellar mass is accreted onto the central object. During this protostellar phase, the inflowing gas must redistribute most of its initial angular momentum outward or else centrifugal forces will balance gravity and prevent accretion, and the growth of the protostellar embryo. This long-standing "angular momentum problem" is quite severe since the gas must reduce its specific angular momentum by ~5 orders of magnitude during this brief accretion phase (<0.1 Myr) (Belloche 2013). Exactly how this is done remains a major unsolved problem in star formation. Early theoretical studies (Terebey *et al.* 1984) suggested that the formation of large (⩾100 au) centrifugally supported disks and protostellar jets (Shu *et al.* 1991) would transfer angular momentum outward, contributing grandly to solving the angular momentum problem by the end of the first collapse phase. However, while the presence of large (radii 100–300 au) accretion disks is well documented in Class I protostars and T-Tauri stars (Watson *et al.* 2007; Duchêne *et al.* 2007; Connelley *et al.* 2008), the progenitors of these protoplanetary disks remain elusive around younger Class 0 protostars (Maury *et al.* 2010; Ohashi *et al.* 2014;

Yen *et al.* 2015a). Whether disks and multiple systems are natural products of the conservation of angular momentum formed during the main accretion phase remains largely debated today from both an observational and a theoretical point of view.

2. The CALYPSO survey

The CALYPSO survey (see http://irfu.cea.fr/Projets/Calypso/), carried out with the IRAM Plateau de Bure (PdBI) interferometer, has been tailored to address this cornerstone question, and ultimately describe how the circumstellar envelope is being accreted onto the central protostellar embryo, also exploring the formation processes of protostellar jets, disks and multiple systems, during the main accretion phase. In the framework of this IRAM large program, we carried out detailed dust continuum and molecular line emission observations of 17 Class 0 protostars, at scales ranging from 50 au to 5000 au, in selected frequency ranges between 1.3 mm and 3.2 mm. Highlights of the CALYPSO observations of the Class 0 protostar NGC1333-IRAS2A (IRAS2A in the following) were presented in Maret *et al.* (2014), Maury *et al.* (2014) and Codella *et al.* (2014).

3. Analysis of the dust continuum emission

Here we briefly present our analysis of the dust continuum emission obtained in the 1mm PdBI bands, namely two independent observations obtained over 4-GHz bandwidths centered respectively at 1.3 mm and 1.4 mm, towards a large fraction of the CALYPSO sample, analyzed as of August 2015 and presented in Table 1.
Class 0 protostars are heavily embedded, envelope-dominated sources. To constrain the protostellar disk properties during this early stage, a precise modeling of envelope emission at all scales is required to discriminate envelope emission from disk emission at small scales. The large-scale envelope structure is recovered thanks to IRAM-30m 1.2-mm continuum observations used as zero-spacing information as well as short-spacing intensity profiles when available (see Motte & André 2001). The IRAM-30m 1.2-mm dust continuum emission maps (Motte & André 2001) tracing protostellar envelopes at large scales $\sim 10-50''$ are shown in the left panels of Figs. 1, 2 and 3, while the central panels show the IRAM-PdBI 1.3-mm dust continuum maps, sensitive to the inner envelope density structure at scales $\sim 0.5-5''$. We examine the flux density versus uv-distance from the PdBI data at both 1.3 mm and 1.4 mm, and use Plummer-like envelope models with $\rho(r)=\rho_0(\frac{R_0}{\sqrt{r^2+R_0^2}})^p$ to reproduce the visibility profiles observed at the two wavelengths (see central panels of Figs. 1, 2 and 3). In the Rayleigh-Jeans approximation and for optically-thin dust emission, if the temperature and density in the envelope follow simple radial power laws $\rho(r)\propto r^{-p}$ and $T(r)\propto r^{-q}$, the emergent dust continuum emission also has a simple power-law form $I(r)\propto r^{-p-q+1}$. In interferometric observations, the visibility distribution then has $V(b)\propto b^{p+q-3}$, solely determined by the power-law indices of the temperature and density profiles in the envelope. We noticed that the PdBI 1.3 mm visibility curves were not satisfactorily reproduced by envelope models, for a few sources in our sample. To check whether this deviation to a single envelope profile traces a circumstellar disk embedded in the envelope, or reflects envelope asymmetries and structure, we also tried to reproduce the intensity profiles with models including the best-fit envelope model and a Gaussian source of variable flux and size, tracing a candidate disk structure. The details of this analysis will be presented in a forthcoming paper (Maury *et al.* in prep.). It is illustrated for three of our CALYPSO sources in Figs. 1 to 3. The right panels show the visibilities fluxes as a function of IRAM-PdBI baseline length. The best-fit Plummer envelope model reproducing the single dish continuum emission profile (at baselines <20 kλ, e.g. tracing the outer envelope properties) is shown as a red curve, for each source. The main results of our analysis, such as best-fit values of the

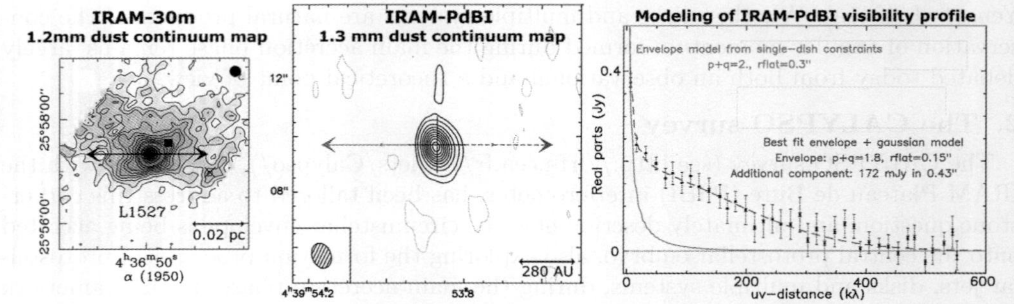

Figure 1. L1527: Dust continuum emission from the IRAM-30 m (map shown in left panel) and PdBI (map shown in central panel, the double-sided arrow marks the outflow axis) is reproduced with a model featuring a Plummer envelope and an additional Gaussian source. The best-fitting model includes a $r \sim 60$ au radius additional Gaussian source at phase center, interpreted as a candidate disk component, and is shown in the right panel as a blue curve overlaid on the PdBI visibility profile at 1.3 mm. *(A color version of this figure is available online.)*

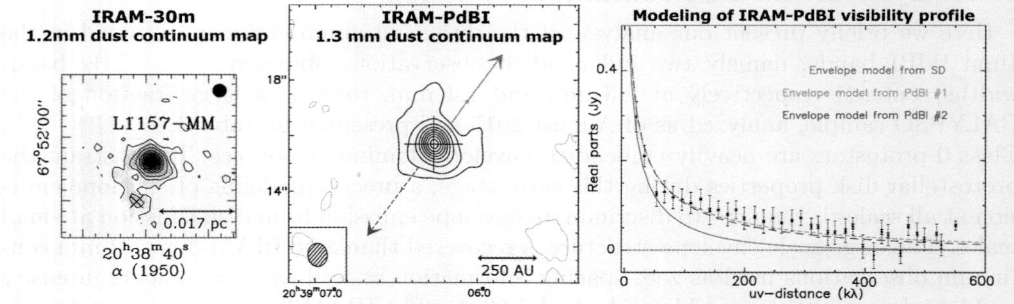

Figure 2. L1157: Dust continuum emission from the IRAM-30 m (map shown in left panel) and PdBI (map shown in central panel, the double-sided arrow marks the outflow axis) is well reproduced with a model only comprising a Plummer envelope. The best model is shown as a blue curve in the right panel : it does not need any additional Gaussian component to reproduce the PdBI 1.3-mm visibility profile apart from the envelope model, therefore suggesting any protostellar disk is $r < 40$ au in this Class 0 source. *(Color version available online.)*

power-law indices for protostellar envelopes and detection of a $r > 50$ au candidate-disk component, are summarized in Table 1.

4. Summary: do most Class 0 protostellar disks have radii < 50 au ?

Among the 12 Class 0 protostars analyzed here, only 3 of them show millimeter dust continuum visibility profiles which may hint to the presence of disk structures with radii > 50 au, and only one source may harbor a candidate disk with radius > 100 au. Being the largest sample of Class 0 protostars analyzed so far, our results confirm the paucity of large Class 0 disks suggested by recent literature (Maury *et al.* 2010, Yen *et al.* 2015, Tobin *et al.* 2015). These upper limits on disk sizes are not easily explained by the standard hydrodynamical description of protostellar collapse, where large (r > 100 au) disk-like structures are expected to form quickly ($t < 10^3$ yr) at the beginning of the protostellar phase, as a consequence of angular momentum conservation during collapse of the large-scale envelope towards the central protostellar embryo (Li *et al.* 2014). One possible explanation for the regulation of disk sizes during the main accretion phase, lies in magnetically-regulated protostellar collapse scenarios, because magnetic braking can significantly contribute to angular momentum transport and hence delay the formation of large disk-structures (Hennebelle & Teyssier 2008). In the sources showing no signs of

Table 1. Preliminary results from the analysis of continuum emission in the CALYPSO sample

Object	α (J2000)	δ (J2000)	Dist. (pc)	M_{env} ($M_\odot$) (1)	$F^{PdBI}_{peak}/F^{SD}_{tot}$ (2)	$p+q$ index (3)	Candidate > 50 au disk ? (4)	Class
L1448-2A	03:25:22.406	+30:45:13.28	220	0.9	0.09-0.10	2-2.3	n	0
L1448-NB	03:25:36.364	+30:45:14.83	220	1	0.1-0.13	2.6-2.8	n	0
L1448-C	03:25:38.878	+30:44:05.32	220	4	0.14-0.18	2.3-2.6	n	0
IRAS2A	03:28:55.575	+31:14:37.05	220	5	0.15-0.2	2.5-2.6	n	0
SVS13-B	03:29:03.075	+31:15:51.71	220	3	0.15-0.2	1.7-2.1	y	0
IRAS4A	03:29:10.531	+31:13:30.96	220	6	0.4-0.6	2.7-2.8	n	0/I
IRAS4B	03:29:12.012	+31:13:08.07	220	3	0.2-0.3	2.7-2.9	y	0
IRAM04191	04:21:56.903	+15:29:46.14	140	1	0.006	1.5-1.6	n	0
L1521-F	04:28:38.936	+26:51:35.11	140	0.7	0.01-0.04	1.4-1.7	n	0
L1527	04:39:53.874	+26:03:09.67	140	1	0.13-0.15	1.8-2.0	y	0/I
L1157	20:39:06.268	+68:02:15.67	250	1.5	0.11-0.13	2.5-2.6	n	0
GF9-2	20:51:30.29.822	+60:18:38.44	200	0.5	<0.04	1.7-1.9	n	0

(1) Envelope masses from the literature. (2) Ratio of the small-scale flux (1.3-mm peak flux computed in PdBI matching-beams maps – synthesized HPBW $\sim$1.5″×1″) to the total envelope flux (as extrapolated from IRAM-30m 1.2-mm or Bolocam 1.1-mm dust continuum maps). (3) Index of the intensity profile, see Maury *et al.* (2014). (4) Indicating whether the best model to reproduce both the IRAM-30m and IRAM-PdBI continuum emission includes a Gaussian source of FWHM>50 au ("y" for yes or "n" for no).

large disks, the role of magnetic fields should be explored in more detail to understand our puzzling results. The kinematics of Class 0 envelopes will have to be explored as well, for example to look for signatures of transitions to Keplerian rotation at small scales in the sources where additional dust continuum structures are needed to explain our PdBI data: the wealth of density and temperature regimes probed by the CALYPSO molecular line observations should allow such investigations in the coming year.

Acknowledgements

This work has received support from the ERC 'ORISTARS' project (ERC Advanced Grant Agreement no. 291294).

References

André, P., Ward-Thompson, D., & Barsony, M., 2000, in *Protostars and Planets IV*, p. 59
Belloche, A., 2013, in *EAS Publications Series* Vol. 62, pp 25
Codella, C., Maury, A. J., Gueth, F., *et al.*, 2014, A&A, 563, L3
Connelley, M. S., Reipurth, B., & Tokunaga, A. T. 2008, ApJ, 135, 2526
Duchêne, G., Bontemps, S., Bouvier, J., *et al.*, 2007, A&A, 476, 229
Hennebelle, P. & Teyssier, R., 2008, A&A, 477, 9
Li Z.-Y., *et al.*, 2014, Protostars and Planets VI, pp 173
Maret, S., Belloche, A., Maury, A. J., *et al.*, 2014, A&A, 563, L1
Maury, A. J., André, Ph. *et al.*, 2010, A&A, 512, A40+
Maury, A. J., Belloche, A., André, Ph. *et al.*, 2014, A&A, 563, L2
Maury, A. J., *et al.*, 2016, in prep., A&A
Motte, F. & André, Ph., 2001, A&A, 365, 440
Ohashi, N., Saigo, K., Aso, Y., *et al.* 2014, ApJ, 796, 131
Terebey, S., Shu, F., & Cassen, P., 1984, ApJ, 286, 529
Tobin, J., *et al.* 2015, ApJ, 805, 125
Shu, F. H., Ruden, S. P. , Lada, C. J., & Lizano, S., 1991, ApJ, 370, 31
Watson, A. M., Stapelfeldt, K. R., *et al*, 2007, in Protostars and Planets V, p. 523
Yen, H.-W., Koch, P. M., Takakuwa, S., *et al.* 2015, ApJ, 799, 193
Yen, H.-W., Takakuwa, S., Koch, P. M., *et al.* 2015, ApJ, 812, 129

From Interstellar Clouds to Star-Forming Galaxies: Universal Processes?
Proceedings IAU Symposium No. 315, 2015
P. Jablonka, Ph. André & F. van der Tak, eds.

doi:10.1017/S1743921316007377

Theory of Protostellar Disk Formation

Zhi-Yun Li[1], Ruben Krasnopolsky[2] and Hsien Shang[2]

[1]Astronomy Department, University of Virginia
email: zl4h@virginia.edu

[2]Academia Sinica, Institute of Astronomy and Astrophysics and Theoretical Institute for Advanced Research in Astrophysics, Taiwan

Abstract. How large, 100-AU scale, rotationally supported disks form around protostars remains unsettled, both observationally and theoretically. In this contribution, we discuss the theoretical difficulties with disk formation in the presence of a dynamically significant magnetic field and their possible resolutions. These difficulties are caused by the concentration of magnetic field lines close to the forming star by protostellar collapse, and the strong magnetic braking associated with the concentrated field. Possible resolutions include magnetic field-rotation axis misalignment, non-ideal MHD effects, and turbulence. The field-rotation misalignment has been shown to promote disk formation, especially when the field is relatively weak and the misalignment angle is relatively large. Non-ideal MHD effects can enable the formation of small disks at early times. How such disks grow at later times remains to be fully quantified. Turbulence has been found to enable disk formation in a number of simulations, but the exact reason for its beneficial effect is debated.

Keywords. stars: formation, accretion disk, magnetic field, turbulence

How disks form is an important, unsolved, problem in star and planet formation. One may expect disks to form as a simple consequence of angular momentum conservation during the hydrodynamic collapse of dense, rotating cores of molecular clouds. However, in the presence of a dynamically significant magnetic field, disk formation becomes more uncertain. We will first discuss the theoretical difficulties with disk formation in magnetized dense cores in the ideal MHD limit. This is followed by a discussion of possible resolutions of these difficulties, including non-ideal MHD effects, field-rotation misalignment, and turbulence.

1. Magnetic Braking Catastrophe in Ideal MHD Limit

The basic difficulty with disk formation in magnetized dense cores can be illustrated most clearly in the strict ideal MHD limit, where the magnetic field lines are perfectly frozen into the core material. In this limit, as a finite amount of mass is accreted onto the central object (the protostar), a finite amount of magnetic flux will be dragged into the object as well. The magnetic flux accumulated at the center forms a magnetic split monopole, with the field lines fanning out radially. As a result, the magnetic field strength increases rapidly with decreasing distance to the center, as $B \propto r^{-2}$. The magnetic energy density, which is proportional to the field strength squared, increases with decreasing radius even more rapidly, as $E_B \propto r^{-4}$. This increase is faster than, for example, the energy density of the accretion flow, which can be estimated approximately from spherical free-fall collapse as $E_{\rm ff} \propto r^{-5/2}$. As the infalling material approaches the central object, it will become completely dominated by the magnetic field sooner or later. The strong magnetic field at small radii is able to remove all of the angular momentum in the collapsing flow, leading to the so-called "magnetic braking catastrophe" for disk formation (Galli *et al.* 2006).

The catastrophic braking of disks in magnetized dense cores in the ideal MHD limit was also found in many numerical as well as semi-analytic calculations. Krasnopolsky & Königl (2002) were the first to show semi-analytically, using the so-called "thin-disk" approximation, that the formation of rotationally supported disks (RSDs hereafter) can be suppressed if the efficiency of magnetic braking is large enough. However, the braking efficiency was parametrized rather than computed self-consistently. Similarly, Dapp & Basu (2010) and Dapp *et al.* (2012) demonstrated that, in the absence of any magnetic diffusivity, a magnetic split-monopole is produced at the center and RSD formation is suppressed, again under the thin-disk approximation.

The catastrophic braking that prevents the formation of RSDs during the protostellar accretion phase has been found in several subsequent 2D and 3D ideal MHD simulations (Tomisaka 1998; Allen *et al.* 2003; Mellon & Li 2008; Hennebelle & Fromang 2008; Duffin & Pudritz 2009; Seifried *et al.* 2012; Santos-Lima *et al.* 2012). Mellon & Li (2008), in particular, formulated the disk formation problem in the same way as Allen *et al.* (2003), by adopting a self-similar rotating, magnetized, singular isothermal toroid as the initial configuration. Although idealized, the adopted initial configuration has the advantage that the subsequent core collapse should remain self-similar. The self-similarity provides a useful check on the correctness of the numerically obtained solution. They found that the disk formation was suppressed by a field as weak as $\lambda = 13.3$, where λ is the mass-to-flux ratio of the dense core in units of the critical value $1/(2\pi\sqrt{G})$.

It should be noted that ideal MHD simulations of core collapse and star/disk formation are intrinsically difficult to converge. This is because the collapse tends to produce a split magnetic monopole near the center, which has a sharp field reversal near the equatorial plane that is prone to (often violent) numerical reconnection (Li *et al.* 2014; Masson *et al.* 2015). When and where the reconnection occurs depends on numerical resolution, which makes it difficult, perhaps not possible, to obtain numerically converged solutions. Nevertheless, both analytic arguments and currently available numerical simulations support the notion that, in the ideal MHD limit, catastrophic braking makes it difficult to form rotationally supported disks in (laminar) dense cores magnetized to a realistic level (with a typical λ of a few; Troland & Crutcher 2008). Additional effects must be included in order to enable disk formation.

2. Non-ideal MHD Effects

In the ideal MHD limit, the flux-freezing condition leads to a magnetic flux problem for the central stellar object, which then results in a magnetic braking catastrophe. Non-ideal MHD effects can in principle break the flux-freezing condition, and that gives a chance to solve the magnetic flux problem, and potentially the magnetic braking catastrophe. The three more well-known non-ideal MHD effects are ambipolar diffusion (AD), the Hall effect, and Ohmic dissipation (see Armitage 2011 for a review). Roughly speaking, in the simplest case of an electron-ion-neutral medium, both ions and electrons are well tied to the magnetic field in the ambipolar diffusion regime. In the Hall regime, electrons remain well tied to the field, but not ions. At the highest densities, both electrons and ions are knocked off the field lines by collisions before they finish a complete gyration; in such a case, Ohmic dissipation dominates. This simple picture is complicated by dust grains, whose size distribution in dense cores is relatively uncertain, but which can become the dominant charge carriers. Under typical cloud conditions, ambipolar diffusion dominates over the other two effects at densities typical of cores (e.g., Nakano *et al.* 2002, Kunz &

Mouschovias 2010). It is the most widely studied non-ideal MHD effect in the context of core formation and evolution in the so-called "standard" picture of low-mass star formation out of magnetically supported clouds (Nakano 1984; Shu *et al.* 1987; Mouschovias & Ciolek 1999). It is the effect that we will first concentrate on.

Ambipolar diffusion enables the magnetic field lines that are tied to the ions to drift relative to the bulk neutral material. In the context of disk formation, its most important effect is to redistribute the magnetic flux that would have been dragged into the central object in the ideal MHD limit to a circumstellar region where the magnetic field strength is greatly enhanced (Li & McKee 1996) solving the problem of the central magnetic flux, but not necessarily the problem of excessive magnetic braking. Indeed, the enhanced circumstellar magnetic field is strong enough to drive a hydromagnetic shock into the protostellar accretion flow (Li & McKee 1996; Ciolek & Königl 1998; Contopoulos *et al.* 1998; Krasnopolsky & Königl 2002; Tassis & Mouschovias 2007; Dapp *et al.* 2012). Krasnopolsky & Königl (2002) showed semi-analytically, using the 1D thin-disk approximation, that disk formation may be suppressed in the strongly magnetized post-shock region if the magnetic braking is efficient enough. The braking efficiency, parametrized in Krasnopolsky & Königl (2002), was computed self-consistently in the 2D (axisymmetric) simulations of Mellon & Li (2009), which were performed under the usual assumption of ion density proportional to the square root of neutral density. 3D simulations of AD were performed by Duffin & Pudritz (2009) using a specially developed, single fluid AMR code (Duffin & Pudritz 2009) as well as by a two fluid SPH code (Hosking & Whitworth 2004). Mellon & Li (2009) found that ambipolar diffusion does not weaken the magnetic braking enough to allow rotationally supported disks to form for realistic levels of cloud core magnetization and cosmic ray ionization rate. In many cases, the magnetic braking is even enhanced. These findings were strengthened by Li *et al.* (2011), who computed the ion density self-consistently using the simplified chemical network of Nakano and collaborators (Nakano *et al.* 2002; Nishi *et al.* 1991) that includes dust grains. We should note that RSD formation may still be possible if the cosmic ray ionization rate can be reduced well below the canonical value of 10^{-17} s^{-1} (Mellon & Li 2009), through for example the magnetic mirroring effect, which may turn a large fraction of the incoming cosmic rays back before they reach the disk-forming region (Padovani *et al.* 2013, 2014). Masson *et al.* (2015) showed that ambipolar diffusion in their simulations lowered the maximum field strength obtained close to the central object compared to the case of ideal MHD, and enabled the formation of rotation-dominated, spiral-like structure of order 50-80 AU in radius for a magnetized, $\lambda \approx 4$, core. Whether this structure can be called a rotationally supported disk or not depends on how the disk is defined, which is somewhat ambiguous. More work is needed to ascertain the role of ambipolar diffusion in disk formation.

As the density increases, the Hall effect tends to become more important. It is less explored than ambipolar diffusion in the star formation literature. A unique feature of this effect is that it can actively increase the angular momentum of a collapsing, magnetized flow through the so-called "Hall spin-up" (Wardle & Ng 1999). In the simplest case of electron-ion-neutral fluid, the spin-up is caused by the current carriers (the electrons) moving in the azimuthal direction, generating a magnetic torque through field twisting; the toroidal current is produced by gravitational collapse, which drags the poloidal field into a pinched, hourglass-like configuration. The Hall spin-up was studied numerically by Krasnopolsky *et al.* (2011) and semi-analytically by Braiding & Wardle (2012a,b). Krasnopolsky *et al.* (2011) showed that a rotationally supported disk can form even in an initially *non-rotating* core, provided that the Hall coefficient is large enough. Interestingly, when the direction of the initial magnetic field in the core is flipped, the disk rotation is reversed. The Hall effect, although dynamically significant, does not appear capable of forming a large (100-AU scale) rotationally supported disk under typical dense core conditions according to Li *et al.* (2011), at least at relatively late times where substantial mass has accreted onto the central stellar object. It does appear to facilitate the formation of a rotationally supported structure during the first core formation at early times (Tsukamoto *et al.* 2015a).

Ohmic dissipation becomes the dominant nonideal MHD effects at high densities (e.g., Nakano *et al.* 2002). It has been investigated by different groups in connection with disk formation. Shu *et al.* (2006) studied semi-analytically the effects of a spatially uniform resistivity on the magnetic field structure during the protostellar mass accretion phase. They found that, close to the central object, the magnetic field decouples from the collapsing material and becomes more or less uniform. They suggested that a rotationally supported disk may form in the decoupled region, especially if the resistivity is higher than the classic (microscopic) value. This suggestion was confirmed by Krasnopolsky *et al.* (2010; see also Santos-Lima *et al.* 2012), who found numerically that a large, 10^2 AU-scale, Keplerian disk can form around a $0.5\,M_\odot$ star, provided that the resistivity is of order $10^{19}\,cm^2\,s^{-1}$ or more; such a resistivity is significantly higher than the classic (microscopic) value over most of the density range relevant to disk formation.

Machida & Matsumoto (2011) and Machida *et al.* (2011) studied disk formation in magnetized cores including only the classic value of resistivity estimated from Nakano *et al.*'s (2002) numerical results. The former study found that a relatively small, 10 AU-scale, rotationally supported disk formed within a few years after the formation of the stellar core. Inside the disk, the density is high enough for magnetic decoupling to occur due to Ohmic dissipation. This work was extended to much later times by Machida *et al.* (2011), who included a central sink region in the simulations. They concluded that the small RSD can grow to large, 10^2-AU size at later times, especially after most of the envelope material has fallen onto the disk and the central object. A caveat, pointed out by Tomida *et al.* (2013; see also Dapp & Basu 2010), is that they used a form of induction equation that is, strictly speaking, inappropriate for the non-constant resistivity adopted in their models; it may generate magnetic monopoles that are subsequently advected away using the Dedner's method (Dedner *et al.* 2002). This deficiency was corrected in Tomida *et al.* (2013), who carried out radiative MHD simulations of magnetized core collapse to a time shortly ($\sim$1 year) after the formation of the second (protostellar) core. They found that the formation of a (small, AU-scale) rotationally supported disk was suppressed by magnetic braking in the ideal MHD limit but was enabled by Ohmic dissipation at this early time; the latter result is in qualitative agreement with Machida & Matsumoto (2011) and Machida *et al.* (2011), although it remains to be seen how the small disks in Tomida *et al.* (2013)'s simulations evolve further in time.

Dapp & Basu (2010) studied the effects of Ohmic dissipation on disk formation semi-analytically, using the "thin-disk" approximation for the mass distribution and an approximate treatment of magnetic braking. The approximations enabled them to follow the formation of both the first and second core. They found that a small, sub-AU, rotationally supported disk was able to form soon after the formation of the second core in the presence of Ohmic dissipation; it was suppressed in the ideal MHD limit, in agreement with the later 3D simulations of Tomida *et al.* (2013). This work was extended by Dapp *et al.* (2012) to include a set of self-consistently computed charge densities from a simplified chemical network and ambipolar diffusion. They showed that their earlier conclusion that a small, sub-AU scale, RSD is formed through Ohmic dissipation holds even in the presence of a realistic level of ambipolar diffusion. This conclusion appears reasonably secure in view of the broad agreement between the semi-analytic work and numerical simulations (see also Tomida *et al.* 2015 and Tsukamoto *et al.* 2015b). When and how such disks grow to the much larger, 10^2 AU-scale, size deserve to be explored more fully.

3. Magnetic Field-Rotation Misalignment

Misalignment between the magnetic field and rotation axis as a way to form large RSDs has been explored extensively by Hennebelle and collaborators (Hennebelle & Ciardi 2009; Ciardi & Hennebelle 2010; Joos *et al.* 2012; see also Machida *et al.* 2006, and Boss & Keiser 2013). The misalignment is expected if the angular momenta of dense cores are generated through turbulent motions (e.g., Burkert & Bodenheimer 2000; Myers *et al.* 2013; Joos *et al.* 2013). Plausible observational evidence for it was recently uncovered by Hull *et al.* (2013) using CARMA, who found that the distribution of the angle between the magnetic field on the 10^3 AU-scale and the bipolar outflow axis (taken as a proxy for the rotation axis) is consistent with being random. If true, it would imply that in half of the sources the two axes are misaligned by an angle greater than 60°. Joos *et al.* (2012) found that such a large misalignment enables the formation of RSDs in moderately magnetized dense cores with a dimensionless mass-to-flux ratio λ of $\sim$ 3–5; RSD formation is suppressed in such cores if the magnetic field and rotation axis are less misaligned. They attributed the disk formation to a reduction in the magnetic braking efficiency induced by large misalignment. In more strongly magnetized cores with $\lambda \lesssim 2$, RSD formation is suppressed independent of the misalignment angle, whereas in very weakly magnetized cores RSDs are formed for all misalignment angles.

Based on the work of Hull *et al.* (2013) and Joos *et al.* (2012), Krumholz *et al.* (2013) estimated that the field-rotation misalignment may enable the formation of large RSDs in $\sim$10–50% of dense cores. If the upper range is correct, the misalignment would go a long way toward solving the problem of excessive magnetic braking in protostellar disk formation.

Li *et al.* (2013) carried out simulations similar to those of Joos *et al.* (2012), except for the initial conditions. They confirmed the qualitative result of Joos *et al.* (2012) that the field-rotation misalignment is conducive to disk formation. In particular, large misalignment weakens the strong outflow in the aligned case and is a key reason behind the formation of RSDs in relatively weakly magnetized cores. For more strongly magnetized cores with $\lambda \lesssim 4$, RSD formation is suppressed independent of the degree of misalignment. This threshold value for the mass-to-flux ratio is about a factor of 2 higher than that obtained by Joos *et al.* (2012). The difference may come, at least in part, from the different initial conditions adopted: uniform density with a uniform magnetic field for Li *et al.* (2013) and a centrally condensed density profile with a nonuniform but unidirectional field for Joos *et al.* (2012); the magnetic braking is expected to be more efficient at a given (high) central density for the former initial configuration, because its field lines would become more pinched, with a longer lever arm for braking. Whether there are other factors that contribute significantly to the above discrepancy remains to be determined.

If the result of Li *et al.* (2013) is correct, then a dense core must have both a large field-rotation misalignment *and* a rather weak magnetic field in order to form a RSD. This dual requirement would make it difficult for the misalignment alone to enable disk formation in the majority of dense cores, which are typically rather strongly magnetized according to Troland & Crutcher (2008, with a median mass-to-flux ratio of $\lambda \sim 2$). In a more recent study, Crutcher *et al.* (2010) argued, based on Bayesian analysis, that a fraction of dense cores could be very weakly magnetized, with a dimensionless mass-to-flux ratio λ well above unity (see Bertram *et al.* 2012 for additional arguments for weak field, including field reversal). However, since the median mass-to-flux ratio remains unchanged for the different distributions of the total field strength assumed in Crutcher's (2012) Bayesian analysis, it is unlikely for the majority of dense cores to have λ much greater than the median value of 2. For example, Li *et al.* (2013) estimated the fraction of dense cores with $\lambda > 4$ at $\sim 25\%$. There is also concern that the random distribution of the field-rotation misalignment angle found by Hull *et al.* (2013) on the 10^3 AU scale may not be representative of the distribution on the larger core scale. Indeed, Chapman *et al.* (2013) found that the field orientation on the core scale (measured using a single-dish telescope) is within $\sim 30°$ of the outflow axis for 3 of the 4 sources in their sample (see also Davidson *et al.* 2011); the larger angle measured in the remaining source may be due to projection effects because its outflow axis lies close to the line-of-sight. If the result of Chapman *et al.* (2013) is robust and if the outflow axis reflects the rotation axis, dense cores with large misalignment between the magnetic and rotation axes would be rare. In such a case, it would be even less likely for the misalignment to be the dominant mechanism for disk formation.

4. Turbulence

Turbulence is a major ingredient for star formation (see reviews by, e.g., Mac Low & Klessen 2004 and McKee & Ostriker 2007). It can generate local angular momentum by shear flows and form highly asymmetric dense cores (see results from Herschel observations, e.g., Men'shchikov *et al.* 2010 and Molinari *et al.* 2010). There is increasing evidence that it also promotes RSD formation. Santos-Lima *et al.* (2012) contrasted the accretion of turbulent and laminar magnetized gas onto a pre-existing central star, and found that a nearly Keplerian disk was formed in the turbulent but not laminar case. The simulations were carried out at a relatively low resolution (with a rather large cell size of 15.6 AU; this was halved, however, in Santos-Lima *et al.* 2013, who found similar results), and turbulence was driven to an rms Mach number of $\sim$4, which may be too large for low-mass cores. Nevertheless, the beneficial effect of turbulence on disk formation is clearly demonstrated. They attributed the disk formation to the turbulence-induced outward diffusion of magnetic flux, which reduces the strength of the magnetic field in the inner, disk-forming, part of the accretion flow. The simulations of Seifried *et al.* (2012b, 2013) also show turbulence allowing disk formation, but these authors propose a different mechanism, in which the turbulence-induced tangling of field lines and strong local shear are mainly responsible for the disk formation: the disorder of the magnetic field weakens its braking efficiency and the shear enhances rotation. However, the connection between the proposed mechanisms and the simulations remains to be fully quantified. It is possible that the results of Seifried *et al.* (2012b) are compatible with magnetic misalignment (Joos *et al.* 2013) or magnetic diffusion (Santos-Lima *et al.* 2013), either due to the mechanism advocated by Santos-Lima *et al.* (2012) or numerical diffusion, especially during the formation of sink particles.

Similarly, Myers *et al.* (2013) also observed formation of a nearly Keplerian disk in their radiative MHD simulation of a turbulent massive ($300\,M_\odot$) core, although they refrained from discussing the origin of the disk in detail since it was not the focus of their investigation. Joos *et al.* (2013) investigated the effects of turbulence of various strengths on disk formation in a core of intermediate mass ($5\,M_\odot$). They found that an initially imposed turbulence has two major effects. It produces an effective diffusivity that enables magnetic flux to diffuse outward, broadly consistent with the picture envisioned in Santos-Lima *et al.* (2012, 2013). It also generates a substantial misalignment between the rotation axis and magnetic field direction (an effect also seen in Seifried *et al.* 2012b and Myers *et al.* 2013). Both of these effects tend to weaken magnetic braking and make disk formation easier. If the turbulence-induced magnetic diffusion is responsible, at least in part, for the disk formation, then numerical effects would be a concern. In the ideal MHD limit, the diffusion presumably comes from turbulence-enhanced reconnections due to finite grid resolution. Indeed, Joos *et al.* (2013) reported that their simulations did not appear to be fully converged, with disk masses differing by a factor up to ~ 2 in higher resolution simulations. The situation is further complicated by numerical algorithms for treating magnetic field evolution, especially those relying on divergence advection and divergence cleaning, which could introduce additional magnetic diffusion and residual magnetic monopoles. To make further progress, it would be useful to determine when and how the reconnections occur and exactly how they lead to the magnetic diffusion that are apparent in the simulations of Joos *et al.* (2013), Santos-Lima *et al.* (2012), Li *et al.* (2014) and perhaps Seifried *et al.* (2012b, 2013). An improvement in our understanding of the detailed structure of the turbulence in star forming cores would also help greatly.

In the simulations of Li *et al.* (2014), disk formation is observed for a turbulent Mach number in the range 0.5–1.0, the disk being persistent for M=1. A strongly warped pseudodisk was observed in the disk-forming cases, and a mechanism is proposed in which this pseudodisk warping is envisioned as favorable to the formation of a flat Keplerian disk inside. This mechanism can perhaps be also extended to the misalignment scenario, which also shows warped pseudodisks surrounding the Keplerian disks (Li *et al.* 2013). These simulations have been performed using a constrained transport code, and therefore are essentially free of magnetic monopoles; they are however not free of magnetic diffusion, especially due to the use of a relatively large spherical hole in their simulation grid. In any case, turbulence has been shown to promote disk formation, but this beneficial effect remains to be full understood and quantified.

In summary, there has been substantial progress in identifying the theoretical difficulties in magnetized disk formation and possible resolutions. Much work remains to be done to determine which of these possibilities, if any, is chosen by Nature.

We thank D. Galli, E. de Gouveia Dal Pino, R. Klessen, S. Lizano, R. Pudritz and R. Santos-Lima for useful discussion. This work was supported in part by NASA NNX10AH30G and NNX14AB38G, and NSF AST1313083.

References

Allen, A., Li, Z.-Y., & Shu, F. H. 2003, *ApJ*, 599, 363
Armitage, P. J. 2011, *ARAA*, 49, 195
Bertram, E., Federrath, C., Banerjee, R., & Klessen, R. S. 2012, *MNRAS*, 420, 3163
Boss, A. P. & Keiser, S. A. 2013, *ApJ*, 764, 136
Braiding, C. R. & Wardle, M. 2012, *MNRAS*, 422, 261
Braiding, C. R. & Wardle, M. 2012, *MNRAS*, 427, 3188
Burkert, A. & Bodenheimer, P. 2000, *ApJ*, 543, 822
Chapman, N. L., Davidson, J. A., Goldsmith, P. F. *et al.* 2013, *ApJ*, 770, 151
Ciardi, A. & Hennebelle, P. 2010, *MNRAS*, 409, L39
Ciolek, G. E. & Königl, A. 1998, *ApJ*, 504, 257
Contopoulos, I., Ciolek, G. E., & Königl, A. 1998, *ApJ*, 504, 247
Crutcher, R. M., Wandelt, B., Heiles, C., Falgarone, E., & Troland, T. H. 2010, *ApJ*, 725, 466
Crutcher, R. M. 2012, *ARAA*, 50, 29
Dapp, W. B. & Basu, S. 2010, *AA*, 521, 56
Dapp, W. B., Basu, S., & Kunz, M. W. 2012, *AA*, 541, A35
Davidson, J. A., Novak, G., Matthews, T. G. *et al.* 2011, *ApJ*, 732, 97

Dedner, A., *et al.* 2002, *Journal of Computational Physics*, 175, 645
Duffin, D. F. & Pudritz, R. E. 2009, *MNRAS*, 391, 1659
Duffin, D. F. & Pudritz, R. E. 2009, *ApJL*, 706, L46
Galli, D., Lizano, S., Shu, F. H., & Allen, A. 2006, *ApJ*, 647, 374
Hennebelle, P. & Ciardi, A. 2009, *AA*, 506, L29
Hennebelle, P. & Fromang, S. 2008, *AA*, 477, 9
Hosking, J. G. & Whitworth, A. P. 2004, *MNRAS*, 347, 1001
Hull, C. L. H., Plambeck, R. L., Bolatto, A. D. *et al.* 2013, *ApJ*, 768, 159
Joos, M., Hennebelle, P., & Ciardi, A. 2012, *AA*, 543, A128
Joos, M., Hennebelle, P., Ciardi, A., & Fromang, S. 2013, *AA*, 554, A17
Krasnopolsky, R. & Königl, A. 2002, *ApJ*, 580, 987
Krasnopolsky, R., Li, Z.-Y., & Shang, H. 2010, *ApJ*, 716, 1541
Krasnopolsky, R., Li, Z.-Y., & Shang, H. 2011, *ApJ*, 733, 54
Krumholz, M. R., Crutcher, R. M., & Hull, C. L. H. 2013, *ApJL*, 767, L11
Kunz, M. W. & Mouschovias, T. C. 2010, *MNRAS*, 408, 322
Li, Z.-Y., Krasnopolsky, R., & Shang, H. 2011, *ApJ*, 738, 180
Li, Z.-Y., Krasnopolsky, R., & Shang, H. 2013, *ApJ*, 774, 82
Li, Z.-Y., Krasnopolsky, R., Shang, H., & Zhao, B. 2014, *ApJ*, 793, 130
Li, Z.-Y. & McKee, C. F. 1996, *ApJ*, 464, 373
Machida, M. N., Inutsuka, S.-i., & Matsumoto, T. 2006, *ApJL*, 649, L129
Machida, M. N., Inutsuka, S.-i., & Matsumoto, T. 2011, *PASJ*, 63, 555
Machida, M. N. & Matsumoto, T. 2011, *MNRAS*, 413, 2767
Mac Low, M.-M. & Klessen, R. S. 2004, *Reviews of Modern Physics*, 76, 125
Masson, J., Chabrier, G., Hennebelle, P., Vaytet, N., & Commerçon, B. 2015, *AA*, in press.
McKee, C. F. & Ostriker, E. C. 2007, *ARAA*, 45, 565
Mellon, R. R. & Li, Z.-Y. 2008, *ApJ*, 681, 1356
Mellon, R. R. & Li, Z.-Y. 2009, *ApJ*, 698, 922
Men'shchikov, A., *et al.* 2010, *AA*, 518, L103
Molinari, S., *et al.* 2010, *AA*, 518, L100
Mouschovias, T. C., & Ciolek, G. E. 1999 in: Magnetic Fields and Star Formation: A Theory Reaching Adulthood, (edited by C. J. Lada and N. D. Kylafis), 305
Myers, A. T., McKee, C. F., Cunningham, A. J., Klein, R. I., & Krumholz, M. R. 2013, *ApJ*, 766, 97
Nakano, T. 1984, *Fundamentals of Cosmic Physics*, 9, 139
Nakano, T., Nishi, R., & Umebayashi, T. 2002, *ApJ*, 573, 199
Nishi, R., Nakano, T., & Umebayashi, T. 1991, *ApJ*, 368, 181
Padovani, M., Hennebelle, P., & Galli, D. 2013, *AA*, 560, A114
Padovani, M., Galli, D., Hennebelle, P., Commerçon, B., & Joos, M. 2014, *AA*, 571, A33
Santos-Lima, R., de Gouveia Dal Pino, E. M., & Lazarian, A. 2012, *ApJ*, 747, 21
Santos-Lima, R., de Gouveia Dal Pino, E. M., & Lazarian, A. 2013, *MNRAS*, 429, 3371
Seifried, D., Pudritz, R. E., Banerjee, R. *et al.* 2012, *MNRAS*, 422, 347
Seifried, D., Banerjee, R., Pudritz, R. E., & Klessen, R. S. 2012, *MNRAS*, 423, L40
Seifried, D., Banerjee, R., Pudritz, R. E. *et al.* 2013, *MNRAS*, 432, 3320
Shu, F. H., Adams, F. C., & Lizano, S. 1987, *ARAA*, 25, 23
Shu, F. H., Galli, D., Lizano, S., & Cai, M. 2006, *ApJ*, 647, 382
Tassis, K. & Mouschovias, T. C. 2007, *ApJ*, 660, 388
Tomida, K., Tomisaka, K., Matsumoto, T., Hori, Y., *et al.* 2013, *ApJ*, 763, 6
Tomida, K., Okuzumi, S., & Machida, M. N. 2015, *ApJ*, 801, 117
Tomisaka, K. 1998, *ApJL*, 502, L163
Troland, T. H. & Crutcher, R. M. 2008, *ApJ*, 680, 457
Tsukamoto, Y., Iwasaki, K., Okuzumi, S., Machida, M. N., & Inutsuka, S. 2015, *ApJ*, 810, L26
Tsukamoto, Y., Iwasaki, K., Okuzumi, S., Machida, M. N., & Inutsuka, S. 2015, *MNRAS*, 452, 278
Wardle, M. & Ng, C. 1999, *MNRAS*, 303, 239

From Interstellar Clouds to Star-Forming Galaxies: Universal Processes?
Proceedings IAU Symposium No. 315, 2015
P. Jablonka, Ph. André & F. van der Tak, eds.

doi:10.1017/S1743921316007389

SMA and ALMA studies of protoplanetary disk formation around low-mass protostars

Shigehisa Takakuwa[1], Nagayoshi Ohashi[1,2], Hsi-Wei Yen[1], Ti-Lin Chou[1], Kazuya Saigo[3], Masao Saito[4], Yusuke Aso[2,5], Yuri Aikawa[6], Shin Koyamatsu[2,5], Masahiro N. Machida[7], Sanemichi Z. Takahashi[8], Kengo Tomida[9] and Kohji Tomisaka[10]

[1]Academia Sinica Institute of Astronomy and Astrophysics, PO Box 23-141, Taipei 10617, Taiwan, email: takakuwa@asiaa.sinica.edu.tw

[2]Subaru Telescope, National Astronomical Observatory of Japan, 650 North A'ohoku Place, Hilo, HI 96720, USA

[3]Department of Physical Science, Graduate School of Science, Osaka Prefecture University, 1-1 Gakuen-cho, Naka-ku, Sakai, Osaka 599-8531, Japan

[4]Nobeyama Radio Observatory, National Astronomical Observatory of Japan, Minamimaki, Minamisaku, Nagano 384-1805, Japan

[5]Department of Astronomy, Graduate School of Science, The University of Tokyo, 7-3-1 Hongo, Bunkyo-ku, Tokyo 113-0033, Japan

[6]Center for Computational Sciences, University of Tsukuba, 1-1-1 Tennoudai, Tsukuba 305-8577, Japan

[7]Department of Earth and Planetary Sciences, Faculty of Sciences, Kyushu University, Fukuoka 812-8581, Japan

[8]Department of Physics, Kyoto University, Oiwake-cho, Kitashirakawa, Sakyo-ku, Kyoto 606-8502, Japan

[9]Department of Earth and Space Science, Osaka University, 1-1 Machikaneyama, Toyonaka, Osaka 560-0043, Japan

[10]National Astronomical Observatory of Japan, Osawa 2-21-1, Mitaka, Tokyo 181-8588, Japan

Abstract. We report our systematic survey observations of protostellar sources with the Sub-Millimeter Array (SMA) and Atacama Large Millimeter/submillimeter Array (ALMA). The purpose of our survey is to investigate formation mechanism of protoplanetary disks, precursors of planetary systems, out of ∼1000 AU-scale protostellar envelopes surrounding the protostars. We found that in the early protostars (B335, NGC1333 IRAS 4B), the envelopes do not show significant rotating motions but infalling motions toward the central protostars. In more evolved protostars (L1527 IRS, L1448-mm, L1551 IRS 5), the envelopes are infalling and rotating with the conserved specific angular momenta (that is, $v_{rot} \propto r^{-1}$). In most evolved sources (L1551 NE, TMC-1A, L1489 IRS) large-scale ($\gtrsim$100 AU) disks in Keplerian rotation or protoplanetary disks are evident. These results demonstrate a systematic evolutionary trend of envelope gas motions toward the disk formation.

Keywords. Protostellar Envelopes, Keplerian Disks, SMA, ALMA

1. Introduction

Keplerian disks are ubiquitous around T-Tauri stars, and are considered to be precursors of planetary systems, "protoplanetary disks". Indeed, recent high-resolution millimeter and submillimeter continuum observations of the disk around HL Tau with ALMA have unveiled presence of seven ringlike gaps in the disk, evidence for presence of planets (ALMA Partnership *et al.* 2015). The formation mechanism of such protoplanetary disks should define the initial conditions of subsequent disk evolution toward planet formation.

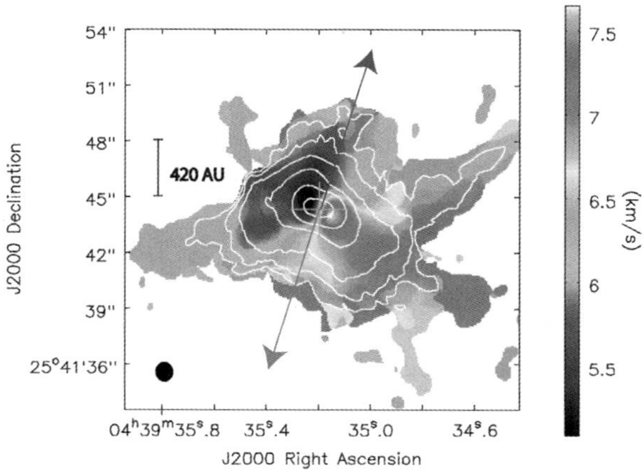

Figure 1. Moment 0 (white contours) and 1 (color) maps of the $C^{18}O$ ($J = 2$–1) emission in TMC-1A observed with ALMA, taken from Aso *et al.* (2015). A cross shows the position of the central protostar (continuum emission peak), and blue and red arrows the direction of the blueshifted and redshifted molecular outflows observed in the ^{12}CO ($J = 2$–1) emission. A filled ellipse at the bottom-left corner denotes the ALMA synthesized beam ($1''.06 \times 0''.90$; P.A.$=-176°$). *(A color version of this figure is available online.)*

Our research team has initiated a large collaborative study of formation mechanism of protoplanetary disks from protostellar envelopes surrounding the protostars with the SMA and ALMA. We have observed eight protostellar sources ranging from early Class 0 to late Class I stages with ALMA and/or SMA, and investigated the gas motions in the envelopes. Below, we will show examples of our results.

2. TMC-1A

TMC-1A is a late Class I protostar ($T_{bol} = 172$ K) located in the Taurus region. Fig. 1 shows moment 0 (white contours) and 1 (color) maps of the $C^{18}O$ ($J = 2$–1) emission in TMC-1A observed with ALMA (Aso *et al.* 2015). There appears a ~1000-AU scale $C^{18}O$ feature elongated approximately perpendicular to the associated molecular outflows (blue and red arrows). This $C^{18}O$ ($J = 2$–1) emission most likely traces the molecular envelope component associated with the protostar. The eastern and western parts of the emission component are blueshifted and redshifted, respectively, and the most blueshifted and redshifted velocities appear close to the protostar.

Fig. 2 compares distributions of the high- and low-velocity blueshifted and redshifted $C^{18}O$ emission in TMC-1A. The high-velocity blueshifted and redshifted $C^{18}O$ emission are located to the east and west of the protostar, respectively. The direction of the velocity gradient is almost perpendicular to the direction of the associated molecular outflow, and there is no clear emission overlap of the blueshifted and redshifted emission. On the other hand, the low-velocity blueshifted and redshifted emission exhibit significant spatial overlaps, and the direction of the overall velocity gradient is tilted with respect to that of the high-velocity emission.

From our protostellar envelope + disk modeling, we found that the high-velocity components can be explained as an r ~100-AU scale Keplerian disk orbiting a protostar with a dynamical mass of 0.68 $M_\odot$, while the low-velocity components an infalling and rotating protostellar envelope. The rotational profile of the envelope follows $v_{rot} \propto r^{-1}$, that is, rotation with the conserved specific angular momentum, and the infalling velocity is ~3 times smaller than free-fall velocity yielded by the dynamical mass of the protostar.

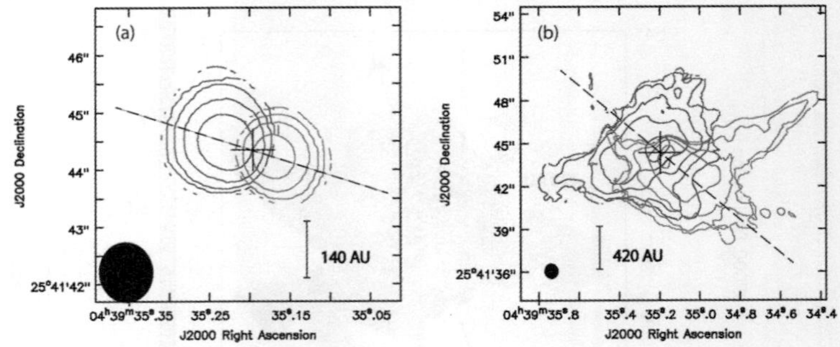

Figure 2. ALMA maps of the blueshifted (blue contours) and redshifted $C^{18}O$ (J = 2–1) emission (red) at (a) high-velocity ($|\Delta V|$ >2.0 km s^{-1}) and (b) low-velocity ($|\Delta V|$ <2.0 km s^{-1}) ranges in TMC-1A, taken from Aso *et al.* (2015). The dashed lines show the directions of the observed velocity gradients. *(A color version of this figure is available online.)*

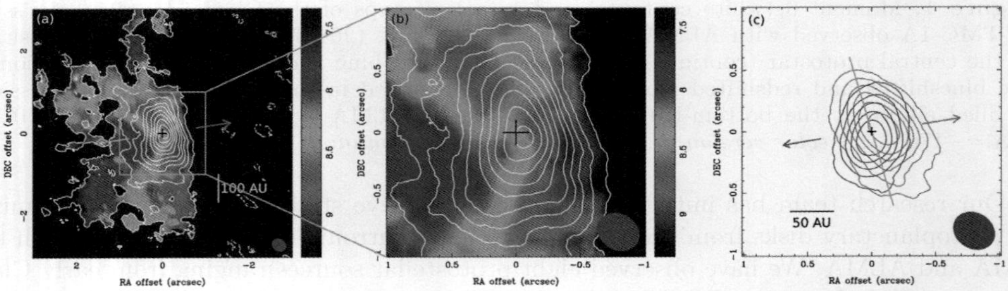

Figure 3. (a) ALMA moment 0 map (contours) overlaid on the moment 1 map (color scale, units in km s^{-1}) of the $C^{18}O$ (J = 2–1) emission in B335. A cross indicates the protostellar position, and blue and red arrows denote the directions of the associated blueshifted and redshifted outflows, respectively. (b) Zoom-up view of the central $2'' \times 2''$ region delineated by the green box in panel (a). (c) Moment 0 map of the high-velocity ($|V|$ >1.2 km s^{-1}) blueshifted and redshifted $C^{18}O$ emission in B335. The green line shows the direction of the major axis of the 1.3 mm continuum emission. These figures are from Yen *et al.* (2015b). *(A color version of this figure is available online.)*

These results indicate that the protoplanetary disk is being formed out of the infalling envelope, where the infalling velocity is decelerated as the material approaches close to the disk.

3. B335

Whereas our ALMA observations of a late-phase protostar TMC-1A found a well-developed Keplerian disk embedded in the rotating and infalling protostellar envelope, there are other protostellar sources without any detectable disk or even envelope rotation. Fig. 3 shows results of our ALMA observations of the Class 0 protostar B335 in the $C^{18}O$ (2–1) emission (Yen *et al.* 2015b). Fig. 3 (a) shows the presence of a ~500-AU scale protostellar envelope elongated perpendicular to the associated molecular outflow. While the protostellar envelope exhibits a conspicuous velocity gradient along the minor axis originated from the infalling motion, there is no clear velocity gradient along the major axis (*i.e.*, rotation) down to the radial scale of $\lesssim$50 AU.

The absence of the envelope rotation implies that the outermost radius of the Keplerian disk is as small as $\lesssim$1–3 AU. These results indicate that in the early Class 0 protostar of B335 the envelope material is actively infalling toward the protostar and the disk

Table 1. Evolution of Envelope Gas Motion toward Protoplanetary-Disk Formation

Protostars	Class	T_{bol} [K]	Envelope	R_{kep} [AU][1]	Our Publ.[2]
B335	0	31	infall+slow rotation	$<$1-3	a,b,c,d,e
NGC1333 IRAS 4B	0	54	infall+slow rotation	$<$5	c,d
L1527 IRS	0-I	59	slow infall+r^{-1} rotation	54	c,d,f
L1448-mm	0-I	69	infall+r^{-1} rotation	$\sim$150	c,d
L1551 NE	I	91	slow infall+slow rotation	300	g,h,i
L1551 IRS 5	I	92	slow infall+r^{-1} rotation	64	j,k
TMC-1A	I	172	slow infall+r^{-1} rotation	100	c,l
L1489 IRS	I-II	238	infall+r^{-1} rotation	$>$300	c,m

Notes:
[1] Estimated outermost radius of the protoplanetary disk in Keplerian rotation.
[2] (a) Yen *et al.* 2010; (b) Yen *et al.* 2011; (c) Yen *et al.* 2013; (d) Yen *et al.* 2015a; (e) Yen *et al.* 2015b; (f) Ohashi *et al.* 2014; (g) Takakuwa *et al.* 2012; (h) Takakuwa *et al.* 2013; (i) Takakuwa *et al.* 2014; (j) Takakuwa *et al.* 2004; (k) Chou *et al.* 2014; (l) Aso *et al.* 2015; (m) Yen *et al.* 2014.

formation around the protostar has not been well proceeded, in contrast with the late-phase Class I protostar TMC-1A.

4. Summary

So far, we have completed ALMA and/or SMA observations and detailed modeling of eight protostellar sources. Table 1 summarizes the observed protostars, their evolutional stages, measured envelope motions and outermost radii of the Keplerian disks, and our publications. Here, "slow rotation" denotes no detectable rotational motion at $r \lesssim 500$ AU, and "slow infall" the infalling velocity $\gtrsim 3$ times smaller than the free-fall velocity. In the early Class 0 protostars of B335 and NGC1333 IRAS 4B, the envelopes do not show clear rotating motions but infalling motions toward the central protostars. In L1527 IRS, L1448-mm, and L1551 IRS 5, the envelopes are infalling and rotating with the conserved specific angular momenta (that is, $v_{rot} \propto r^{-1}$), while the radii of the inner Keplerian disks are still small ($r \lesssim 100$ AU). In the most evolved sources of L1551 NE, TMC-1A and L1489 IRS, large-scale disks in Keplerian rotation or protoplanetary disks are present. These results demonstrate the systematic evolutionary trend of the envelope motion toward the disk formation. In the early-stage protostars the envelopes have just started collapsing and only a small amount of the envelope angular momentum is carried in, and thus the disk formation is not yet initiated. As more envelope materials fall in, more angular momenta are carried toward the center, initiating the disk formation. In the most evolved protostellar sources disk formation has been well developed.

References

ALMA Partnership, Brogan, C. L., Pérez, L. M., Hunter, T. R., *et al.* 2015, *ApJ*, 808, 3
Aso, Y., Ohashi, N., Saigo, K., *et al.* 2015, *ApJ*, 812, 27
Chou, T.-L., Takakuwa, S., Yen, H.-W., Ohashi, N., & Ho, P. T. P. 2014, *ApJ*, 796, 70
Ohashi, N., Saigo, K., Aso, Y., *et al.* 2014, *ApJ*, 796, 131
Takakuwa, S., Ohashi, N., Ho, P. T. P., *et al.* 2004, *ApJ* (Letters), 616, L15
Takakuwa, S., Saito, M., Lim, J., *et al.* 2012, *ApJ*, 754, 52
Takakuwa, S., Saito, M., Lim, J., & Saigo, K. 2013, *ApJ*, 776, 51
Takakuwa, S., Saito, M., Saigo, K., *et al.* 2014, *ApJ*, 796, 1
Yen, H.-W., Takakuwa, S., & Ohashi, N. 2010, *ApJ*, 710, 1786
Yen, H.-W., Takakuwa, S., & Ohashi, N. 2011, *ApJ*, 742, 57
Yen, H.-W., Takakuwa, S., Ohashi, N., & Ho, P. T. P. 2013, *ApJ*, 772, 22
Yen, H.-W., Takakuwa, S., Ohashi, N., *et al.* 2014, *ApJ*, 793, 1
Yen, H.-W., Koch, P. M., Takakuwa, S., *et al.* 2015a, *ApJ*, 799, 193
Yen, H.-W., Takakuwa, S., Koch, P. M., *et al.* 2015b, *ApJ*, 812, 129

From Interstellar Clouds to Star-Forming Galaxies: Universal Processes?
Proceedings IAU Symposium No. 315, 2015
P. Jablonka, Ph. André & F. van der Tak, eds.

doi:10.1017/S1743921316007390

Formation of star clusters: Models and simulations

Philip C. Myers
Harvard-Smithsonian Center for Astrophysics
60 Garden Street, Cambridge, MA 02138 USA
email: pmyers@cfa.harvard.edu

Abstract. Recent progress is reviewed in describing the physical processes of star cluster formation, from the viewpoint of theoretical models and numerical simulations. In many studies, supersonic turbulent flows structure the gas into a network of clumps, filaments, and cores. Self-gravity drives star formation, moderated by turbulence, magnetic fields, and stellar feedback. Recent models and simulations describe the core mass function, the protostar mass function, and the star-formation rate. Several simulations indicate similar protostar birthrates, a mass advantage for firstborn protostars, and accretion rates that are superthermal and nearly constant. Recently observed core-filament systems with sonic line widths and parsec lengths present challenges for turbulent models.

Keywords. Molecular clouds, star formation, star clusters

1. Introduction

Embedded clusters are parsec-scale concentrations of dense gas and very young stars. On time scales of a few Myr, their dense cores form protostars, their protostars gain mass, and the circumstellar disks around their pre-main sequence stars begin to form planets. These regions provide the best laboratories to observe the processes of star formation. Recent observational reviews include Longmore *et al.* (2014), Luhman (2012), Portegies Zwart *et al.* (2010), Allen *et al.* (2007), and Zinnecker & Yorke (2007).

The processes of star formation in clusters have also been the subject of an increasing number of theoretical and numerical studies. Large-scale numerical simulations have greatly extended the reach of analytical models. Recent reviews of cluster theories and simulations address the initial mass function (Offner *et al.* 2014), formation of massive stars (Tan *et al.* 2014), the role of stellar feedback (Krumholz *et al.* 2014), and theoretical aspects of cluster formation (McKee & Ostriker 2007).

This review discusses recent developments that shed light on the key questions:

(1) Is cluster formation *slow or fast*? Can cluster formation be described by quasi-equilibrium models where turbulence and magnetic forces oppose gravity over many initial free-fall times (Tan *et al.* 2006; Huff & Stahler 2007)? Or is a more dynamic picture needed, where supersonic turbulent flows and stellar feedback act over only a few free-fall times (Ballesteros-Paredes *et al.* 1999; Elmegreen 2000; Bonnell *et al.* 2001)?

(2) Do massive stars in clusters form *first or last*? Do the most massive stars start accreting before lower-mass stars, in the deepest parts of the potential well, as in some simulations (Bonnell *et al.* 1997; Wang *et al.* 2010), or do they form later, when their winds and feedback disperse dense gas and shut off further star formation (Herbig 1962)?

(3) Do cluster protostars gain mass at a rate expected from models of *"collapse" or "accretion"*? The standard model of isolated low-mass star formation (Shu *et al.* 1987) relies on the constant accretion rate due to collapse of a singular isothermal sphere

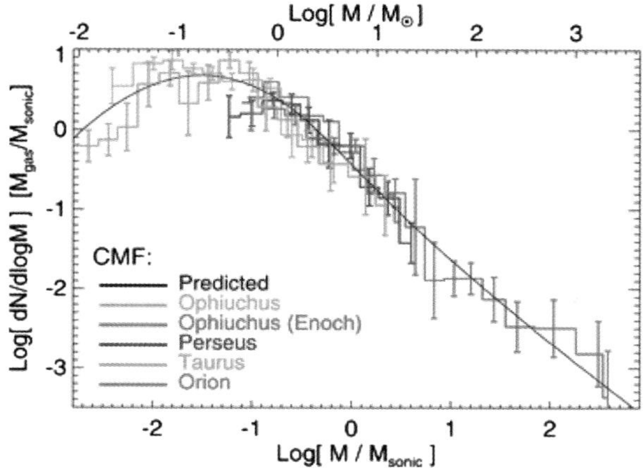

Figure 1. Predicted core mass function, based on a log-normal distribution of density fluctuations in a turbulent medium, which at each spatial scale are dense enough to be gravitationally bound (Hopkins 2012). The distribution agrees well with observed estimates of the core mass function. *(A color version of this figure is available online.)*

(SIS – Shu 1977), while more massive stars are expected to gain mass more rapidly, by gravitational accretion, whose rate increases with protostar mass (Zinnecker 1982; Bonnell *et al.* 1997).

2. Analytic models

Here, three analytic models are described, which match theoretical predictions to observational data on young clusters and regions of star formation.

Accelerating star formation. The ages of pre-main sequence (PMS) stars in Orion, based on photometric and spectroscopic observations, and on models of PMS evolution, indicate that most PMS stars were born in the last 1–2 Myr, but their age distribution has a tail extending to ages of ∼10 Myr (Palla & Stahler 2000). This distribution is well fit by a quasi-equilibrium model of spherical clump contraction against the slow dissipation of turbulent support due to colliding shocks (Tan *et al.* 2006; Huff & Stahler 2007). Then the protostar births "accelerate" as the clump collapses.

Core mass function (CMF). In these models, the CMF is the mass distribution of gravitationally bound density fluctuations on all spatial scales, where a fluctuation is bound if its mass equals the Jeans mass corresponding to the thermal and turbulent velocity dispersion. These models require proper counting of small bound condensations that lie inside larger bound condensations, the "cloud-in-cloud problem." Fig. 1 shows a good match of observed and predicted CMFs. However, it is based on a relatively large turbulent Mach number, $M = 30$ (Hopkins 2012).

Star-formation rate. An important consequence of the core mass function is the rate at which cores form stars. This regional property can be expressed as the gain rate of the mass surface density of stars $(d/dt)\sum_{\rm stars}$ as a function of the mass surface density of gas $\sum_{\rm gas}$. The empirical Kennicutt–Schmidt (KS) relation, $(d/dt)\sum_{\rm stars} \propto \sum_{\rm gas}^{p}$, where $p \sim 1.4$, applies over a wide range of galactic and extragalactic regions (Schmidt 1959; Kennicutt 1998). The CMF models of Hennebelle & Chabrier (2008) predict KS relations for Milky Way clouds of various sizes and for extragalactic clouds (Chabrier *et al.* 2014). These models match observed KS data with significant scatter at low surface density.

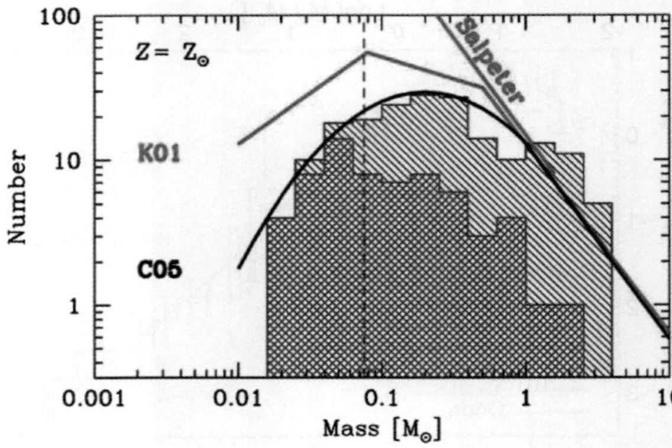

Figure 2. Mass functions of stars that have stopped accreting (heavy shading) and that are still accreting at the end of the simulation (light shading). The lines indicate the IMF distributions of Salpeter (1955), Kroupa (2001), and Chabrier (2005). Adapted from Bate (2014).

3. Numerical simulations

Analytic models give physical insight into broad properties of cluster formation but cannot account for the complex structure and evolution of embedded clusters. In contrast, numerical simulations of forming clusters have improved in resolution, computational power, and run time. Now one can vary the initial gas configuration, the strength and geometry of the associated magnetic field, and properties of stellar feedback (e.g., Federrath 2015; Körtgen & Banerjee 2015; Bate 2014; Myers *et al.* 2014).

Methods. Many simulations compare results with stellar and gas properties, including the distributions of stellar masses and multiplicity, and the rate of star formation. The typical protocluster has initial properties of several hundred $M_{\odot}$ and several hundred Jeans masses in a 1-pc clump, with turbulent hydrodynamic and Alfvénic Mach numbers $\sim$10. The turbulent motions stir the initial gas into a clumpy, filamentary network.

The turbulent motions may be generated by a large-scale "turbulence driver" that causes the gas to have velocity dispersion σ increasing with size scale R, as in the Larson relation $\sigma \sim R^q$, where $q \approx \frac{1}{2}$ (Federrath 2015). Alternately, colliding supersonic flows can form a turbulent, filamentary layer (Vázquez-Semadeni *et al.* 2006).

The evolving clump is followed for about a free-fall time corresponding to its initial mean density. Its densest gas collapses into sink cells, which are taken as protostars if the simulation has sufficient resolution. The evolution is terminated when the mass in protostars has reached a fraction $\sim$ 10% of the total mass, as in observed clusters.

Matching the IMF. Simulations of turbulent fragmentation, using a smooth particle hydrodynamics code with resolution $\sim$20 AU, make enough protostars to compare their mass distribution with the IMF, as the metallicity of the gas is varied (Bate 2014). These simulations include radiative heating from their recently formed protostars. This heating inhibits the formation of low-mass stars and brown dwarfs, and thereby provides mass distributions in relatively good accord with the IMF (Bate 2012).

In these studies, it was found that the protostar mass functions do not vary significantly as the metallicity of the gas changes from 0.01 to 3 times the solar value. Fig. 2 shows the protostar mass distribution for solar metallicity. This figure also shows that the mass function arising from the simulation agrees approximately with estimates of the field star IMF, within statistical uncertainty.

Matching observed star-formation rates. To be realistic, such simulations must also match the rate of star formation estimated from observations of young clusters. The earliest hydrodynamic simulations made protostars at too great a rate, because their initial clumps were too dense and because they did not include enough physical processes that oppose gravity and disperse star-forming gas. Recent simulations using the FLASH adaptive mesh refinement code found that the rate of star formation is reduced by a factor 2–3 from that due to gravity alone, as the effects of turbulent stirring, magnetic fields and outflow feedback are sequentially included (Federrath 2015). This rate reduction was realized by running each simulation until it formed 20% of the initial mass into sink cells. In combination, these effects increase the star-forming life of a young cluster by an order of magnitude, bringing the mean star-formation rate to within a few times the typical observational estimate of a few $M_\odot$ pc^{-2} Myr^{-1}. These results extend and confirm the pioneering study of Wang *et al.* (2010).

4. Protostar accretion histories

As numerical simulations of cluster formation have become more realistic, it has become possible to study the mass accretion histories $m(t)$ of their sink cells, which are identified as young stellar objects (YSOs). Accretion histories are useful to understand the protostar birth rate, the gain rate of protostar mass, and the nature of accretion stopping. These processes are essential parts of how protostars gain their mass, but observations cannot sample protostar masses over a sufficiently long time span.

A recent study compares the accretion histories of YSOs in four hydrodynamic simulations carried out by groups using different codes and initial conditions (Myers *et al.* 2016). The simulations, by Bate (2012), Krumholz *et al.* (2012), Girichidis *et al.* (2011), and Wang *et al.* (2010), have initial clumps with 500–1200 $M_\odot$ log mean density (in cm^{-3}) of 4–6 and turbulent Mach number 3–15. Each simulation produces 100–200 YSOs over 0.5–1.3 initial free-fall times.

In each simulation, the accretion histories allow a clear distinction between two kinds of YSO. At the end time of the simulation, some YSOs have rising $m(t)$ while others have flat $m(t)$. A rising $m(t)$ is interpreted as due to a protostar (PS) that is still accreting, while a flat $m(t)$ is interpreted as due to a pre-main sequence star (PMS), which has essentially stopped accreting.

Birth history. Each simulation shows a continuous history of protostar births, at a mean rate ~ 100 births per free-fall time. Some birth histories have a nearly constant rate, while others show a gradual acceleration or a "burst" as a new subcluster is born. Some simulations resemble very young clusters, where PS outnumber PMS stars, while others resemble more evolved clusters, where PMS stars outnumber PS.

Early-birth mass advantage. Most of the simulations also show a tendency for early-born stars to become relatively massive. This early-birth mass advantage appears both as an excess of early-born stars with high final mass and as a deficit of early-born stars with low final mass, as first noted by Wang *et al.* (2010) and by Bate (2012).

Mass accretion rates. In these simulations, the form of the average PS accretion history, $\langle m(t) \rangle$, is remarkably linear. Individual protostar accretion histories $m(t)$ typically fluctuate about a rising trend. But when ~ 10 histories having similar start times are averaged together, the fluctuations are reduced, and the resulting average history can be fit with a simple linear function, as shown in Fig. 3.

The constant superthermal mass accretion rates in Fig. 3 differ from those expected in well-known models. The rates exceed that of the collapsing SIS by more than an order of magnitude. Such an increase could arise from collapse of a configuration whose initial density exceeds that of the SIS by a large factor (Shu 1977), but such an initial state is highly unlikely. The factor of increase is closer to the factor 47 expected for collapse

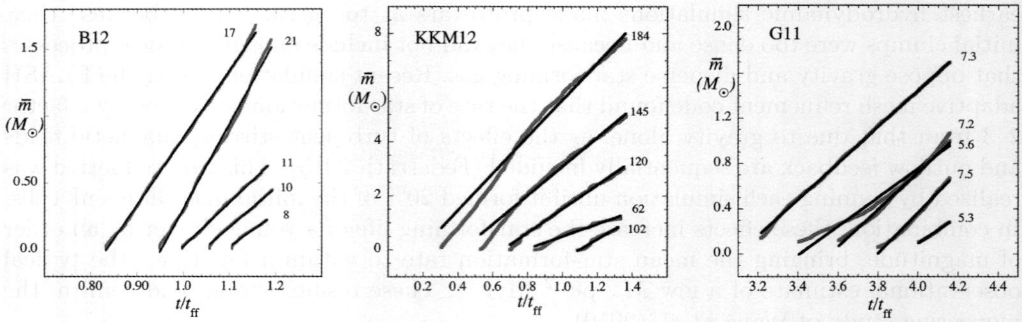

Figure 3. Mean accretion histories $\langle m(t) \rangle$ of groups of ~ 10 protostars having similar start times, in each of three turbulent fragmentation simulations (red curves). Black lines are linear fits to each $\langle m(t) \rangle$. The corresponding mass accretion rate is shown in units of the rate for a collapsing singular isothermal sphere (SIS). The rates greatly exceed the SIS rate (Myers *et al.* 2016). *(A color version of this figure is available online.)*

of an initially uniform isothermal sphere (Larson 1969, Penston 1969), but star-forming cores are observed to be centrally condensed. The constant rate also differs from the increasing rate expected for mass-dependent accretion models such as Bondi accretion (Bondi 1952).

The protostar mass accretion rates reported here resemble those in a numerical study of turbulent fragmentation, where shock formation of a massive filament drives supersonic flows into and along the filament (Banerjee *et al.* 2006).

5. The end of turbulence

It has been known for many years that some dense regions have spectral lines whose widths are trans-sonic or subsonic (Rydbeck *et al.* 1977; Myers 1983). In these regions, the large-scale supersonic motions that structure gas into filaments are relatively negligible. "Dense core" gas dominated by thermal motions is often considered an initial condition for isolated low-mass star formation (André *et al.* 2014).

Regions dominated by thermal motions were found to have typical size 0.1 pc based on maps of NH_3 line emission, a size similar to the thermal Jeans length for 10 K gas (Myers & Benson 1983). The origin of such low-turbulence gas in large-scale regions of greater turbulence has been attributed to stagnation zones in colliding flows (Ballesteros-Paredes *et al.* 1999; Mac Low & Klessen 2004). Their 0.1-pc size scale regions is much greater than the mpc size scale for true dissipation of subsonic turbulence (Falgarone *et al.* 2009).

An important recent finding is that sonic motions are not confined only to 0.1-pc cores. Instead, sonic motions are now known to dominate $C^{18}O$ line widths in filamentary core environments in Taurus and in Musca. The filaments widths are again close to 0.1 pc, but their lengths can be many pc (Hacar & Tafalla 2011; Hacar *et al.* 2013; Kainulainen *et al.* 2016; Fig. 4).

When such filaments have sufficient mass per length, they harbor dense cores of similar width and similar subsonic motions. The cores and their host filaments also share similar line center velocities, suggesting that they have together decoupled from the turbulent motions that formed them (Tafalla & Hacar 2015). Matching these findings presents a challenge and an opportunity for turbulent simulations.

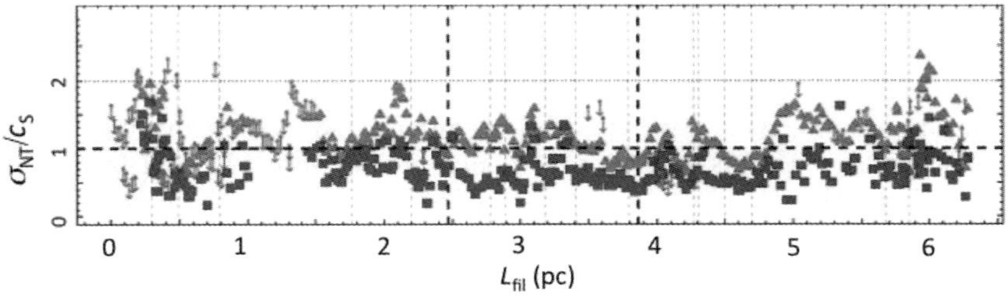

Figure 4. A sonic filament in Musca, extending over ~6 pc. The red and blue points indicate the ratio of opacity-corrected nonthermal velocity dispersion to sound speed, for lines of ^{13}CO and $C^{18}O$, respectively. Adapted from Kainulainen *et al.* (2016). *(A color version of this figure is available online.)*

6. Conclusion

The results discussed in Sections 2–5 suggest some possible answers to the questions posed in the introduction.

(1) Is cluster formation *slow or fast*? Cluster observations and simulations of turbulent fragmentation indicate that embedded clusters have a significant population of stars younger than 1 Myr. A quasi-equilibrium evolution has not been ruled out, but the majority of recent studies match embedded cluster properties with a "fast" dynamical picture involving supersonic flows.

(2) Do massive stars in clusters form *first or last*? Several turbulent simulations indicate that the most massive stars start their accretion early in the star-forming life of the cluster and have relatively long accretion durations. In this picture, massive stars are born early, but their powerful winds and radiation feedback do not arise until late in the cluster history, when they disperse dense gas and terminate further star formation.

(3) Do cluster protostars gain mass at a rate expected from models of *"collapse" or "accretion"*? In several simulations of hydrodynamic turbulence, the average $\langle m(t) \rangle$ of protostar accretion histories having similar start times shows protostar mass increasing at a nearly constant rate. This rate exceeds that of a collapsing singular isothermal sphere by one to two orders of magnitude, increasing with the level of initial supersonic turbulence. This rate does not increase with mass, in contrast with mass-dependent accretion models. Thus the typical gain rate is a constant, as in SIS collapse, but its magnitude is much greater than in SIS collapse models and is more typical of accretion models.

7. Summary

The main points of this review are:

Analytic models of cluster formation. Quasi-static contraction of a cluster-forming clump matches the star-formation rate in Orion as a function of time, over ~10 Myr (Tan *et al.* 2006; Huff & Stahler 2007). The core mass function (CMF) observed in nearby star-forming clouds is well matched by a model where cores are a subset of lognormal density fluctuations at all spatial scales, provided their density exceeds a turbulent Jeans criterion (Hennebelle & Chabrier 2008; Hopkins 2012). The Kennicutt–Schmidt increase of star-formation rate surface density with gas surface density in galactic and extragalactic clouds is approximated by an extension of the CMF model (Chabrier *et al.* 2014).

Simulations of cluster formation by turbulent fragmentation. Evolution of a 1-pc clump having 500 $M_\odot$, supersonic hydrodynamic turbulence, and hundreds of Jeans masses produces a filamentary network that makes more than 100 sink cells in an initial free-fall time, some still accreting and others that have stopped accreting. Their mass distribution matches the peak of the IMF and approximates its shape (Bate 2012, 2014). Another study addresses the low efficiency of star formation by evolving four initially identical clumps, until each has converted 20% of its mass into stars. The clumps collapse (1) with gravity alone, (2) with gravity and turbulence, (3) with gravity, turbulence, and magnetic fields, and (4) with gravity, turbulence, magnetic fields, and protostar outflow jets. The star-formation time increases by an order of magnitude from (1) to (4), bringing the resulting star-formation rate to within a factor of a few of the typical observed rate of a few $M_{\odot}$ pc^{-2} Myr^{-1} (Federrath 2015; see also Wang *et al.* 2010).

Accretion histories of cluster protostars. Comparison of three simulations of cluster formation by hydrodynamic turbulent fragmentation indicates that protostars are born at a typical rate of a few hundred per initial free-fall time. Those born early tend to have greater final masses, and their masses tend to increase linearly with time. Their mass accretion rate exceeds that due to a collapsing SIS by a factor 10–100, increasing with the initial turbulence of the simulation. Such a constant, superthermal accretion rate differs in magnitude from SIS collapse, and its constant magnitude differs from mass-dependent accretion models such as Bondi accretion, where the rate increases with mass and with time (Myers *et al.* 2015).

Extended regions of low turbulence. Recent observations indicate that dense gas with subsonic turbulence is not confined only to dense cores with size ~0.1 pc but also is seen in the filamentary environments of some cores, according to observations in lines of $C^{18}O$ (Hacar & Tafalla 2011; Kainulainen *et al.* 2016). These filaments, in the nearby Taurus and Musca regions, extend up to several pc. The cores and filaments in these systems have similar spatial structures in their line centers and widths, suggesting that they have together decoupled from their turbulent environments. This result presents a challenge and opportunity to models of cloud turbulence.

References

Allen, L., Megeath, S., Gutermuth, R., Myers, P., Wolk, S., *et al.* 2007, in: B. Reipurth, D. Jewitt, & K. Keil (eds.), *Protostars and Planets V* (Tucson: U. of Arizona Press), p. 361

André, P., di Francesco, J., Ward-Thompson, D., Inutsuka, S., Pudritz, R., & Pineda, J. 2014, in: H. Beuther, R. Klessen, C. Dullemond, & T. Henning (eds.), *Protostars and Planets VI*, (Tucson: U. of Arizona Press), p. 27

Ballesteros-Paredes, J., Harmann, L., & Vázquez-Semadeni, E. 1999, *ApJ*, 527, 285

Banerjee, R., Pudritz, R., & Anderson, D. 2006, *MNRAS*, 373, 1091

Bate, M. 2012, *MNRAS*, 419, 3115

Bate, M. 2014, *MNRAS*, 442, 285

Bondi, H. 1952, *MNRAS*, 112, 195

Bonnell, I., Bate, M., Clarke, C., & Pringle, J. 1997, *MNRAS*, 285, 201

Bonnell, I., Bate, M., Clarke, C., & Pringle, J. 2001, *MNRAS*, 323, 785

Chabrier, G. 2005, in: E. Corbelli, F. Palla, & H. Zinnecker (eds.), *The Initial Mass Function 50 Years Later*, Astrophysics and Space Science Library, v. 327 (Dordrecht: Springer), p. 41

Chabrier, G., Hennebelle, P., & Charlot, S. 2014, *ApJ*, 796, 75

Elmegreen, B. 2000, *ApJ*, 530, 277

Falgarone, E., Pety, J., & Hily-Blant, P. 2009, *A&A*, 507, 355

Federrath, C. 2015, *MNRAS*, 450, 4035

Girichidis, P., Federrath, C., Banerjee, R., & Klessen, R. 2011, *MNRAS*, 413, 2741

Hacar, A. & Tafalla, M. 2011, *A&A*, 533, 34

Hacar, A., Tafalla, M., Kauffmann, J., & Kovács, A. 2013, *A&A*, 554, A55

Hennebelle, P. & Chabrier, G. 2008, *ApJ*, 684, 395

Herbig, G. 1962, *ApJ*, 135, 73

Hopkins, P. 2012, *MNRAS*, 423, 2037

Huff, E. & Stahler, S. 2007, *ApJ*, 666, 281
Kainulainen, J., Hacar, A., Alves, J., Beuther, H., Bouy, H., & Tafalla, M. 2016, *A&A*, 586, A27
Kennicutt, R. 1998, *ApJ*, 498, 541
Körtgen, B. & Banerjee, R. 2015, *MNRAS*, 451, 334
Kroupa, P. 2001, *MNRAS*, 322, 231
Krumholz, M., Klein, R., & McKee, C. 2012, *ApJ*, 754, 71
Krumholz, M., Bate, M., Arce, H., Dale, J., Gutermuth, R., Klein, R., Li, Z-Y., Nakamura, F., & Zhang, Q. 2014, in: H. Beuther, R. Klessen, C. Dullemond, & T. Henning (eds.), *Protostars and Planets VI*, (Tucson: U. of Arizona Press), p. 243
Larson, R. 1969, *MNRAS*, 145, 271
Longmore, S., Kruijssen, J., Bastian, N., Bally, J., Rathborne, J., Testi, L., Stolte, A., Dale, J., Bressert, E., & Alves, J. 2014, in: H. Beuther, R. Klessen, C. Dullemond, & T. Henning (eds.), *Protostars and Planets VI*, (Tucson: U. of Arizona Press), p. 291
Luhman, K. 2012, *ARAA*, 50, 65
Mac Low, M. & Klessen, R. 2004, *RvMP*, 76, 125
McKee, C. & Ostriker, E. 2007, *ARAA*, 45, 565
Myers, A., Klein, R., Krumholz, R., & McKee, C. 2014, *MNRAS*, 439, 342
Myers, P. 1983, *ApJ*, 270, 105
Myers, P. & Benson, P. 1983, *ApJ*, 266, 309
Myers, P., Bate, M., Elmegreen, B., Federrath, C., Girichidis, P., Klein, R., Klessen, R., Krumholz, M., Li, Z-Y., McKee, C., Offner, S., & Ridge, N. 2016, in preparation
Offner, S., Clark, P., Hennebelle, P., Bastian, N., Bate, M., Hopkins, P., Moreaux, E., & Whitworth, A. 2014, in: H. Beuther, R. Klessen, C. Dullemond, & T. Henning (eds.), *Protostars and Planets VI*, (Tucson: U. of Arizona Press), p. 53
Palla, F. & Stahler, S. 2000, *ApJ*, 540, 255
Penston, M. 1969, *MNRAS*, 144, 425
Portegies Zwart, S., McMillan, S., & Gieles, M. 2010, *ARAA*, 48, 431
Rydbeck, O., Sume, A., Hjalmarson, A., Elldér, J., Rönnäng, B., & Kollberg, E. 1977, *ApJL*, 215, 35
Salpeter, E. 1955, *ApJ*, 121, 161
Schmidt, M. 1959, *ApJ*, 129, 243
Shu, F. 1977, *ApJ*, 214, 488
Shu, F., Adams, F., & Lizano, S. 1987, *ARAA*, 25, 23
Tafalla, M. & Hacar, A. 2015, *A&A*, 574, 104
Tan, J., Krumholz, M., & McKee, C. 2006, *ApJ*, 641, 121
Tan, J., Beltrán, M., Caselli, P., Fontani, F., Fuente, A., Krumholz, M., McKee, C., & Stolte, A. 2014, in: H. Beuther, R. Klessen, C. Dullemond, & T. Henning (eds.), *Protostars and Planets VI*, (Tucson: U. of Arizona Press), p. 149
Vázquez-Semadeni, E., Ryu, D., Passot, T., González, R., & Gazol, A. 2006, *ApJ*, 643, 245
Wang, P., Li, Z-Y., Abel, T., & Nakamura, F. 2010, *ApJ*, 709, 27
Zinnecker, H. 1982, *Ann. NYAS*, 395, 226
Zinnecker, H. & Yorke, H. 2007, *ARAA*, 45, 481

From Interstellar Clouds to Star-Forming Galaxies: Universal Processes?
Proceedings IAU Symposium No. 315, 2015
P. Jablonka, Ph. André & F. van der Tak, eds.

doi:10.1017/S1743921316007407

The ALMA view of the Antennae galaxy collision: How galaxy interaction triggers the formation of super star clusters

Cinthya N. Herrera[1]†, François Boulanger[2], Edith G. Falgarone[3], Guillaume Pineau des Forêts[2], Santiago García-Burillo[4], Daisuke Iono[1] and Pierre Guillard[5]

[1]National Astronomical Observatory of Japan,
2-21-1 Osawa, Mitaka, Japan
email: herrera@iram.fr (present address), diono@nao.ac.jp

[2]Institut d'Astrophyisque Spatiale
Université Paris Sud, Batiment 121, 91405 Orsay, France
email: francois.boulanger@ias.u-psud.fr, guillaume.pineaudesforets@ias.u-psud.fr

[3]Ecole Normale Superieure, LERMA
24 rue Lhomond, 75231 Paris, France
email: edith.falgarone@ens.fr

[4]Observatorio Astronómico Nacional
Aptdo 1143, 28800 Alcalá de Henares, Spain
email: burillo@oan.es

[5]Sorbonne Universités, UPMC Univ Paris 6 et CNRS, UMR 7095,
Institut d'Astrophysique de Paris 98 bis bd Arago, 75014 Paris, France
email: guillard@iap.fr

Abstract. The Antennae galaxies are a spectacular example of a burst of star formation triggered by the encounter of two galaxies, being an ideal source to understand how the dynamics of galaxy mergers drives star formation. We present archive ALMA CO(3−2) and VLT near-IR H_2 spectro-imaging observations, and new ALMA ^{13}CO(2−1) and dust continuum observations, at ~50 pc resolution. Combining tracers of density and velocity structure of the gas and its energetics, we demonstrate that star formation involves a complex interplay of merger-driven gas dynamics and turbulence, and the dissipation of the gas kinetic energy. We focus on a compact, bright H_2 source, associated with cold molecular gas and dust continuum emission, located where the velocity gradient in the interaction region is observed to be the largest. The characteristics of this source suggest that we are witnessing the formation, initiated by turbulent dissipation, of a cloud massive enough ($\sim 4\times10^6 M_\odot$) to form a super star cluster within 1 Myr.

Keywords. ISM; galaxies: starbursts, interaction, star clusters

1. Introduction

Galaxy mergers and interactions are a common phenomenon in the Universe. They play a large role in determining the morphologies of galaxies and the properties of their stellar populations along the Hubble sequence. In mergers, most of massive star formation is observed in Super Star Clusters (SSCs), compact (few parsec size) and massive ($>10^4$ $M_\odot$) star clusters. Our understanding on how the galaxy interaction triggers the formation of SSCs relies on theoretical arguments with only limited observational

† Present address: IRAM, 300 Rue de la Piscine, 38406 Saint Martin d'Hères, France.

support. The formation of SSCs must involve a complex interplay of merger-driven gas dynamics, turbulence fed by the galaxy interaction, and dissipation of the kinetic energy of the gas. Hydrodynamical simulations suggest that massive complexes of cold gas, i.e. *Super Giant Molecular Complexes* (SGMCs), form where merger driven gas flows trigger compression, cooling and gravitational fragmentation (Teyssier *et al.* 2010). Within SGMCs, a hierarchy of structures must be present including *pre-cluster clouds* (PCCs), i.e. dense and compact concentrations of molecular gas that are sufficiently massive to form SSCs (Weidner *et al.* 2010).

We address this topic from an observer's standpoint, by studying the density and velocity structure of the gas and its energetics, which are essential observational information.

2. Molecular gas in the Antennae overlap region

The Antennae (at 22 Mpc) is a merger of two gas-rich spiral galaxies hosting thousands of SSCs (Whitmore *et al.* 1995). The most massive ($>10^5$ $M_\odot$) and youngest (<10 Myr) SSCs are located where the two disks interact – overlap region – and are associated with large amount of molecular gas fragmented into SGMCs (Wilson *et al.* 2000). Recent ALMA observations show that SGMCs have abundant substructures and complex velocity components (Herrera *et al.* 2012; hereafter H12; Whitmore *et al.* 2014). The large surface density of the CO molecular gas across the overlap region and its fragmentation into these massive clouds are evidence for cooling and gravitational fragmentation of the diffuse gas compressed in the galaxy collision (Herrera *et al.* 2011, hereafter H11).

Nature of the warm molecular gas emission. SINFONI/VLT observations revealed extended near-IR H_2 emission associated with SGMCs (Fig. 1a). The cold (CO) and warm (H_2) molecular gas show similar morphologies and kinematics. Extinction-free diagnostics indicate that H_2 is mainly excited by collisions. Most of the H_2 emission traces the dissipation of the mechanical energy driven by the galaxy interaction. SGMCs would form through gas accretion and energy dissipation within converging flows driven by the galaxy interaction. The observed H_2 luminosities of SGMCs can be accounted for by the loss of turbulent kinetic energy associated with the formation of gravitationally bound gas within convergent gas flows driven by the galaxy interaction (H11, H12).

Excitation of the cold molecular gas. ^{13}CO(2−1) observations obtained with ALMA during Cycle 2, at $0''.4$ angular resolution ($\sim$40 pc), trace the densest parts of the ^{12}CO molecular clouds (see Fig. 1 a). Fig. 1b shows the CO(3−2)/CO(2−1) ratio, a tracer of the gas excitation. The ratio changes between SGMCs, as seen in lower resolution observations (e.g., Ueda *et al.* 2012), and, as revealed now by ALMA, within individual SGMCs. In particular, SGMC 2 shows the largest variations. The gas can be heated by UV radiation (probably in SGMC 4/5 which has an embedded SSC, Herrera *et al.* 2015) and/or the dissipation of the gas turbulent kinetic energy.

In the overlap region, single dish observations measured a ^{12}CO/^{13}CO(2−1) ratio between 14 and 28 at 1.5 kpc scales (Schulz *et al.* 2007), as also observed in other active star forming/interacting galaxies (e.g., Aalto *et al.* 1995). ALMA unveils ratios up to $\sim$50 at 100 pc scales (Fig. 1b, towards PCC 1, see §3). Physical and chemical modelling of these ALMA data will be presented in a future paper. Different CO transitions and isotopes may have different filling factors at 100 pc scales, and gas abundances of ^{12}CO and ^{13}CO may not be the canonical ones. For instance, towards the PCC 1 cloud, the CO(3−2)/CO(2−1) ratio is 0.5 and ^{12}CO/^{13}CO ratio is $\sim$50.

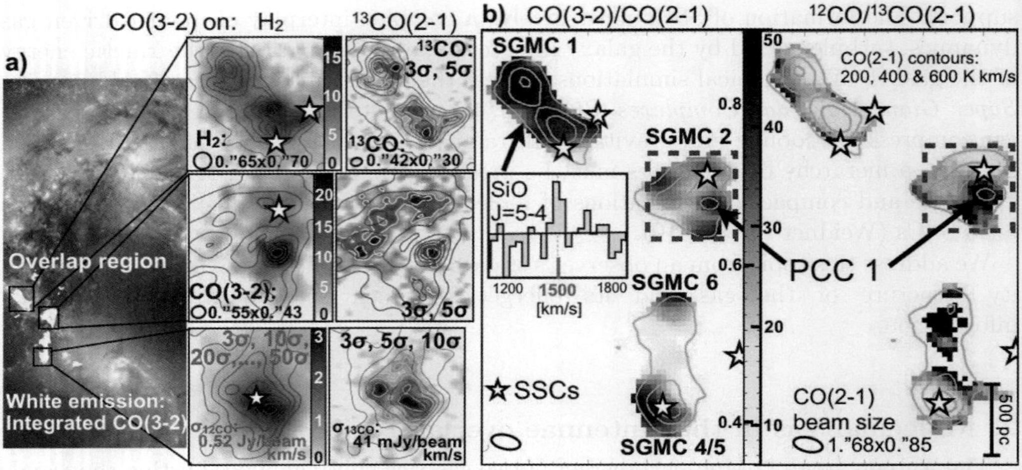

Figure 1. a) Integrated CO(3−2) emission from ALMA overlaid on an HST image of the Antennae. We zoom on the main SGMCs to show the distribution of the H_2, CO(3−2) and ^{13}CO(2−1) integrated intensity lines. Red and white contours are the ^{13}CO(2−1) and CO(3−2) lines, respectively. H_2 color bars are in units of 10^{-17} erg s^{-1} cm^{-1}. **b)** CO integrated intensity ratios: CO(3−2) to CO(2−1) and ^{12}CO to ^{13}CO(2−1), respectively. Data were convolved to a common resolution, the CO(2−1) beam size. We highlight SGMC 2 and PCC 1 within it. The spectrum is the SiO profile of SGMC 1, after stacking each individual spectra to 1500 km s^{-1} based on the H_2 velocity map. The most massive and youngest SSCs are marked with stars. *(A color version of this figure is available online.)*

The large variations of the CO line ratios reflect the complexity of physical processes in the Antennae overlap region. We are still missing more information of other CO transitions, specially its isotopes, to well constrain the gas properties at GMC scales.

3. A pre-cluster cloud in the Antennae merger

Our near-IR data obtained with the SINFONI spectro-imager on the VLT revealed a compact ($\sim$0.″6, $\sim$50 pc) source not associated with any known SSC (hereafter PCC 1; H11). With CO(3−2) ALMA observations we identified the CO counterpart of this PCC. It has the largest H_2/CO emission ratio and coincides with the highest velocity shear in all of the overlap region (Fig. 2), i.e. the shortest dissipation timescale $\sim 1\,$Myr (H12).

Herrera *et al.* (2011) proposed that the PCC 1 may be a massive cloud on its way to form a SSC within the next few Myr as its turbulent energy dissipates. Within this framework, the H_2 lines trace turbulent energy that is being dissipated and radiated away, as the SGMC complexes and the PCC within, form. Based on dust continuum emission at 1.3 mm of our Cycle 2 ALMA observations, for PCC 1 we derive a mass of $\sim 4 \times 10^6$ M$_\odot$, which is one order of magnitude smaller than the virial mass estimated from the CO line-width and continuum size. Our mass estimate agrees with that obtained by Johnson *et al.* (2015) using the 870 μm dust continuum emission. Johnson *et al.* argued that for the cloud to be confined it must be subject to a high ($P/k_B \gtrsim 10^8$ K cm^{-3}) external pressure, which may be dominated by ram pressure.

Our ^{13}CO(2−1) and dust continuum emission observations do not resolve the internal structure of PCC 1. After beam deconvolution, we find 40 ± 3 and 49 ± 6 pc, respectively. The detection of non-resolved HCO + (4−3) and HCN(4−3) ($n_{\rm crit} \sim 10^6, 10^7$ cm^{-3})

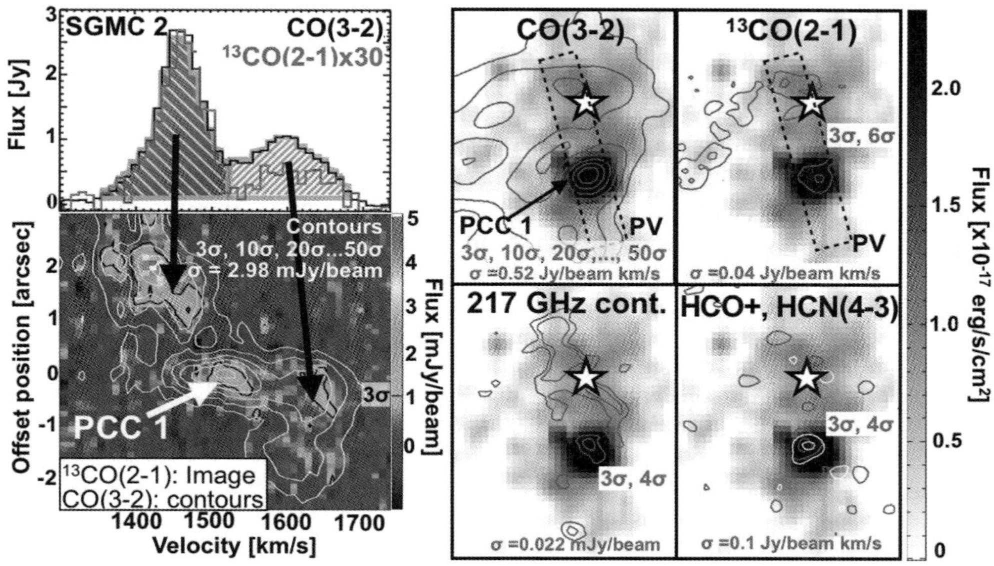

Figure 2. Top-left panel shows the total CO(3−2) and ^{13}CO(2−1) emission from SGMC 2. At the bottom, the position-velocity (PV) diagram across SGMC 2. PCC 1 is identified in CO and ^{13}CO, and seems to be located at the interface of blue and red shifted gas. The right 4 panels show the H_2 integrated emission in SGMC 2 in comparison to CO(3−2), ^{13}CO(2−1), 1.3 mm continuum, HCO+(4−3) and HCN(4−3), in contours. PCC 1 is detected in all of these tracers. *(A color version of this figure is available online.)*

emission towards PCC 1 (Fig. 2) may indicate the presence of a gravitationally bound dense core within the source.

References

Aalto, S., Booth, R. S., Black, J. H., & Johansson, L. E. B., 1995, *A&A*, 300, 369

Flower, D. R. & Pineau Des Forêts, G 2010, *MNRAS*, 406, 1745

Herrera, C. N., Boulanger, F., & Nesvadba, N. P. H. 2011, *A&A*, 534, A138

Herrera, C. N., Boulanger, F., Nesvadba, N. P. H., & Falgarone, E. 2012, *A&A*, 538, L9

Herrera, C. N & Boulanger, F. 2015, *Proceedings IAU Symposium 316*, Article ID IAU-15-IAUS316-0210: *Stellar Feedback From a Massive Super Star Cluster in the Antennae Merger*

Johnson, K. E., Leroy, A. K., Indebetouw, R., Brogan, C. L.,Whitmore, B. C., Hibbard, J., Sheth, K., & Evans, A. S. 2015, *ApJ*, 806, 35

Schulz, A., Henkel, C., Muders, D., Mao, R. Q, Röllig, M., & Mauersberger, R. 2007, *A&A*,466, 467

Teyssier, R., Chapon, D., & Bournaud, F. 2010, *ApJ*, 720, 149

Ueda, J., Iono, D., Petitpas, G., Yun, M. S., Ho, P. T. P., Kawabe, R., Mao, R.-Q., Martín, S., Matsushita, S., Peck, A. B., Tamura, Y., Wang, J., Wang, Z., Wilson, C. D., & Zhang, Q., 2012, *ApJ*, 745, 65

Weidner, C., Bonnell, I. A., & Zinnecker, H. 2010, *ApJ*, 724, 1503

Whitmore, B. C., Brogan, C., Chandar, R., Evans, A., Hibbard, J., Johnson, K., Leroy, A., Privon, G., Remijan, A., & Sheth, K. 2014, *ApJ*, 795, 156

Whitmore, B. C. & Schweizer, F. 1995, *AJ*, 109, 960

Wilson, C., Scoville, N., Madden, S., & Charmandaris, V. 2000, *ApJ*, 542, 120

From Interstellar Clouds to Star-Forming Galaxies: Universal Processes?
Proceedings IAU Symposium No. 315, 2015
P. Jablonka, Ph. André & F. van der Tak, eds.

doi:10.1017/S1743921316007419

A multi-wavelength classification system for the evolution of star clusters

Bradley C. Whitmore[1], Crystal Brogan[2], Rupali Chandar[3], Aaron Evans[2,4], John Hibbard[2], Kelsey Johnson[4], Adam Leroy[2], George Privon[4], Anthony Remijan[2] and Kartik Sheth[2]

[1]Space Telescope Science Institute, 3700 San Martin Drive, Baltimore, MD, 21218, USA
[2]National Radio Astrophysical Observatory, Charlottesville, VA, 22903, USA
[3]Department of Physics & Astronomy, The University of Toledo, Toledo, OH, 43606, USA
[4]Department of Astronomy, University of Virginia, Charlottesville, VA 22904, USA
email: whitmore@stsci.edu

Abstract. The availability of high spatial resolution molecular gas observations from ALMA, and similar resolution observations in the radio continuum using the VLA, is providing the opportunity to make comparisons with specific features seen in optical observations more directly than in the past. Using our ALMA observations of the Antennae galaxies as a springboard, we have compared the locations of small-scale CO (3–2) features with a variety of multi-wavelength observations, in particular optical and near-infrared imaging using both broad (UBVI) and narrow-band data (H_α and Pa_β) taken with the HST, and radio (3.6 cm) continuum observations taken with the VLA. This comparison leads to the development of an evolutionary classification system which provides a framework for studying the sequence of star cluster formation and evolution, from diffuse Giant Molecular Clouds (GMCs), to proto, embedded, emerging, young, and intermediate/old star clusters. Using this evolutionary framework, we estimate the maximum age range of clusters formed in a single GMC is approximately 10 Myr. This suggests that the molecular gas is removed over this timescale, resulting in the cessation of star formation and the destruction of the GMC within a radius of about 200 pc.

1. Introduction

In the past, the difference in spatial resolution for optical observations (e.g., 0.1″ with the *Hubble Space Telescope* = HST) and radio observations (typically a few ″s) has been a limiting factor for many studies. With the advent of ALMA (the *Atacama Large Millimeter/submillimeter Array*), and the *Very Large Array* (VLA), we have entered a new era where it is possible to study all three components with comparable resolution.

Figure 1 shows an illustrative example from the Antennae galaxies, as described in more detail in Whitmore *et al.* (2014). The background images are HST observations; the green contours show CO (3-2) observations from ALMA; the yellow contours are radio continuum (3.6 cm) observations from the VLA, and the orange shows age and extinction estimates for several star clusters based on HST observations from Whitmore *et al.* (2010). Evolutionary stages, as discussed below, are included in blue.

As shown in Figure 1, regions with diffuse CO emission (i.e., GMCs) have neither optical clusters or radio continuum emission associated with them. The youngest star clusters, with ages around 1 Myr and Av extinctions greater than about 1 mag, have both CO and radio continuum emission. For the older clusters, with ages greater than about 5 Myr, there is no CO or radio continuum emission. Hence, region W10-3 provides a clear story line of how clusters form and evolve in the Antennae galaxies.

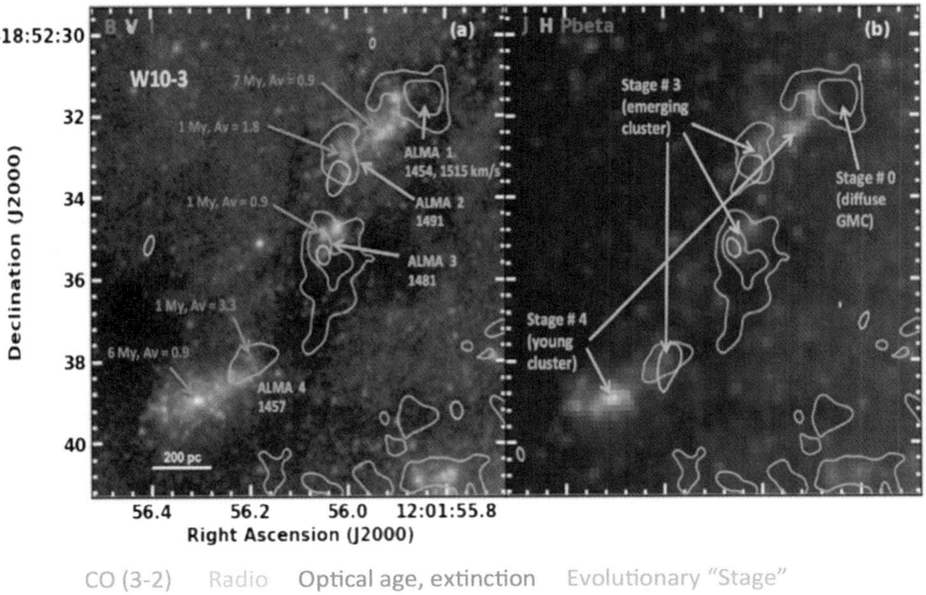

Figure 1. Correspondence between CO knots (green), 3.6 cm radio emission (yellow) and optical star clusters (orange) in W10-3 (from Whitmore *et al.* 2014). *(A color version of this figure is available online.)*

2. A Multi-Wavelength Classification System

Other regions in the Antennae show similar correlations as those found in W10-3. Based on these results, and expanding on earlier schemes discussed in Johnson (2002, 2005) and Whitmore *et al.* (2011), we developed a classification system for the evolution of star clusters from GMCs to old globular clusters, as outlined in Figure 2.

Below we provide a brief description; see Whitmore *et al.* (2014) for more details.

Stage 0 (**diffuse giant molecular clouds**) [turbulent equilibrium] - Knots of diffuse CO emission are detected, but no radio, or optical/near-IR emission is observed.

Stage 1 (**protocluster**, $<$0.1 Myr) [high pressure, gravitational collapse] - Compact CO emission is detected, but no radio or optical/near-IR emission is observed.

Stage 2 (**embedded cluster**, 0.1 - 1 Myr) [onset of star formation] - Thermal radio emission is detected along with CO emission. No optical/near-IR continuum is observed, but weak line emission may be present if an O star has formed. The very recently formed cluster is deeply embedded in its natal gas during this stage.

Stage 3 (**emerging cluster**, 1 - 3 Myr) [removal of gas and dust] - The very young cluster is observed primarily in the radio, optical/near-IR emission lines, and faintly in the optical continuum. The cluster is moderately extincted by dust, with A_V typically $>$1 mag.

Stage 4 (**young cluster**, 3 Myr - 10 Myr) [ISM feedback] - The cluster is increasingly observable in the optical and in optical/near-IR emission lines (the latter typically in the form of bubbles) due to the removal of much of the natal gas and dust by feedback. CO emission is gone, but weak radio continuum emission is still observed in many cases. The A_V values have dropped to $<$1 mag.

Stage 5 (**intermediate/old clusters**, $>$10 Myr) [spectral dimming, evaporation] - The cluster is observed in the optical/near-IR, although it has faded, and ionized gas is no longer observed since the massive stars have evolved into stellar remnants.

cluster formation and evolution stage	CO	Radio	Optical/ near-IR continuum	Optical/ near-IR line emission (e.g., Ha, Pa, Pb, ...)	Notes
Stage 0 (diffuse GMC)	Yes, diffuse	---	---	---	Turbulent equilibrium
Stage 1 (protocluster)	Yes, compact, bright	---	---	---	High pressure, grav. collapse, onset of star formation (< 0.1 Myr)
Stage 2 (embedded cluster)	Yes, compact > diffuse, bright	Yes, bright	---	Possible, compact, faint	Increasing star formation (0.1 - 1 Myr)
Stage 3 (emerging cluster)	Yes, diffuse, faint	Yes, moderate	Yes, faint, (Av > 1)	Possible, compact > bubble, bright	Removal of gas and dust (1 - 3 Myr)
Stage 4 (young cluster)	---	Yes, faint	Yes, bright, (Av < 1)	Possible, bubble, moderate	ISM feedback (3 -10 Myr)
Stage 5 (intermediate/old cluster)	---	---	Yes, moderate	---	Spectral dimming, evaporation (> 10 Myr)

Figure 2. Matrix showing the definition of the various stages of cluster formation and evolution used in the classification framework (from Whitmore *et al.* 2014).

3. Other Regions in the Antennae - the "Long-Thin Filament" and the "Firecracker"

Figure 3 shows a more populated region in the Antennae that contains all six evolutionary stages from the matrix in Figure 2. The region has been nicknamed the Long-Thin Filament due to the presence of a 3 kpc CO filamentary structure with an internal velocity dispersion less than 10 km/s (Whitmore *et al.* 2014).

Another region discussed in Whitmore *et al.* (2014) is Super Giant Molecular Cloud 2 (SGMC2), first defined in Wilson *et al.* (2000). A particularly interesting feature is a compact (radius less than 24 pc), high mass (greater than 5×10^6 $M_\odot$), region with intense compact CO emission but no radio emission. This would be a Stage 1 (protocluster), as defined in Figure 2. See Johnson *et al.* (2015) for a detailed discussion of this system, which has been nicknamed the Firecracker.

4. The Timescale for GMC Destruction

An example of the utility of the classification system is provided by estimating the destruction time of GMCs once star formation has started. This is determined by counting the relative fractions of objects in each evolutionary stage. The resulting estimate for the maximum age range of clusters in a GMC after star formation has started is less than 10 Myr, after which time the GMC has been destroyed within a radius of about 200 pc, probably by feedback from the cluster (Whitmore *et al.* 2014). Note that this timescale is consistent with region W10-3 shown in Figure 1, since no CO emission is associated with the older (6 and 7 Myr) clusters. Similar calculations have been made by Kawamura *et al.* (2009) and by Miura *et al.* (2012), resulting in estimates of somewhat longer GMC destruction time scales (i.e., 20 40 Myr).

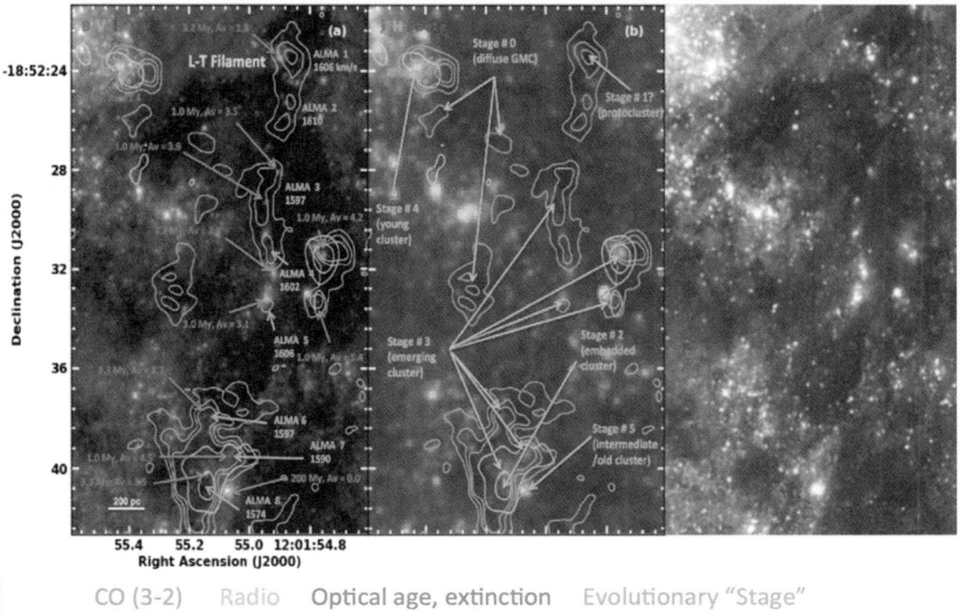

Figure 3. Correspondence between CO (3–2) knots, 3.6 cm radio emission and optical star clusters in the Long-Thin (L-T) Filament. The same color scheme as defined in Figure 1 is used (from Whitmore *et al.* (2014)). *(A color version of this figure is available online.)*

5. Summary

1. The availability of similar resolution CO (ALMA), radio continuum (VLA), and optical/near-IR (HST) makes it easier to compare features and find patterns.

2. This leads to the development of a multi-wavelength evolutionary classification system ranging from diffuse GMCs (stage 0) to old globular clusters (stage 5).

3. Two particularly interesting objects in the Antennae are the Long-Thin Filament (a 3 kpc long CO emission filament with a velocity dispersion less than 10 km/s) and the Firecracker (a candidate protocluster with a radius less than 24 pc and mass greater than 5×10^6 $M_\odot$).

4. Using this classification system leads to an estimate of less than 10 Myr for the destruction time of a giant molecular cloud due to feedback from the young star clusters.

References

Johnson, K. E. 2002, *Science*, 297, 776

Johnson, K. E. 2005, *IAUS*, 227, 413

Johnson, K. E. Leroy, A. K., Indebetouw, R., Brogan, C. L., Whitmore, B. C., Hibbard, J., Sheth, K., & Evans, A. S. 2015, *ApJ*, 806, 35

Kawamura, A., Mizuno, Y., Minamidani, T., Filipovic M. D., Staveley-Smith, L., Kim, S., Mizuno, N., Onishi, T., Mizuno, A., & Fukui, Y. 2009, *ApJS*, 184, 1

Miura, R. E. *et al.* 2012, 761, 37

Whitmore, B. C., Chandar, R., Schweizer, F., Rothberg, B., Leitherer, C., Rieke, M., Rieke, G., Blair, W. P., Mengel, S., & Alonso-Herrero, A. 2010, *AJ*, 140, 75

Whitmore, B. C. *et al.* 2011, *ApJ*, 729, 78

Whitmore, B. C., Brogan, C., Chandar, R., Evans, A., Hibbard, J., Johnson, K., Leroy, A., Privon, G., Remijan, A., & Sheth, K. 2014, *ApJ*, 795, 156

Wilson, C. D., Scoville, N., Madden, S. C., & Charmandaris, V. 2000, *ApJ*, 542, 120

From Interstellar Clouds to Star-Forming Galaxies: Universal Processes?
Proceedings IAU Symposium No. 315, 2015
P. Jablonka, Ph. André & F. van der Tak, eds.

doi:10.1017/S1743921316007420

HOBYS and W43-HERO: Two more steps toward a Galaxy-wide understanding of high-mass star formation

Frédérique Motte[1,2], Sylvain Bontemps[3] and Jérémy Tigé[4]

[1]Institut de Planétologie et d'Astrophysique de Grenoble (IPAG), Univ. Grenoble Alpes - CNRS/INSU, BP 53, 38041 Grenoble Cedex 9, France
email: frederique.motte@ujf-grenoble.fr

[2]Laboratoire AIM Paris-Saclay, CEA/IRFU - CNRS/INSU - Univ. Paris Diderot, Service d'Astrophysique, Bât. 709, CEA-Saclay, F-91191 Gif-sur-Yvette Cedex, France

[3]OASU/LAB-UMR 5804, CNRS - Université Bordeaux 1, 33270 Floirac, France

[4]Laboratoire dAstrophysique de Marseille (LAM), Aix Marseille Université - CNRS, UMR 7326, 13388, Marseille, France

Abstract. The *Herschel*/HOBYS key program allows to statistically study the formation of $10 - 20\ M_\odot$ stars. The IRAM/W43-HERO large program is itself dedicated to the much more extreme W43 molecular complex, which forms stars up to $50\ M_\odot$. Both reveal high-density cloud filaments of several pc^3, which are forming clusters of OB-type stars. Given their activity, these so-called mini-starburst cloud ridges could be seen as "miniature and instant models" of starburst galaxies. Both surveys also strongly suggest that high-mass prestellar cores do not exist, in agreement with the dynamical formation of cloud ridges. The HOBYS and W43 surveys are necessary steps towards Galaxy-wide studies of high-mass star formation.

Keywords. stars: formation, ISM: structure, star clusters, starburst

1. Introduction

Despite their low numbers, massive stars (OB-type, $>10\ M_\odot$ on the main sequence) produce much of the luminosity of galaxies and are the main driver of their evolution. Both their large impact on their environment and how their formation is linked to natal clouds remain poorly understood (see reviews by Elmegreen 2011; Krumholz 2014).

High-mass star formation scenarios currently undergo a change of paradigm, in which this process is no longer quasi-static but simultaneously evolves with both cloud and cluster formation. Theoretically, OB stars could either form through (1) a powerful accretion driven by a high degree of turbulence (e.g., McKee & Tan 2002; Hosokawa & Omukai 2009) or (2) colliding flows initiated by competitive accretion or cloud formation (e.g., Bonnell & Bate 2006; Hartmann *et al.* 2012). From an observational point of view, gravitational streamers and shearing motions have been reported from the cloud to the protostellar scales in a few high-mass star-forming regions (on $10 - 0.1$ pc scales, Schneider *et al.* 2010; Csengeri *et al.* 2011a). We hereafter show that these pioneering studies are now comforted by sensitive *Herschel* images, cloud kinematic studies, and (sub)millimeter interferometric images, favoring the second family of models.

"HOBYS, the *Herschel* imaging survey of OB Young Stellar objects" is a key program, which is exclusively dedicated to high-mass star formation (Motte, Zavago, Bontemps *et al.* 2010; see http://hobys-herschel.cea.fr). It uses the SPIRE and PACS cameras of the *Herschel* spatial observatory to image essentially all of the regions forming OB-type stars at distances less than 3 kpc from the Sun ($d = 0.7 - 3.2$ kpc). HOBYS images

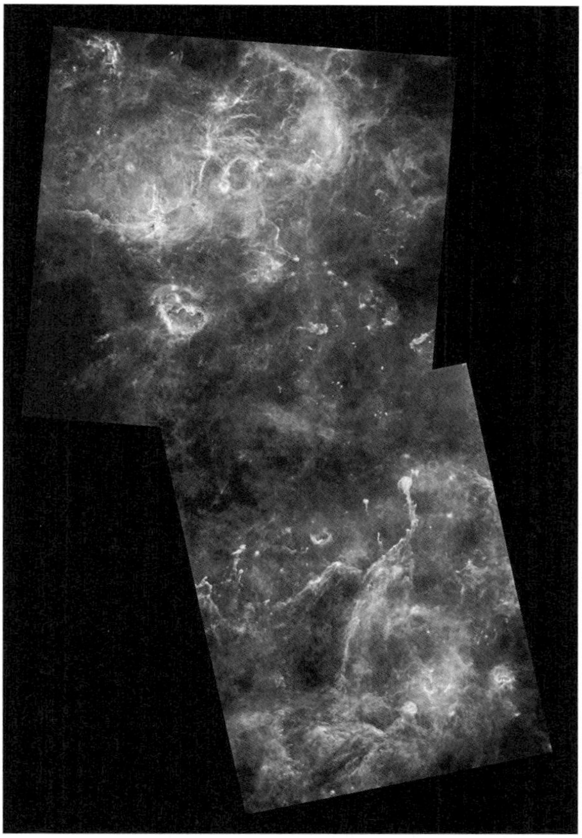

Figure 1. Composite 3-color *Herschel* image of the Cygnus X molecular complex with red=250 μm, green=160 μm, and blue=70 μm (taken from Hennemann *et al.* 2012 and Schneider *et al.* in prep.). The mosaic performed by the HOBYS key program approximately covers a $5^\circ \times 2.5^\circ$ or 120 pc $\times$ 60 pc area. The diffuse blue emission at the center is an H II region powered by the massive Cyg OB2 cluster while earlier stage star-forming sites are seen as red filaments. *(A color version of this figure is available online.)*

revealed networks of filaments and clusters of burgeoning YSOs (see Fig. 1). The W43-HERO large program is dedicated to the "Origins of molecular clouds and star formation in W43". This extreme molecular complex is located at 5.5 kpc from the Sun, at the tip of the long bar of the Milky Way. This IRAM 30 m survey is complemented by IRAM Plateau de Bure, *Herschel*, and ALMA imagings. Among the HOBYS and W43-HERO highlights are the discovery of "mini-starburst ridges" (see Sect. 2) and the recurrent inability to find good prestellar core candidates (see Sect. 3).

2. Ridges & hubs: Hyper-massive clumps forming clusters of high-mass stars

2.1. *Discovery of hyper-massive clumps*

One of the most unexpected results of the *Herschel* mission was in revealing the importance and ubiquity of interstellar filaments (e.g., André *et al.* 2010; Hill *et al.* 2011). They were known from molecular line surveys but could, for the first time, be 1) quantitatively studied thanks to sensitive column density and temperature images and 2)

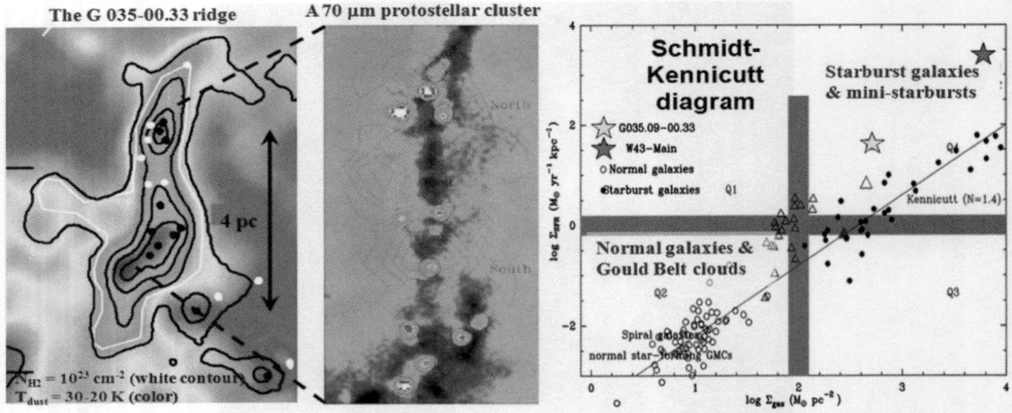

Figure 2. The G035-00.33 ridge and its cluster of high-mass protostars (Left and Center). Ridges are high-density clumps with large star formation activity (Right, see yellow and red star markers). On a Schmidt-Kennicutt diagram, they qualify as mini-starburst regions. Figures extracted from Nguyen Luong *et al.* (2011a) and (in prep.). *(A color version of this figure is available online.)*

directly compared with the protostellar population (see, e.g., Figs. 2 Left-Center). The HOBYS group has developed a definite expertise in building clean and reliable column density maps, in identifying and analyzing cloud filaments, in properly extracting protostars from their surroundings and characterizing them (e.g., Hill *et al.* 2012; Hennemann *et al.* 2012; Men'shchikov *et al.* 2012; Tigé *et al.* 2016). Within the high-mass molecular complexes imaged by HOBYS, we have discovered high-density dominating elongated clumps, which are the preferential sites to form massive stars and which we quote as ridges (see Figs. 2 Left-Center). The existence of ridges is predicted by dynamical models of cloud formation such as colliding flow simulations (e.g., Heitsch & Hartmann 2008; Federrath *et al.* 2010). The densest ridges should coincide with the precursors of "young massive clusters" that will become bound stellar clusters (e.g., Ginsburg *et al.* 2012).

In more details, Hill *et al.* (2011) studied the Vela C molecular cloud complex and defined the central ~3 pc-long filament with column density above 10^{23} cm^{-2} as a ridge. They showed that the Vela C ridge represents a gravitational well that is dominating and shaping its surroundings, in marked contrast to other more typical filaments (e.g., Arzoumanian *et al.* 2011). In parallel, Hennemann *et al.* (2012) analyzed the structure of the DR21 ridge (see Fig. 3 Left), previously identified by us as hosting a cluster of ~20 high-mass protostars (Motte *et al.* 2007; Bontemps *et al.* 2010). In agreement with the kinematical study and simulations of Schneider *et al.* (2010, see Fig. 3 Right), Hennemann *et al.* (2012) proposed the DR21 ridge is fed by and formed through the merging of several sub-filaments already forming stars. A handful of other $\sim 10^4\ M_\odot$ ridges were recently identified in *Herschel*/HOBYS and Hi-GAL images (Nguyen Luong *et al.* 2013; Hill *et al.* in prep.).

More spherical and less massive, $\sim 10^3\ M_\odot$, clumps were also found to form stellar clusters at the junction of filaments (Schneider *et al.* 2012; Didelon *et al.* 2015). These clumps recall the so-called "hubs" discovered in a few extreme infrared dark clouds (e.g., Peretto *et al.* 2013) and predicted by numerical simulations (e.g., Dale & Bonnell 2011). We propose to generalize the concept of ridges and hubs into that of "hyper-massive clumps", which could either be spherical or filamentary and either form a couple or a rich cluster of high-mass stars.

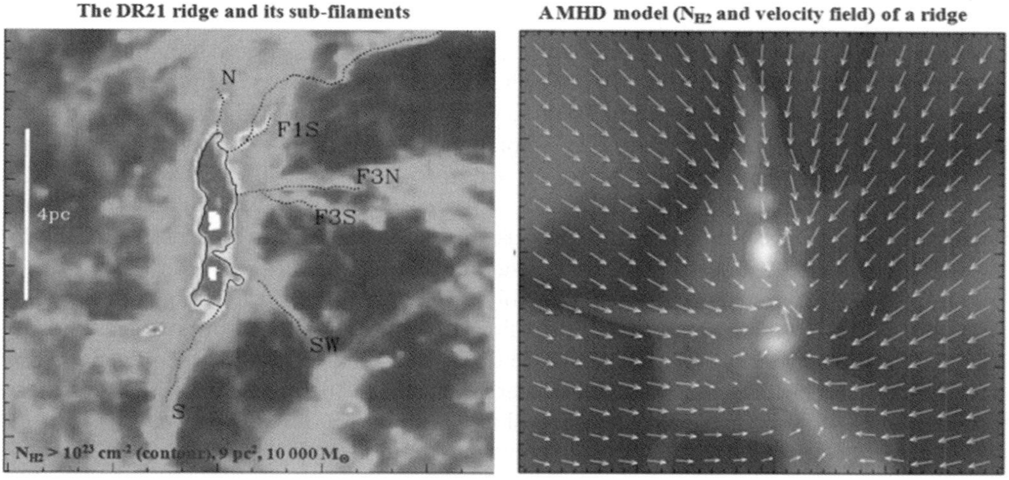

Figure 3. The DR21 ridge and its feeding sub-filaments network (Left), compared to numerical MHD simulations (Right). Panels extracted from Hennemann *et al.* (2012) and Schneider *et al.* (2010). *(A color version of this figure is available online.)*

2.2. *Dynamical processes at the origin of hyper-massive clouds*

Based on *Herschel* and kinematic studies (Schneider *et al.* 2010; Hill *et al.* 2011; Hennemann *et al.* 2012; Nguyen Luong *et al.* 2011a, 2013), we proposed that most ridges/hubs could have been formed by dynamical scenarios such as converging flows (see Fig. 3 Right). Minier *et al.* (2013) themselves used numerical simulations to show that the ionization of a very nearby H II region/OB cluster could be the main process responsible for shaping some ridges like Vela C (see his Fig. 5). In any case, whatever the origin of the additional pressure, arising from colliding flows or ionization, it supersedes the thermal and micro-turbulence pressure. Those dynamical scenarios for cloud and then subsequent star formation are part of the second family of models shortly described in Sect. 1. They are also consistent with the observational lack of high-mass prestellar cores (e.g., Motte *et al.* 2007, 2010; Russeil *et al.* 2010; Tigé *et al.* 2016).

In parallel to these *Herschel* structural studies, we have investigated the kinematics of three ridges identified as such from HOBYS data (Hennemann *et al.* 2012; Nguyen Luong *et al.* 2013). A long-term survey of Cygnus X and an IRAM large program on W43 (W43-HERO, Motte, Schilke *et al.*, see http://www. iram-institute.org/EN/content-page-292-7-158-240-292-0.html) has shown that the DR21, W43-MM1, and W43-MM2 ridges are all undergoing global collapse on the ridge scale with supersonic inward velocities ($1-2$ km s^{-1} over $1-10$ pc^2; Schneider *et al.* 2010; Motte *et al.* in prep.). Such infall motions, sometimes called gravitational focusing, are expected both in colliding flows and ionization compression models (e.g., Vázquez-Semadeni *et al.* 2005; Tremblin *et al.* 2012). We started studying the inner kinematic structure of these ridges, which suggests they are braids/bundles of filaments or gas layers (Louvet *et al.* in prep.; Bontemps *et al.* in prep.). We have also studied the shocks associated with gas shears created by sub-filaments braiding, using SiO imaging and detailed shock modeling (Nguyen Luong *et al.* 2013; Duarte-Cabral *et al.* 2014; Louvet *et al.* 2016). Such extreme kinematics is observed by other groups at the clump scale (e.g. Galván-Madrid *et al.* 2010; Henshaw *et al.* 2014) and is reminiscent of gas flows and shocks observed at the protostellar scale by Csengeri *et al.* (2011a-b).

2.3. *Galactic mini-starburst events*

Extreme characteristics of ridges and hubs, in terms of density and kinematics, could lead to an atypical star formation activity. Star formation efficiency is indeed predicted to increase with gas density (Krumholz & McKee 2005; Hennebelle & Chabrier 2011; Padoan & Nordlund 2011). The clear accumulation of massive protostars we observed along ridges (see, e.g., Fig. 2 Center) leads, on $1-10$ pc^2 areas, to star formation rate densities worthy of starburst galaxies (see Fig. 2 Right). For this reason, W43-main, G035.39-00.33, and W43-MM1 qualify as mini-starburst regions, i.e. "miniature and instantaneous models of starburst galaxies" (Motte *et al.* 2003; Nguyen Luong 2011a; Louvet *et al.* 2014).

These mini-starburst events most probably follow the formation of the ridges, which were proposed to develop through colliding flows. Indeed, short mini-bursts of star formation are to be expected after a fast episode of cloud formation (e.g., Vázquez-Semadeni *et al.* 2008) or, equivalently, for a cloud under compressive turbulent forcing (Federrath & Klessen 2012). In this scenario, star formation should gradually settle within ridges. We should thus measure very different SFR levels depending on the evolutionary status of the ridge. Louvet *et al.* (2014) measured SFRs within subregions of the W43-MM1 ridge and found a clear correlation of SFR with cloud density or, equivalently, SFR with the evolution of cloud concentration. Moreover, SFR is smaller within the eastern part of the ridge, where the shocks associated with more recent cloud formation are the strongest (Louvet *et al.* 2014, 2016). Therefore, the dynamical ridge formation may well be followed by series of intense bursts of star formation.

3. A ten years-long quest for high-mass prestellar cores

The dynamical picture presented above is consistent with the fact that observers have tried for ten years, in vain, to identify good candidate high-mass prestellar cores (Motte *et al.* 2007, 2010; Russeil *et al.* 2010; Tigé *et al.* 2016). The quest for high-mass prestellar cores started with ground-based millimeter images and continued with sets of *Herschel* images at far-infrared to submillimeter wavelengths. Given the $1-3$ kpc distance of the nearest high-mass star-forming complexes, both surveys are unable to resolve individual high-mass prestellar cores, assumed to be individual mass reservoirs of $\sim$0.02 pc sizes and $>10^6$ cm^{-3} densities. They however do resolve their parental massive dense cores (MDCs, a few 0.1 pc). While millimeter surveys failed to find starless MDCs (Motte *et al.* 2007, Russeil *et al.* 2010), *Herschel* studies, thanks to their careful dust temperature measurements, did identify tens of starless MDC candidates (Tigé *et al.* 2016).

However, several follow-ups do suggest that most starless MDCs may not contain fragments dense and massive enough to be high-mass prestellar cores. A handful of the NGC 6334 starless MDCs, observed at higher-resolution proved indeed not to be centrally concentrated but to mostly contain filaments and diffuse gas (Tigé *et al.* 2016). Similarly, the starless MDC candidates of the W43-MM1 ridge observed with ALMA disperse into diffuse structures and a few low-mass cores at the 0.02 pc scale (project Motte *et al.*). When located in the mass versus radius diagram of Fig. 4, starless MDCs are expected to follow the $M(<r) \propto r^2$ relation, which links a sample of starless clumps and their prestellar sub-fragments/cores (Butler & Tan 2012; Tan *et al.* 2013). This mass concentration is much lower than the one generally found for protostellar envelopes and ridges, which have $\rho \propto r^{-2}$ density structures (e.g., Motte & André 2001; Didelon *et al.* 2015) and thus $M(<r) \propto r$ mass concentration. At 0.02 pc scales, NGC 6334 starless MDCs of Tigé *et al.* (2016) and all starless fragments found by Tan *et al.* (2013), including C1-S, should merely host low- to intermediate-mass prestellar cores.

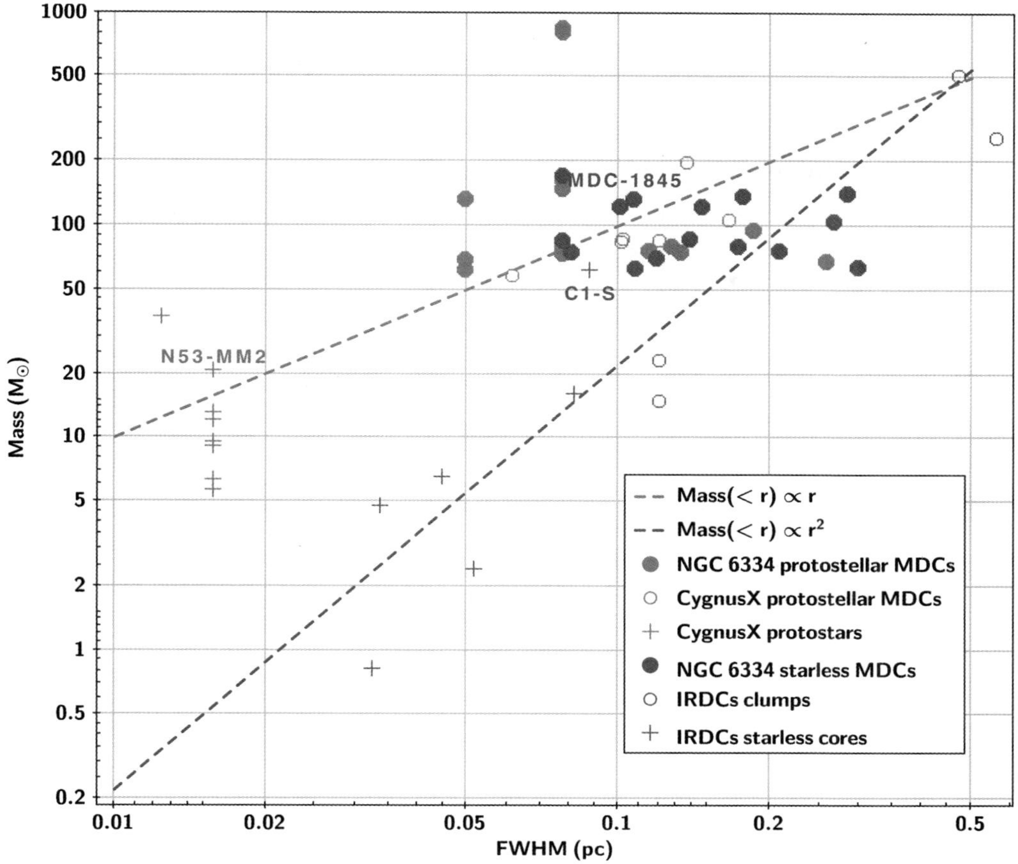

Figure 4. Prediction of gas concentration within starless MDC candidates of NGC 6334 (filled blue circles). Given that they are expected to follow the $M(< r) \propto r^2$ relation (yellow dashed line), linking other starless clumps and their prestellar sub-fragments/cores (open blue symbols), NGC 6334 starless MDCs should merely host low- to intermediate-mass prestellar cores. Figure extracted from Tigé *et al.* (2016). *(A color version is available online.)*

Pre-stellar cores could also be hosted within IR-quiet (young) protostellar MDCs. An interferometric study of similar objects in Cygnus X however showed that most if not all massive sub-fragments at 0.02 pc scales are associated with protostars (Duarte-Cabral *et al.* 2013).

After ten years of research, we therefore still lack a good candidate high-mass prestellar core. Several interferometric studies toward large samples of MDCs are ongoing. We are then at the dawn of finally proving that high-mass prestellar cores do or do not exist.

4. Conclusion

The dynamical picture of Sect. 2 and the lack of high-mass prestellar cores presented in Sect. 3 are consistent. Indeed, the high-mass equivalent of prestellar cores may be replaced by low-mass protostars within hyper-massive clumps. The latter may thus constitute the pc-scale "gas reservoir", from which gas is accreted onto 0.02 pc-scale protostellar envelopes and in turn onto protostellar embryos. As a consequence, protostars will grow

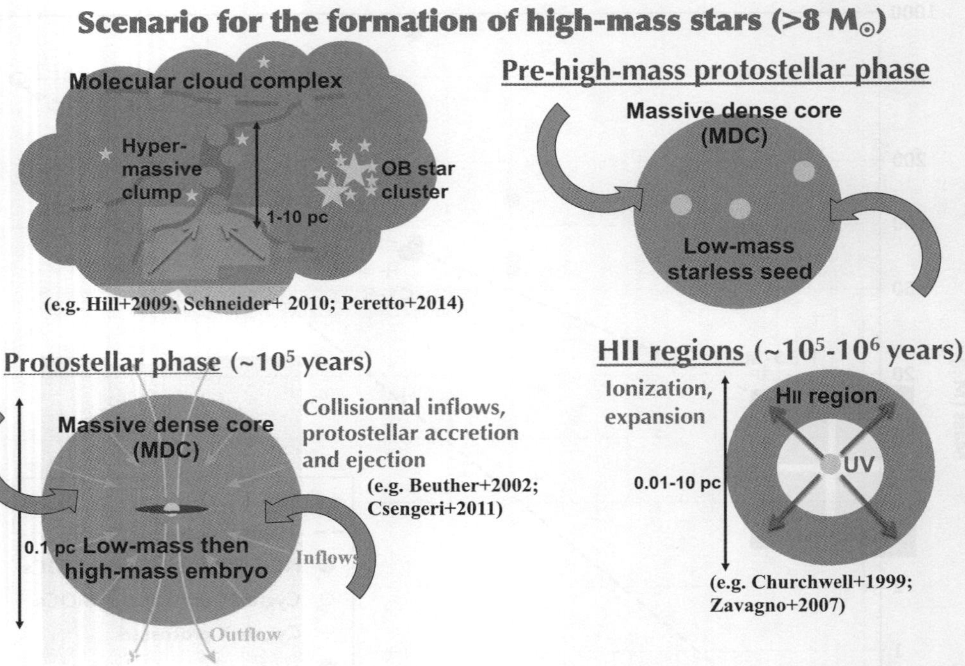

Figure 5. Schematic evolutionary diagram proposed for the simultaneous formation of an hyper--massive filament and one of its high-mass protostars. *(A color version of this figure is available online.)*

in mass at the same time as hyper-massive clumps will form and get denser. Figure 5 illustrates the evolutionary scheme we propose for the formation of high-mass stars.

In this context, it will be very interesting to investigate, in massive protostellar clusters, the origin of the initial mass function of stars (IMF, e.g., Kroupa 2001). Indeed, while low-mass stars in the Gould Belt clouds accrete their final mass from well-defined prestellar cores with a mass distribution mimicking the IMF (Motte *et al.* 1998, Könyves *et al.* 2015), (high-mass) stars within hyper-massive clumps should accrete gas from pc-scales, through sporadic flows, probably favoring the formation of high-mass stars during $10^5 - 10^6$ yrs bursts. An ALMA study of the W43-MM1 ridge is ongoing but many more high-resolution studies will be necessary to solve this issue.

References

André, Ph., Men'shchikov, A., Bontemps, *et al.* 2010, *A&A*, 518, L102+
Arzoumanian, D., André, Ph., Didelon, P., *et al.* 2011, *A&A*, 529, L6
Bonnell, I. A. & Bate, M. R. 2006, *MNRAS*, 370, 488
Bontemps, S., Motte, F., Csengeri, T., & Schneider, N. 2010, *A&A*, 524, A18
Butler, M. J. & Tan, J. C. 2012, *ApJ*, 754, 5
Csengeri, T., Bontemps, S., Schneider, N., Motte, F., & Dib, S. 2011a, *A&A*, 527, A135
Csengeri, T., Bontemps, S., Schneider, N., *et al.* 2011b, *ApJ*, 740, L5
Dale, J. E. & Bonnell, I. 2011, *MNRAS*, 414, 321
Didelon, P., Motte, F., Tremblin, P. *et al.* 2015, *A&A*, 584, A4
Duarte-Cabral, A., Bontemps, S., Motte, F., *et al.* 2013, *A&A*, 558, A125
Duarte-Cabral, A., Bontemps, S., Motte, F., *et al.* 2014, *A&A*, 570, A1
Elmegreen, B. 2011, in *Star Formation in the Local Universe*, EAS Pub. Series, 51, 45

Federrath, C. & Klessen, R. S. 2012, *ApJ*, 761, 156
Federrath, C., Roman-Duval, J., Klessen, R. S., Schmidt, W., & Mac Low, M.-M. 2010, *A&A*, 512, A81
Galván-Madrid, R., Zhang, Q., Keto, E., *et al.* 2010, *ApJ*, 725, 17
Ginsburg, A., Bressert, E., Bally, J., & Battersby, C. 2012, *ApJ*, 758, L29
Hartmann, L., Ballesteros-Paredes, J., & Heitsch, F. 2012, *MNRAS*, 420, 1457
Heitsch, F. & Hartmann, L. 2008, *ApJ*, 689, 290
Hennebelle, P. & Chabrier, G. 2011, *ApJ*, 743, L29
Hennemann, M., Motte, F., Schneider, N., *et al.* 2012, *A&A*, 543, L3
Henshaw, J. D., Caselli, P., Fontani, F., Jimez-Serra, I., & Tan, J. C. 2014, *MNRAS*, 440, 2860
Hill, T., Motte, F., Didelon, P., *et al.* 2011, *A&A*, 533, A9
Hill, T., Motte, F., Didelon, P., *et al.* 2012, *A&A*, 542, A114
Hosokawa, T. & Omukai, K. 2009, *ApJ*, 691, 823
Könyves, V., André, Ph., Men'shchikov, A., *et al.* 2015, *A&A*, 584, A91
Kroupa, P. 2001, *MNRAS*, 322, 231
Krumholz, M. 2014, *Physics Reports*, 539, Iss. 2, 49-134
Krumholz, M. R. & McKee, C. F. 2005, *ApJ*, 630, 250
Louvet, F., Motte, F., Hennebelle, P., *et al.* 2014, *A&A*, 570, A15+
Louvet, F., Motte, F., Gusdorf, A., *et al.* 2016, soon to be submitted to *A&A*
McKee, C. F. & Tan, J. C. 2002, *Nature*, 416, 59
Menshchikov, A., André, P., Didelon, P., *et al.* 2012, *A&A*, 542, A81
Minier, V., Tremblin, P., Hill, T., *et al.* 2013, *A&A*, 550, A50
Motte, F. & André, Ph. 2001, *A&A*, 365, 440
Motte, F., André, Ph., & Neri, R. 1998, *A&A*, 336, 150
Motte, F., Bontemps, S., Schilke, P., *et al.* 2007, *A&A*, 476, 1243+
Motte, F., Schilke, P., & Lis, D. C. 2003, *ApJ*, 582, 277
Motte, F., Zavagno, A., Bontemps, S., *et al.* 2010, *A&A*, 518, L77+
Nguyen Luong, Q., Motte, F., Carlhoff, P., *et al.* 2013, *ApJ*, 775, 88
Nguyen Luong, Q., Motte, F., Hennemann, M., *et al.* 2011a, *A&A*, 535, A76
Nguyen Luong, Q., Motte, F., Schuller, F., *et al.* 2011b, *A&A*, 529, A41+
Padoan, P. & Nordlund, A. 2011, *ApJ*, 730, 40
Peretto, N., Fuller, G. A., Duarte-Cabral, A., *et al.* 2013, *A&A*, 555, A112
Russeil, D., Zavagno, A., Motte, F., *et al.* 2010, *A&A*, 515, A55+
Schneider, N., Csengeri, T., Bontemps, S., *et al.* 2010a, *A&A*, 520, A49+
Schneider, N., Csengeri, T., Hennemann, M., *et al.* 2012, *A&A*, 540, L11
Tan, J. C., Kong, S., Butler, M. J., Caselli, P., & Fontani, F. 2013, *ApJ*, 779, 96
Tigé, J., Motte, F., Russeil, *et al.* 2016, soon to be submitted to *A&A*
Tremblin, P., Audit, E., Minier, V., Schmidt, W., & Schneider, N. 2012, *A&A*, 546, A33
Vázquez-Semadeni, E., Kim, J., & Ballesteros-Paredes, J. 2005, *ApJ*, 630, L49
Vázquez-Semadeni, E., González, R. F., Ballesteros-Paredes, J., Gazol, A., & Kim, J. 2008, *MNRAS*, 390, 769

From Interstellar Clouds to Star-Forming Galaxies: Universal Processes?
Proceedings IAU Symposium No. 315, 2015
P. Jablonka, Ph. André & F. van der Tak, eds.

doi:10.1017/S1743921316007432

Comparison of Low-Mass and High-Mass Star Formation

Jonathan C. Tan

Depts. of Astronomy & Physics, University of Florida, Gainesville, FL 32611, USA
email: jctan.astro@gmail.com

Abstract. I review theoretical models of star formation and how they apply across the stellar mass spectrum. Several distinct theories are under active study for massive star formation, especially *Turbulent Core Accretion*, *Competitive Accretion* and *Protostellar Mergers*, leading to distinct observational predictions. These include the types of initial conditions, the structure of infall envelopes, disks and outflows, and the relation of massive star formation to star cluster formation. Even for Core Accretion models, there are several major uncertainties related to the timescale of collapse, the relative importance of different processes for preventing fragmentation in massive cores, and the nature of disks and outflows. I end by discussing some recent observational results that are helping to improve our understanding of these processes.

1. Introduction

A universal theory of star formation that successfully describes the birth of low- and high-mass stars is a long-sought goal. In particular, large efforts have been made to see if massive stars form in a similar way to low-mass stars, i.e., via the *Core Accretion* paradigm, starting with pre-stellar cores (PSCs) that then collapse to form a single or binary protostar, with most matter accreting via a rotationally-supported disk that also helps to launch protostellar outflows (e.g., Shu *et al.* 1987; Inutsuka 2012). However, alternatives to this mechanism include *Competitive Accretion* (Bonnell et al. 2001; Wang *et al.* 2010) and *Protostellar Mergers* (Bonnell et al. 1998; Bally & Zinnecker 2005; Moeckel & Clarke 2011). Both have been proposed for massive star formation in crowded regions of protostellar clusters, the most common, perhaps near-universal, sites where massive stars form. However, these regions are also relevant for the bulk of star formation in galaxies and, in fact, Competitive Accretion has been invoked to explain most of the range of the stellar initial mass function (IMF) $\gtrsim 1\,M_\odot$ (e.g., Bate 2012). Here I discuss latest work that seeks to elucidate the formation mechanism of massive stars, including expected similarities and differences with low-mass protostars and role of environment.

2. Massive Star Formation Environments

Massive stars have dominated the universe since its earliest epochs. The first, "Pop III" stars, forming in individual dark matter minihalos from pure H & He gas, are thought to have been $\gtrsim 10 - 100\,M_\odot$, so that they then initiated reionization and chemical enrichment of their surroundings (e.g., Bromm 2013). Continued radiative, mechanical and chemical feedback from massive stars is then theorized to have shaped the formation and evolution of galaxies (e.g., Vogelsberger et al. 2014). Observationally, the light seen from distant galaxies is dominated by that from massive stars (e.g., Sobral *et al.* 2015). This is true also in regions of nearby galaxies with active star formation, where the clustered nature of the process is readily apparent (e.g., Whitmore *et al.* 2014). There is a wide range of mass surface densities of these young star clusters from $\Sigma \sim 0.1$–$30\,\mathrm{g\,cm^{-2}}$

and a narrower range of sizes of $\sim$ 1–10 pc (Tan et al. 2014 [T14]). Driven by these star clusters and dissolving OB associations, feedback on the interstellar medium is powered via radiation pressure, stellar winds, ionization and supernovae. Such feedback is likely to play a prominent role in the regulation of star formation rates (e.g., Walch *et al.* 2015).

Large fractions of all stars, including massive stars, form in clusters (e.g., de Wit *et al.* 2005; Gutermuth *et al.* 2009) and the distribution of star formation is much more clustered than that of molecular gas. This means that most stars and planetary systems may have been strongly influenced by massive star feedback at the time of their birth.

Another environment where massive stars form is galactic center regions, including in dense circumnuclear starburst disks, where mean disk mass surface densities can range from $\Sigma \sim 0.1$–$100\,\mathrm{g\,cm^{-2}}$ (Downes & Solomon 1998; Wilson et al. 2014). Our Galactic center has a large population of young ($\lesssim 6$ Myr), relatively massive stars, some orbiting in a disk-like structure around the supermassive black hole (e.g., Lu et al. 2014). The growth of supermassive black holes may be regulated by such star formation activity.

3. The Physics of Star Formation

The physics of star formation involves a competition between self-gravity of gas clumps (i.e., structures that fragment into star clusters) and cores (i.e., structures that collapse to a central rotationally-supported disk that forms a single star or small-N multiple) and processes that resist collapse, i.e., various kinds of pressure forces, such as thermal, turbulent and magnetic. The evolution of these pressures needs to be followed by considering heating and cooling processes, generation and decay of turbulence, and generation and diffusion of B-fields. Once the clump or core contains stars, then support against collapse may be provided by mechanical feedback (i.e., protostellar outflows or stellar winds), radiation pressure (acting mostly on dust), or enhanced thermal pressure from photoionized regions. The chemical evolution of the gas and dust needs to be followed during collapse, especially as this helps set the trace ionization fraction that is important for coupling B-fields to the mostly neutral gas. Rotational support becomes important in protostellar disks, but theoretical prediction of their scale is quite uncertain due to the dominant role of magnetic braking in transferring angular momentum (Li *et al.* 2014). Fragmentation of gas in the clump and in the core's disk will depend on the evolution of the local pressure and rotation/shear support contributions. To accurately model star formation, a wide range of spatial and temporal scales extending down to those of the protostellar surface need to be followed. Unlike the cosmological Pop III case, numerical simulations of "local" star formation face the additional challenge of uncertain choices for initial conditions: e.g., how close is the initial clump to virial and pressure equilibrium?

Due to the complexity of this highly nonlinear and high dynamic range problem, which remains beyond full computational tractability, there are many open questions. What causes a certain region of molecular cloud to form stars, i.e., does it typically occur due to an external trigger (e.g., converging flows, cloud collisions or stellar feedback) or via spontaneous gravitationally instability (e.g., as a cloud evolves and loses its earlier level of internal pressure support)? This question also relates to the specification of the initial conditions. Then, what is the accretion mechanism of the protostar, i.e., is most of the mass already organized into a self-gravitating PSC or is it later accreted competitively to the protostar from a previously unbound state from the larger-scale clump? Are protostellar mergers important? Are protostellar interactions that may disturb cores and disks important? What is the timescale of individual star formation: e.g., does a core collapse on a timescale that is similar to or much slower than the local free-fall time (for convenience defined with reference to that of a uniform density sphere: $t_{\rm ff} \equiv (3\pi/[32G\rho])^{1/2}$)?

Similarly, is the clump undergoing rapid global collapse to form a star cluster (Elmegreen 2000; Hartmann & Burkert 2007) or is the process much slower allowing the clump to be in quasi equilibrium (Tan et al. 2006), perhaps regulated by outflow-driven turbulence (Nakamura & Li 2014)? A consequence of these different possibilities is the range of age spreads present in young star clusters, again relative to the local clump free-fall time (Da Rio *et al.* 2014). Are massive stars the first to start forming in a cluster, becoming massive because of this head start (Wang *et al.* 2010; Bate 2012), or are their individual formation times short compared to the duration of cluster formation (McKee & Tan 2002), allowing contemporaneous formation with lower-mass stars that is potentially independent of the evolutionary stage of the protocluster? What processes shape the stellar IMF and initial binary properties? How do these quantities vary with environment?

3.1. *Turbulent Core Accretion*

A variety of Core Accretion models for massive star formation have been proposed, extending the basic concepts of low-mass star formation models (see T14 for a review). The Turbulent Core Accretion model (McKee & Tan 2003 [MT03]) recognizes that thermal pressure is unimportant for supporting massive PSCs (i.e., they are much more massive than the local Jeans or Bonnor-Ebert mass), so such support must come from some combination of nonthermal means, i.e., turbulence and/or magnetic fields. The next key assumption is that the pressure of the clump environment, which is $P_{\rm cl} \simeq G\Sigma_{\rm cl}^2$ for a self-gravitating cloud, sets the surface boundary condition of the core of a given mass. PSCs are modeled as singular polytropic spheres not too far from virial and hydrostatic equilibrium. Then the radius is given by $R_c = 0.057 M_{c,60}^{1/2} \Sigma_{\rm cl,1}^{-1/2}$ pc, where $M_{c,60}$ is the core mass normalized by 60 $M_\odot$ and $\Sigma_{\rm cl,1}$ is the clump mass surface density normalized by $1\,{\rm g\,cm^{-2}}$. This is the degree to which the core mass must be concentrated in order to become gravitationally unstable. Note that no assumption has been made about whether the PSC has taken a short or long time to form from the clump, although achieving approximate pressure and virial equilibrium is expected to take at least $\sim 1 t_{\rm ff}$. In principle, it is possible to estimate the age of a given PSC via comparison of predictions of astrochemical models, especially of deuterated species (e.g., Kong *et al.* 2015a), with observed abundances (§4.1). MT03 assumed cores and clumps have an internal radial density gradient $\rho \propto r^{-k_\rho}$, with $k_\rho \simeq 1.5$ set empirically (and also consistent with later studies of Infrared Dark Clouds (IRDCs) clumps/cores; Butler & Tan 2012 [BT12]).

The core is then assumed to undergo inside-out collapse at a rate comparable with that of local free-fall collapse, i.e., $\dot{m}_{*d} = 9.3 \times 10^{-4} \epsilon_{*d} (M_{c,60}\Sigma_{\rm cl,1})^{3/4} (M_{*d}/M_c)^{1/2}\, M_\odot\,{\rm yr}^{-1}$, where $\dot{m}_{*d}$ is the rate of increase of the mass of the protostar and its disk, ϵ_{*d} is the current efficiency of the infall rate with respect to uninhibited collapse (values ~ 0.5 are expected due to protostellar outflow feedback, e.g., Zhang *et al.* 2014 [ZTH14]) and M_{*d} is the idealized collapsed mass supplied to the central disk in the no-feedback limit. Note, this accretion rate estimate can be applied to stars forming from cores of all masses, with model assumptions only beginning to break down close to the Bonner-Ebert mass, $M_{\rm BE} = 0.050(T/20{\rm K})^2 \Sigma_{\rm cl,1}^{-1}\, M_\odot$. The timescale for star formation is $t_{*f} = 1.3 \times 10^5 M_{c,60}^{1/4} \Sigma_{\rm cl,1}^{-3/4}$ yr, which has a very weak dependence on core mass. This timescale is similar to the clump's free-fall time. Thus it is important to know if the clump is undergoing rapid, free-fall collapse, or whether star cluster formation is a slower, more drawn-out affair.

The question of what, if anything, prevents fragmentation of massive PSCs is crucial, since this physics may play a decisive role in shaping the IMF via the (pre-stellar) core mass function (PS)CMF. In the Turbulent Fragmentation model of Padoan & Nordlund (2002), a power law spectrum of super-Alfvénic turbulence is assumed and the size of dense cores is associated with the thickness of post-shock gas, so cores are born in a very dynamic fashion. This model has been compared to numerical simulations by Padoan *et al.* (2007; see also Collins *et al.* 2011; Hennebelle *et al.* 2011), who point out the potential importance of B-fields in setting the CMF. Hennebelle & Chabrier (2008) presented a theory of the CMF based on the Press-Schecter statistical formalism applied to supersonic turbulence. In their model, more massive cores derive more support from turbulence, but also require lower density initial conditions to help set their larger "turbulent Jeans mass," than lower-mass cores. Alternatively, Kunz & Mouschovias (2009) proposed a dominant role for B-fields, with the CMF being set by ambipolar diffusion. Motivated by their studies of massive, cold IRDC cores that contain $\gtrsim 100\ M_{\rm BE}$, BT12 proposed B-field support was the dominant mechanism inhibiting fragmentation, so that $M_c \simeq M_B = 38(B_c/{\rm mG})^3(n_{\rm H,c}/10^6{\rm cm}^{-3})^{-2}\ M_\odot$, the magnetic critical mass (Bertoldi & McKee 1992). One should note that massive starless cores, like massive stars, will be rare objects within the clump, and this rarity may be set by being in the high B-field-strength tail of the distribution. Krumholz & McKee (2008) proposed radiative heating from surrounding lower-mass protostars prevents fragmentation of massive PSCs, which requires high accretion luminosities and thus high $\dot{m}_{*d}$ and so high $\Sigma_{\rm cl}$, $\gtrsim 1\ {\rm g\ cm}^{-2}$. This model predicts massive star formation requires the presence of an already forming protocluster and that massive protostars would not form from cold, dense cores within IRDCs.

Many numerical simulations of the collapse of massive, turbulent "cores" have been carried out. Dobbs *et al.* (2005) presented hydrodynamics-only simulations, finding extensive fragmentation into stars with masses of only $\sim 0.1\ M_\odot$, and thus argued for Competitive Accretion. Krumholz *et al.* (2007) included dust reprocessed radiative feedback that heated the gas, raised the Jeans mass and thus reduced fragmentation. These simulations also showed radiation pressure does not prevent the formation of massive stars, given the high optical depths of the accretion flows. The strong influence of B-fields is illustrated in the results of Peters et al. (2011), who simulated a $100\ M_\odot$ core with a $10\ \mu$G field that suffered extensive fragmentation, and those of Seifried *et al.* (2011) and Myers *et al.* (2013), who included $\sim$mG B-fields and found very limited fragmentation.

The analytic MT03 Turbulent Core model of collapse of massive cores in the limit of no fragmentation has been coupled to semi-analytic models of accretion disk structure (Zhang & Tan 2011), disk-wind protostellar outflows (Zhang *et al.* 2013b) and a 1D numerical protostellar evolution code (ZTH14). These models have been used as inputs for continuum radiative transfer simulations to predict multiwavelength images and spectral energy distributions (SEDs). Lower-mass cores were studied by Zhang & Tan (2015). Effects of varying the main environmental variable, $\Sigma_{\rm cl}$, have been explored. Density and temperature structures around example low- & high-mass protostars are shown in Fig. 1.

Model predictions include the strong rise of fluxes at $\lambda < 100\ \mu$m as protostars grow in mass and luminosity and open up outflow cavities. Intensity profiles along the outflow axis are typically very asymmetric at $\lambda \lesssim 40\ \mu$m, with the near-facing outflow cavity being much brighter than the far-facing one. This asymmetry decreases for wavelengths $\gtrsim 70\,\mu$m at which the core envelope is becoming optically thin. Protostellar cores in high $\Sigma_{\rm cl}$, i.e., high pressure, environments are smaller, denser, have higher accretion rates and luminosities, and are thus warmer, than cores of the same mass in low $\Sigma_{\rm cl}$ regions. High-mass envelopes and disks have similar overall morphologies and densities as their lower-mass counterparts, but, given the more luminous central protostar, are warmer.

In general, the advantages of the ZTH14 models for massive star formation are that they include: detailed, self-consistent physical models for the core structure and protostellar evolution; gas, as well as dust, opacities; and core properties that respond to clump environment, so typical massive cores are much smaller and denser and collapse with higher accretion rates compared to those of other models (e.g., Indebetouw *et al.* 2006;

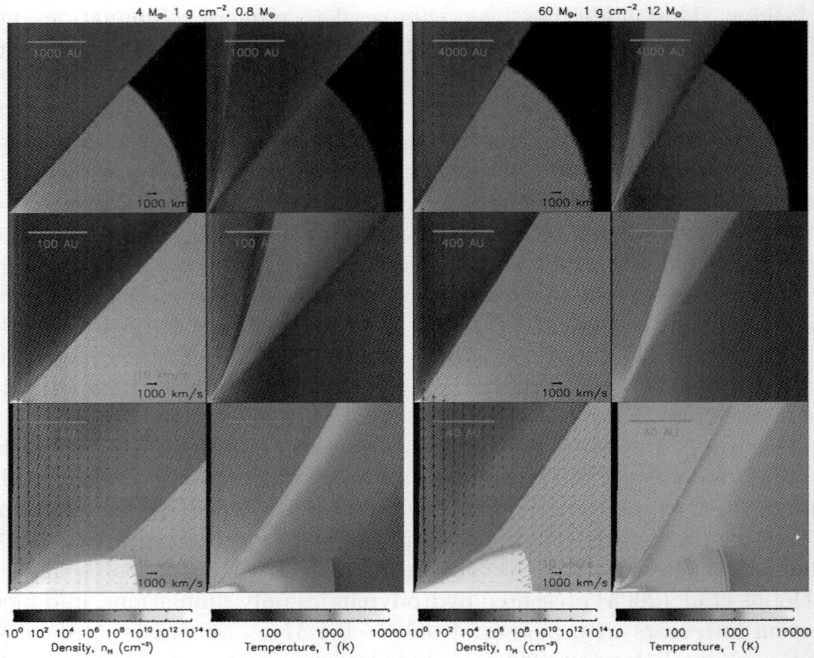

Figure 1. Comparison of low- (Zhang & Tan 2015) and high-mass (ZTH14) protostars forming via Turbulent Core Accretion. Left 6 panels show slices through a 4 $M_\odot$ core in a 1 g cm^{-2} clump that has collapsed to form a protostar with 0.8 $M_\odot$ (i.e., 20% of initial core mass), which is located in the bottom left corner of each panel. Left column shows H nuclei number density, $n_{\rm H}$, and inflow/outflow velocities (note different arrow scales). Right column shows temperature. The view zooms from scale of the core (top row), to inner infall envelope (middle), to accretion disk (bottom). Right 6 panels show the equivalent figures for a 60 $M_\odot$ core in the same clump environment that has collapsed to form a 12 $M_\odot$ protostar (also 20% of initial M_c). Note linear scales in each row are $\simeq 4\times$ larger than the low-mass case. While overall morphologies and absolute densities are quite similar, temperatures are significantly greater in the high-mass case, given its much larger luminosity of $5 \times 10^4\ L_\odot$ (cf., 300 $L_\odot$ of the low-mass protostar).

Robitaille *et al.* 2006). However, they are still highly idealized in having: axisymmetric, smooth density structures (cf., Indebetouw *et al.* 2006); a simplified treatment of the interface between outflow cavity and infall envelope; and do not treat binary protostars.

An important difference of massive protostars compared to low-mass ones is that they can become significant sources of FUV and EUV radiation, since their accretion timescales are generally longer than their Kelvin-Helmholz contraction timescales. EUV photons should first lead to ionization of the protostellar outflows, i.e., initially in a phase of an "Outflow-Confined HII Region" (Tan & McKee 2003; Tanaka *et al.* 2015). Here confinement is mostly in lateral directions, i.e., the disk and infall envelope are shielded. However, ionization soon becomes unconfined in directions along the outflow, which thus leads to a very elongated photoionized region that may appear as a cm continuum "radio jet." Later, the ionizing flux becomes strong enough to begin to have an effect on the accretion flow, e.g., by photoevaporation, although the presence of dust can significantly reduce the effectiveness of this feedback (Tanaka *et al.*, in prep.), in contrast to the Population III case (McKee & Tan 2008; Hosokawa et al. 2011). If massive protostars do not have strong, magnetocentrifugally-driven protostellar outflows, then their HII region structures may be "Accretion-Confined" (Keto 2007; Peters et al. 2010).

3.2. *Competitive Accretion and Protostellar Mergers*

Competitive Accretion (Bonnell *et al.* 2001; Wang *et al.* 2010; Smith et al. 2011) involves protostars first forming from low-mass cores, with masses typically set by the Bonnor-Ebert mass, i.e., $\ll 1\,M_\odot$ in $\Sigma_{\rm cl} \sim 1\,{\rm g\,cm^{-2}}$ environments. The protostars then continue to accrete gas from the clump and for intermediate-mass and high-mass stars this clump-fed, Bondi-Hoyle mode of accretion plays the dominant role in setting the final mass.

A significant difference between Competitive Accretion and Turbulent Core Accretion is the former's much lower accretion rates to massive protostars if the clump is not undergoing rapid global collapse, i.e., if the star formation efficiency per free-fall time, $\epsilon_{\rm ff}$, of the protocluster is $\lesssim 0.1$. This is seen in the simulation of Wang *et al.* (2010), which includes outflow feedback that helps stabilize the clump. The most massive star in the simulation reaches $46.4\ M_\odot$ in 1 Myr, with an average accretion rate of only $4.6 \times 10^{-5}\,M_\odot\,{\rm yr}^{-1}$. In absence of this feedback and in absence of B-fields that can help support massive PSCs, rapid global collapse and fragmentation of the clump is seen, leading to faster rates of Competitive Accretion (Bonnell *et al.* 2001; Smith *et al.* 2011).

For massive protostars forming via Competitive Accretion, the environment is required to be that near the center of a dense protocluster, which is also crowded with lower-mass protostars. Dynamical harassment of the protostar and its accretion envelope and disk is much more severe. This will limit the size of disks and lead to more rapid changes in their orientations, which will also be reflected in the orientations of any associated protostellar outflows. Simple bipolar outflows that maintain a fairly constant orientation would not be expected in these models, also because of presence of multiple overlapping, randomly-aligned outflows from the surrounding lower-mass protostars. Such morphologies are also expected to some extent for massive protostars forming via Turbulent Core Accretion due to turbulence in the core, small-N multiple formation in the core, and neighboring protostars, so it is a question of the degree to which relatively ordered morphologies are preserved in the two scenarios, which still needs to be quantified in simulations.

Protostellar mergers (Bonnell *et al.* 1998), including via hardening of binaries (Bonnell & Bate 2005) have been proposed as a massive star formation mechanism that operates in dense protocluster centers. For collisional growth to be rapid compared to cluster formation or massive stellar evolution timescales of $\sim$ 1–10 Myr, requires extreme stellar densities $\gtrsim 10^8\ {\rm pc^{-3}}$, equivalent to $n_{\rm H} \gtrsim 3 \times 10^9\ {\rm cm^{-3}}$ (e.g., Moeckel & Clarke 2011). Efficient growth by mergers leads to runaway growth of one or two extreme objects, rather than a smoothly filled upper IMF. For these reasons, mergers are generally considered to be unimportant in typical massive star-forming environments, although a merger has been invoked to explain activity of the Orion KL protostar (Bally & Zinnecker 2005).

4. Observational Constraints

Here we discuss just a few recent examples of observations that test massive star formation theories and compare to low-mass cases (see T14 for a more extensive review).

4.1. *Pre-Stellar Cores*

One of the best studied "low-mass" PSCs is L1544 (Caselli & Ceccarelli 2012). However, this core actually has $\sim 8\,M_\odot$, and its slow, subsonic infall ($\lesssim$10% of free-fall), suggests B-fields play a significant role in its dynamics (Keto *et al.* 2015).

Searches for more massive PSCs in higher $\Sigma_{\rm cl}$ environments have focussed on IRDCs. Tan *et al.* (2013), following up a sub-sample of the BT12 MIR extinction map peaks with *ALMA*, identified six cores via $N_2D^+(3-2)$ emission, with the most massive being C1-S with $\sim 60\,M_\odot$. The observed line-widths are on average 80% of that predicted by the fiducial Turbulent Core model that assumes an Alfvén Mach number, $m_A = 1$. However, for C1-S the observed velocity dispersion is only 40% of this level. Thus virial equilibrium would require a stronger B-field of ~ 1 mG (i.e., $m_A = 0.3$). This field strength, given the core density of $n_{\rm H,c} \sim 6 \times 10^5\ {\rm cm}^{-3}$, also yields a magnetic critical mass close to C1-S's observed mass, which could help explain the core's limited fragmentation. Such B-field strengths are similar to those predicted using the empirical relation $B \simeq n_{\rm H}^{0.65}\ \mu$G (for $n_{\rm H} > 300\ {\rm cm}^{-3}$) (Crutcher *et al.* 2010) and are also similar to values inferred in some massive protostars (§4.2). The deuteration fraction, $D_{\rm frac}^{\rm N2H+} \equiv [{\rm N_2D^+}]/[{\rm N_2H^+}]$, has been measured in C1-S to be 0.2–0.7 (Kong *et al.* 2015b), several orders of magnitude greater than the cosmic [D]/[H]. By comparison with chemodynamical models, Kong *et al.* (2015b) conclude it is likely that C1-S is contracting at a relatively slow rate, ∼1/10th of free-fall, so as to have had enough time to reach this level of deuteration. This result needs confirmation by direct measurement of infall speeds, but this is challenging given the core is embedded in a much more massive and kinematically complex clump environment.

C1-S provides physical and chemical evidence for the existence of a massive, monolithic, centrally-concentrated, potentially-virialized pre-stellar or early-stage core. G11.92-0.61-MM2 is another candidate (Cyganowski *et al.* 2014), but based only on mm continuum emission & absence of line emission. The inferred mass of $\sim 30 M_\odot$ within ~ 1000 AU, i.e., $n_{\rm H} \gtrsim 10^9\ {\rm cm}^{-3}$, yet without lines being seen from a lower density envelope, is surprising.

4.2. *Protostellar Cores*

Nearby low-mass protostellar cores are being studied in unprecedented detail with *ALMA*, including detection of rotating infall envelopes, bipolar outflows, and potentially Keplerian disks on scales $\lesssim 100$ AU (e.g., Codella *et al.* 2014). For the typically much more distant massive protostars, there are clear examples of collimated outflows (e.g., Beuther *et al.* 2002; Duarte-Cabral et al. 2013) and "rotating toroids" (e.g., Beltrán *et al.* 2011; Sánchez-Monge *et al.* 2013). Infall has been detected in a number of sources (e.g., Wyrowski *et al.* 2012). However, unambiguous detection of disks remains challenging, which is not unexpected if diameters are $\lesssim 1000$ AU, i.e., $\lesssim 0.5$" at 2 kpc. Girart et al. (2009) & Q. Zhang *et al.* (2014) have inferred "hour-glass" morphologies of B-fields around some massive protostars, with field strengths ~ 1 mG and concluded they play a dynamically important role in massive star formation. Ionized, collimated outflows traced as radio continuum "jets" have been seen in some massive protostars (e.g., Gibb *et al.* 2003; Guzmán *et al.* 2014), although the relative importance of shock- versus photo-ionization remains to be established. The example of G35.20-0.74N contains all of the above elements and detailed radiative transfer models based on Turbulent Core Accretion have been fit to both its SED and resolved multiwavelength images that probe outflow axis intensity asymmetries to derive a protostellar mass of $m_* \sim 20$–$34\ M_\odot$ embedded in a core with $M_c \simeq 240\ M_\odot$ in a clump with $\Sigma_{\rm cl} \simeq 0.4$–$1\ {\rm g\ cm}^{-2}$ (Zhang *et al.* 2013b).

However, some massive protostellar cores appear much more disordered, with the most famous example being the Orion KL "hot core," which is thought to be powered by a massive protostar detected as radio "source I." There is an apparently "explosive" outflow from the region, which has been interpreted as a signature of a protostellar merger (Bally & Zinnecker 2005). From X-ray images, Rivilla et al. (2013) inferred a protostellar density of $\sim 10^6\ {\rm pc}^{-3}$ in this region, still $\sim 100\times$ less than that required for efficient mass growth by mergers (§3.2). However, a single merger event involving a large disk-aided capture cross-section of ~ 100 AU may be expected to occur in $\lesssim 10^5$ yr. An alternative interpretation of Orion KL proposes that the explosive outflow is caused by tidal harassment of source I's accretion disk, leading to enhanced accretion and thus outflow, by a passing runaway star, the BN Object (Chatterjee & Tan 2012). In this case, the interaction can be regarded as a moderate perturbation of the Core Accretion model. Future proper motion studies of source I and BN will help distinguish these scenarios.

4.3. *Protocluster Clumps*

Fragmentation of clumps and filaments is under active study (e.g., Beuther *et al.* 2013; Zhang *et al.* 2015), but interpretation of the detected structures is hampered by lack of observational probes of their B-fields. As reviewed by T14, infall times, $t_{\rm infall} \equiv M_{\rm cl}/\dot{M}_{\rm infall}$, relative to local free-fall time, $t_{\rm ff}$, have been measured in several clumps, including IRDCs, with typical observed ratios of ~ 10, indicating slow, quasi equilibrium collapse. This suggests a dynamically important role for B-fields, consistent with recent observations of two IRDCs by Pillai *et al.* (2015), and/or stabilization of collapse by protostellar outflow feedback, as modeled by Nakamura & Li (2014). As discussed in §3, this question of the dynamical state of the star-forming clump environment, including nature of injected turbulence and distribution of B-field strengths, is crucial for its effect on fragmentation to the CMF and thus the formation mechanism of intermediate- and high-mass stars.

The quasi equilibrium cluster formation model (Tan *et al.* 2006) predicts age spreads of stars $\sim 10 t_{\rm ff}$ in systems that achieve final star formation efficiencies of $\sim 50\%$, consistent with Da Rio *et al.*'s (2014) analysis of the Orion Nebula Cluster (ONC). It also implies there is significant time during star cluster formation for additional gas supply via infall from the clump surroundings and for dynamical evolution of the stellar population leading to mass segregation, so present-day locations of massive stars, such as $\theta^1 C$ in the ONC, are not likely to have been where they formed. Improved observations of stellar kinematics (e.g., Foster *et al.* 2015; Cottaar *et al.* 2015) have the potential to test these theories.

References

Bally, J. & Zinnecker, H. 2005, *AJ*, 129, 2281
Bate, M. R. 2012, *MNRAS*, 419, 3115
Bertoldi, F. & McKee, C. F. 1992, *ApJ*, 395, 140
Beltrán, M. T. *et al.* 2011, *A&A*, 525, 151
Beuther, H., Schilke P. *et al.* 2002, *A&A*, 383, 892
Beuther, H., Linz, H., Tackenberg, J. *et al.* 2013, *A&A*, 553, 115
Bonnell, I. A. & Bate, M. R. 2005, *MNRAS*, 362, 915
Bonnell, I. A., Bate, M. R., & Zinnecker, H. 1998, *MNRAS*, 298, 93
Bonnell, I. A., Clarke, C. J., Bate, M. R., & Pringle, J. E. 2001, *MNRAS*, 324, 573
Bromm V., 2013, *Rep. Prog. Phys.*, 76, 11
Butler M. J. & Tan J. C., 2012, *ApJ*, 754, 5 [BT12]
Caselli, P. & Ceccarelli, C. 2012, *A&AR*, 20, 56
Chatterjee, S. & Tan, J. C. 2012, *ApJ*, 754, 152
Codella, C., Cabrit, S., Gueth, F. *et al.* 2014, *A&A*, 568, L5
Collins, D., Padoan, P., Norman, M. L., & Xu, H. 2011, *ApJ*, 731, 59
Cottaar, M., Covey, K. R., Foster, J. B. *et al.* 2015, *ApJ*, 807, 1
Crutcher, R. M., Wandelt, B., Heiles, C., *et al.* 2010, *ApJ*, 725, 466
Cyganowski, C. J., Brogan, C. L., Hunter, T. R. *et al.* *ApJL*, 796, 2
Da Rio, N., Tan, J. C., & Jaehnig, K. 2014, *ApJ*, 795, 55
de Wit W. J., Testi L. *et al.* 2005, *A&A*, 437, 247
Dobbs, C. L. *et al.* 2005, *MNRAS*, 360, 2
Downes, D. & Solomon, P. M. 1998, *ApJ*, 507, 615
Duarte-Cabral, A. *et al.* 2013, *A&A*, 558, 125
Elmegreen, B. G. 2000, *ApJ*, 530, 277
Foster, J. B., Cottaar, M., Covey, K. R. *et al.* 2015, *ApJ*, 799, 136
Gibb, A. G. *et al.* 2003, *MNRAS*, 339, 198

Girart, J. M., Beltrán, M. T., Zhang, Q. *et al.* 2009, *Science*, 324, 1408
Gutermuth, R. A., Megeath, S. T., Myers, P. C., *et al.* 2009, *ApJS*, 184, 18
Guzmán, A. E., Garay, G., Rodríguez, L. F., *et al.* 2014, *ApJ*, 796, 117
Hartmann, L. & Burkert, A. 2007, *ApJ*, 654, 988
Hennebelle, P. & Chabrier, G. 2008, *ApJ*, 684, 395
Hennebelle, P., Commerçon, B., Joos, M. *et al.* 2011, *A&A*, 528, 72
Hosokawa, T., Omukai, K., Yoshida, N., & Yorke, H. W. 2011, *Science*, 334, 1250
Indebetouw, R., Whitney, B. A., Johnson, K. E., & Wood, K. 2006, *ApJ*, 636, 362
Inutsuka, S. 2012, *PTEP*, 2012, 01A307
Keto, E. 2007, *ApJ*, 666, 976
Keto, E., Caselli, P., & Rawlings, J. 2015, *ApJ*, 446, 3731
Kong, S., Caselli, P., Tan, J. C. *et al.* 2015a, *ApJ*, 804, 98
Kong, S., Tan, J. C., Caselli, P. *et al.* 2015b, *ApJ*, submitted (arXiv:1509.08684)
Krumholz, M. R. & McKee, C. F. 2008, *Nature*, 451, 1082
Krumholz, M. R., Klein, R. I. *et al.* 2007, *ApJ*, 656, 959
Kunz, M. W. & Mouschovias, T. Ch. 2009, *MNRAS*, 399, L94
Li, Z.-Y., Banerjee, R. *et al.* 2014, *Protostars & Planets VI*, eds. Beuther *et al.*, p173
Lu, J. R., Ghez, A. M., Morris, M. R. *et al.* 2014, *IAUS*, 303, 211
McKee C. F. & Tan J. C., 2002, *Nature*, 416, 59
McKee C. F. & Tan J. C., 2003, *ApJ*, 585, 850 [MT03]
McKee C. F. & Tan J. C., 2008, *ApJ*, 681, 771
Myers A. T., McKee C. F. *et al.* 2013, *ApJ*, 766, 97
Moeckel, N. & Clarke, C. J. 2011, *MNRAS*, 410, 2799
Nakamura, F. & Li, Z.-Y. 2014, *ApJ*, 783, 115
Padoan, P. & Nordlund, A. 2002, *ApJ*, 576, 870
Padoan, P., Nordlund, A., Kritsuk, A. G. *et al.* 2007, *ApJ*, 661, 972
Peters, T., Banerjee, R., Klessen, R. S., & Mac Low, M.-M. 2011, *ApJ*, 729, 72
Peters, T., Mac-Low, M.-M., Banerjee, R. *et al.* 2010, *ApJ*, 719, 831
Pillai, T., Kauffmann, J., Tan, J. C. *et al.* 2015, *ApJ*, 799, 74
Rivilla, V., Martín-Pintado, J., Jiménez-Serra, I., & Rodríguez-Franco, A. 2013, *A&A*, 554, 48
Robitaille, T. P., Whitney, B. A., Indebetouw, R. *et al.* 2006, *ApJS*, 167, 256
Sánchez-Monge, Á., Cesaroni, R., Beltrán, M. T. *et al.* 2013, *A&A*, L10
Seifried, D., Banerjee, R., Klessen R. S. *et al.* 2011, *MNRAS*, 417, 1054
Shu, F. H., Adams, F. C., & Lizano, S. 1987, *ARA&A*, 25, 23
Smith, R. J., Glover, S. C. O., Bonnell, I. A. *et al.* 2011, *MNRAS*, 411, 1354
Sobral, D., Matthee, J., Darvish, B. *et al.* 2015, *ApJ*, 808, 139
Tan, J. C., Beltrán, M. *et al.* 2014, *Protostars & Planets VI*, eds. Beuther *et al.*, p149 [T14]
Tan, J. C., Kong, S., Caselli, P. *et al.* 2013, *ApJ*, 779, 96
Tan, J. C., Krumholz, M. R., & McKee, C. F. 2006, *ApJL*, 641, L121
Tan, J. C. & McKee, C. F. 2003, astro-ph/0309139.
Tanaka, K. E. I., Tan, J. C., & Zhang, Y. 2015, *ApJ*, submitted (arXiv:1509.06754)
Vogelsberger, M., Genel, S., Springel, V. *et al.* 2014, *MNRAS*, 444, 1518
Walch, S., Girichidis, P., Naab, T. *et al.* 2015, *MNRAS*, 454, 238
Wang, P., Li, Z.-Y., Abel, T., & Nakamura, F. 2010, *ApJ*, 709, 27
Whitmore, B. C., Chandra, R., Bowers, A. S. *et al.* 2014, *AJ*, 147, 78
Wilson, C. D., Rangwala, N., Glenn, J. *et al.* 2014, *ApJ*, 789, 36
Wyrowski, F., Güsten, R., Menten, K. M. *et al.* 2012, *A&A*, 542, L15
Zhang, Q., Qiu, K., Girart, J. M. *et al.* 2014, *ApJ*, 792, 116
Zhang, Q., Wang, K., Lu, X., & Jiménez-Serra, I. 2015, *ApJ*, 804, 141
Zhang, Y. & Tan, J. C. 2011, *ApJ*, 733, 55
Zhang, Y. & Tan, J. C. 2015, *ApJL*, 802, L15
Zhang, Y., Tan, J. C., & Hosokawa, T. 2014, *ApJ*, 788, 166 [ZTH14]
Zhang, Y., Tan, J. C., De Buizer *et al.* 2013b, *ApJ*, 767, 58
Zhang, Y., Tan, J. C. & McKee 2013a, *ApJ*, 766, 86

From Interstellar Clouds to Star-Forming Galaxies: Universal Processes?
Proceedings IAU Symposium No. 315, 2015
P. Jablonka, Ph. André & F. van der Tak, eds.

doi:10.1017/S1743921316007444

Central Molecular Zone of the Milky Way: Star Formation in an extreme Environment†

Jens Kauffmann

Max–Planck–Institut für Radioastronomie,
Auf dem Hügel 69, D–53121 Bonn, Germany
email: jens.kauffmann@gmail.com

Abstract. The Central Molecular Zone (CMZ; inner ∼ 100 pc) hosts some of the most dense and massive molecular clouds of the Milky Way. These clouds might serve as local templates for dense clouds seen in nearby starburst galaxies or in the early universe. The clouds have a striking feature: they form stars at a very slow pace, considering their mass and high average density. Here we use interferometer data from ALMA and the SMA to show that this slow star formation is a consequence of the cloud density structure: CMZ clouds have a very flat density structure. They might, for example, exceed the average density of the Orion A molecular cloud by an order of magnitude on spatial scales ∼ 5 pc, but CMZ "cores" of ∼ 0.1 pc radius have masses and densities lower than what is found in the Orion KL region. This absence of highest–density gas probably explains the suppression of star formation. The clouds are relatively turbulent, and ALMA observations of H_2CO and SiO indicate that the turbulence is induced by high–velocity shocks. We speculate that these shocks might prevent the formation of high–mass cores. It has been argued that the state of CMZ clouds depends on their position along the orbit around Sgr A*. Our incomplete data indicate no evolution in the density structure, and only a modest evolution in star formation activity per unit mass.

Keywords. stars: formation, ISM: clouds, Galaxy: center

1. Introduction: Dense Clouds with little Star Formation

The Central Molecular Zone (CMZ) — i.e. the inner ∼ 100 pc of our Galaxy — is a star–forming (SF) environment with very extreme physical properties. The CMZ produces 5 to 10% of the Milky Way's infrared and Lyman continuum luminosity, and ∼ 10% of its total molecular gas reside here. The molecular clouds have unusually high average densities $\gtrsim 10^4$ cm^{-3}, they are closely packed, and they are subject to an average pressure (from X–ray data) of order 10^6 K cm^{-3} (Morris & Serabyn 1996, for all of the above). Using dust emission maps from Herschel, Molinari *et al.* (2011) recently showed that most of the gas in the CMZ resides on an orbit of about 100 pc radius that has a complex geometry (Kruijssen *et al.* 2014).

It is important to study SF under these extreme conditions. First, here we have a unique chance to explore an extreme point in the SF parameter space. Comprehensive models should describe SF both in the CMZ as well as the solar neighborhood. Second, the CMZ clouds might serve as local templates for the types of clouds seen in nearby starbursts and the early universe. For example, the CMZ clouds have high average H_2 column densities $\sim 10^{23}$ cm^{-2}. These resemble values found in nearby starburst galaxies like the Antennae and NGC 253.

One particularly striking feature of CMZ molecular clouds is that star formation is suppressed. For example, Guesten & Downes (1983) found that very few H_2O masers

† with key contributions from T. Pillai, P. Goldsmith, Q. Zhang, X. Lu, A. Ginsburg, K. Immer, and A. Guzman

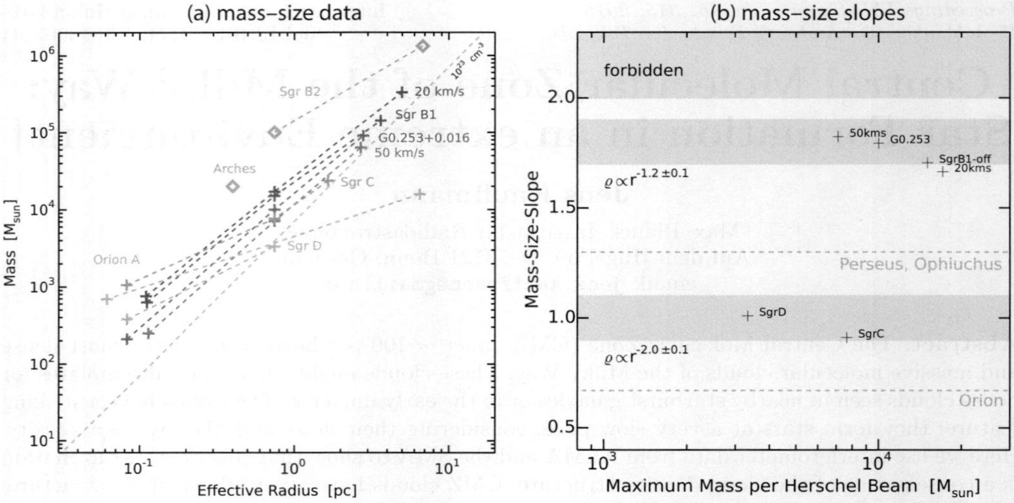

Figure 1. Mass–size analysis for CMZ clouds. *(A Color version is available online.)*

reside in the CMZ, relative to the high mass of dense gas. Longmore *et al.* (2012) provide an updated discussion of this trend: averaged over the entire CMZ star formation is suppressed by at least an order of magnitude, when compared to the solar neighborhood. Lis *et al.* (1994) and Kauffmann *et al.* (2013) demonstrate that SF is also suppressed on the scale of individual clouds.

In Kauffmann *et al.* (2013) we demonstrated that in one CMZ cloud, G0.253+0.016, the absence of significant SF can be explained by the absence of high–mass cores of ~ 0.1 pc size that could form stars efficiently. Here we show that this is a general feature of CMZ molecular clouds (Sec. 2) that is potentially induced by high–velocity shocks (Sec. 3). We explore the SF activity within the CMZ (Sec. 4) and find little evolution of cloud density structure and SF activity along the orbit of CMZ clouds (Sec. 5).

2. Cloud Density Structure

We conduct a mass–size analysis for the CMZ clouds to explore the density structure of our target clouds. The cloud structure on spatial scales $\gtrsim 1$ pc is investigated using Herschel–based dust emission maps. Smaller spatial scales are probed using dust continuum emission maps from the SMA that were obtained at about 280 GHz frequency. Specifically we observed the following targets: Sgr C, 20 $\mathrm{km\,s^{-1}}$ cloud, 50 $\mathrm{km\,s^{-1}}$ cloud, G0.253+0.016, Sgr B1–off, and Sgr D. The maps are used to essentially obtain masses within apertures of varying size which we define via contours of constant column density (see Kauffmann *et al.* 2010a for details). For reference we characterize the Orion A molecular cloud in the same fashion.

The results of these measurements are shown in Fig. 1(a). On spatial scales ~ 5 pc, the masses — and therefore densities — of CMZ molecular clouds exceed the one of Orion A by an order of magnitude and more. However, CMZ cloud fragments of size ~ 0.1 pc are typically less massive than the column density peak associated with Orion KL. This means that, compared to Orion, CMZ molecular clouds are less efficient in concentrating a large fraction of the total mass into structures of small spatial scale.

Figure 1(b) provides an alternative view at the data. Here we present the mass–size slopes, $b = \mathrm{d}\ln(m)/\mathrm{d}\ln(r)$, as a function of the maximum mass per Herschel beam.

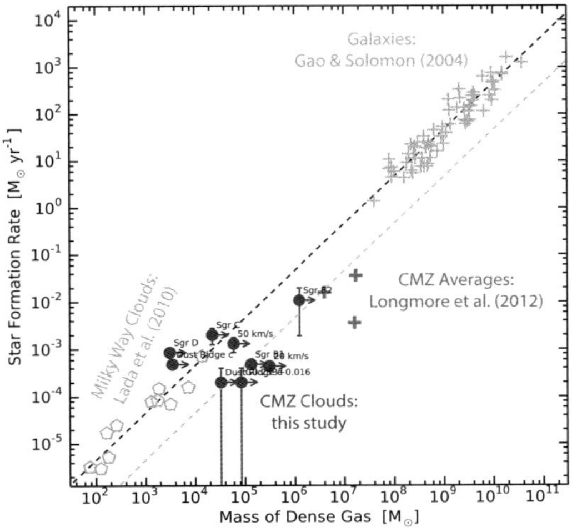

Figure 2. SF rate of molecular clouds as a function of the mass of dense gas. Note that many CMZ molecular clouds fall below the line supposedly describing regular Milky Way clouds and galaxies. *(A color version of this figure is available online.)*

In Kauffmann *et al.* (2010b) we show that $\varrho \propto r^{3-b}$ holds for singular power–law density profiles. The density laws indicated in Fig. 1(b) reveal that CMZ molecular clouds have unusually flat density profiles, compared to SF regions in the solar neighborhood (Kauffmann *et al.* 2010b).

In Kauffmann *et al.* (2013) we already revealed such a density profile in the cloud G0.253+0.016. Now we can assume that these flat density gradients are a general feature of many but not all CMZ molecular clouds. This absence of dense cores that could form stars likely explains the suppression of SF.

3. CMZ Shocks revealed by ALMA

In Kauffmann *et al.* (2013) we speculated that the formation of high–mass cores might be prevented by high–velocity shocks. We use ALMA observations of G0.253+0.016 to explore this picture near 220 GHz frequency. Emission from SiO is usually considered to be a tracer of regions where gas streams collide at relative velocities $\gtrsim 20\ \mathrm{km\,s^{-1}}$. Our data reveal thin SiO arcs in G0.253+0.016, which we take to be shock fronts pervading the cloud. Further, ratios between H_2CO lines indicate that the gas temperature varies between about 40 K and $\geqslant$ 200 K within G0.253+0.016. These temperature variations *within* a given cloud are most easily explained by shock heating which will induce a spatially dependent heating term.

We take this as further evidence that shocks shape CMZ molecular clouds. Numerical simulations must now show to what extent such shocks might help to suppress the fragmentation of clouds into high–mass cores. Also, it is not clear how the strong and numerous shocks in the CMZ are induced.

4. Star Formation Activity

Lada *et al.* (2010) suggest that about $5\,M_{\odot}$ of gas residing at visual extinctions $A_V \geqslant$ 7 mag are needed to form one young stellar object. This is an important observation since this trend might also explain the supposed extragalactic correlation between dense gas and SF proposed by Gao & Solomon (2004), as indicated in Fig. 2. We have compiled SF rates to explore this trend in the CMZ. Based on SF rates from radio–continuum emission revealing HII regions, and dense gas masses derived from Herschel dust emission data,

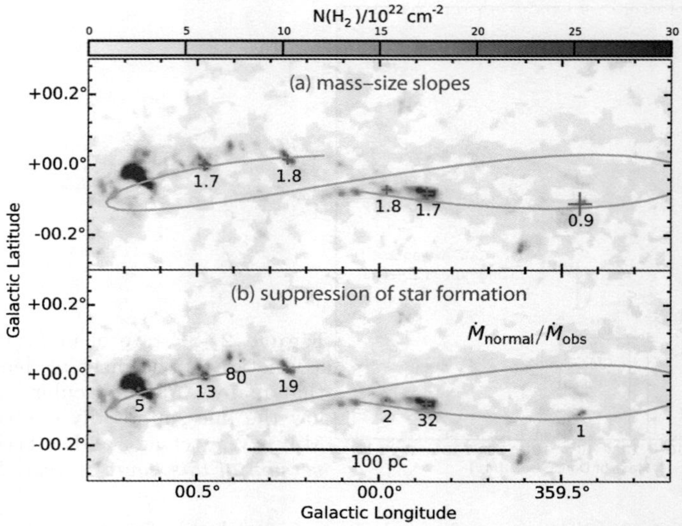

Figure 3. The *orange line* indicates the orbit proposed by (Kruijssen *et al.* 2014). No or only modest evolution is evident in density structure and SF activity. *(A color version of this figure is available online.)*

many CMZ molecular clouds have lower SF rates per unit dense gas than observed by Lada *et al.* (2010) and Gao & Solomon (2004), as shown in Fig. 2.

5. Evolution along Orbit

Longmore *et al.* (2013) suggest that clouds evolve along the orbit identified by Molinari *et al.* (2011) and Kruijssen *et al.* (2014), for example due to compression events triggered by passages close to Sgr A*. The data collected here can be used to evaluate that picture.

Figure 3(a) presents the mass–size slopes derived from our SMA maps and Herschel data. No evolution is seen in the data. However, we caution that the current data set only covers a fraction of the CMZ. Figure 3(b) indicates the factor by which the observed SF rate in a CMZ cloud, $\dot{M}_{\rm obs}$ is below the SF rate expected for Milky Way clouds based on the Lada *et al.* (2010) analysis, $\dot{M}_{\rm normal}$. A modest and non–monotonous increase in SF activity might be evident at positive longitudes.

References

Gao, Y. & Solomon, P. M. 2004, *ApJ*, 606, 271
Guesten, R. & Downes, D. 1983, *A&A*, 117, 343
Kauffmann, J., Pillai, T., Shetty, R., Myers, P. C., & Goodman, A. A. 2010a, *ApJ*, 712, 1137
—. 2010b, *ApJ*, 716, 433
Kauffmann, J., Pillai, T., & Zhang, Q. 2013, *ApJ*, 765, L35
Kruijssen, J. M. D., Dale, J. E., & Longmore, S. N. 2014, *MNRAS*, 447, 1059
Lada, C. J., Lombardi, M., & Alves, J. a. F. 2010, *ApJ*, 724, 687
Lis, D. C., Menten, K. M., Serabyn, E., & Zylka, R. 1994, *ApJ*, 423, L39
Longmore, S. N., Bally, J., Testi, L., *et al.* 2012, *MNRAS*, 429, 987
Longmore, S. N., Kruijssen, J. M. D., Bally, J., *et al.* 2013, *MNRAS*, 433, L15
Molinari, S., Bally, J., Noriega-Crespo, A., *et al.* 2011, *ApJ*, 735, L33
Morris, M. & Serabyn, E. 1996, *ARAA*, 34, 645

From Interstellar Clouds to Star-Forming Galaxies: Universal Processes?
Proceedings IAU Symposium No. 315, 2015
P. Jablonka, Ph. André & F. van der Tak, eds.

doi:10.1017/S1743921316007468

Star formation rates, laws, thresholds

Stefanie Walch

I. Physics Institute, University of Cologne
Zülpicher Str. 77, 50937 Cologne
email: walch@ph1.uni-koeln.de

Abstract. On scales of $\gtrsim$ several hundred parsec, there are relatively well defined observed relations between the star formation rate surface density of a galaxy and its gas content. I present a short summary of the so-called 'star formation laws' and discuss the most established models to explain their origin. Nevertheless, none of the models can explain the offset of these relations, which requires a low star formation efficiency within the total and/or molecular gas. Stellar feedback is proposed to limit the star formation efficiency locally. I discuss the available feedback mechanisms and show recent simulation results on the evolution of the multi-phase interstellar medium under the influence of gravity, cooling, feedback-driven outflows and heating.

Keywords. stars: formation, ISM: supernova remnants, ISM: general, ISM: kinematics and dynamics, ISM: structure, turbulence, MHD, galaxies: ISM

1. The galactic 'star formation law'

The most established observed 'star formation (SF) law' in galaxies is the Kennicutt-Schmidt relation (Schmidt, 1959; Kennicutt, 1989; Kennicutt, 1998), which relates the average star formation rate (SFR) surface density ($\Sigma_{\rm SFR}$ in units of $[{\rm M}_\odot\,{\rm yr}^{-1}\,{\rm kpc}^{-2}]$) to the total gas surface density ($\Sigma_{\rm Gas}$ in units of $[{\rm M}_\odot\,{\rm kpc}^{-2}]$) of a galaxy using a power law $\Sigma_{\rm SFR} \propto \Sigma_{\rm Gas}^{n_{\rm KS}}$. Schmidt (1959) originally suggested $n_{\rm KS} \approx 2$, while Kennicutt (1998) determines $n_{\rm KS} \approx 1.4$. A power law index of $n_{\rm KS} = 1.5$ can be readily explained with a dynamical model of star formation on galactic scales. This can be formulated in two ways: intrinsically three-dimensional – per free-fall time of the denser, star-forming molecular clouds, $\tau_{\rm ff} \propto (G\rho_{\rm Gas})^{-1/2}$; or intrinsically two-dimensional – per galaxy dynamical time, $\tau_{\rm dyn}$, which can be calculated from the Toomre criterion (Elmegreen-Silk relation; Elmegreen, 1997; Silk, 1997; see e.g. Kennicutt, 1998, and Genzel *et al.*, 2015 for an application). The first argument goes as follows (e.g. Krumholz *et al.*, 2005, Krumholz & Tan, 2007, Leroy *et al.*, 2008):

$$\Sigma_{\rm SFR} = \epsilon_{\rm SF}\,\Sigma_{\rm Gas} \times {\rm dynamical\,rate}, \tag{1.1}$$

where the dynamical rate is $1/\tau_{\rm ff}$ in this case, and $\epsilon_{\rm SF}$ is the star formation efficiency. Assuming a constant gas scale height we have $\Sigma_{\rm Gas} \propto \rho_{\rm Gas}$ and therefore $n_{\rm KS} = 1.5$. These global dependencies have been verified in numerous numerical studies that mostly include simple hydrodynamics and gas cooling (e.g. Tasker *et al.*, 2006; Harfst *et al.*, 2006; Wada & Norman, 2007).

However, a constant scale height is not a good assumption for all star-forming galaxies (e.g. in dIrr's, where $H \propto 1/\Sigma_{\rm Gas}$ with scale height H, which implies $\rho \propto \Sigma_{\rm Gas}^2$; Elmegreen & Hunter, 2015). Therefore, the simple Toomre analysis fails in these galaxies while models based on the volumetric density (e.g. Krumholz & Matzner, 2009) work well. In addition, in radially resolved galaxies the observed slopes seem to vary with galactocentric radius, from linear in the inner discs to quadratic in the outer discs (Barnes *et al.* 2012). Elmegreen (2015) propose disc flaring as a theoretical explanation for a steeper SF relation in outer discs.

Interestingly, the gas depletion time, $\tau_{\rm dep}$, in a normal star forming galaxy is quite long, around 2 Gyr (Genzel *et al.* 2010). This implies that the star formation efficiency in this model is low with $\epsilon_{\rm SF} = \tau_{\rm ff}/\tau_{\rm dep} \sim 1-3\%$. The question remains why star formation is so inefficient. This might be caused by super-sonic turbulence that prevails within molecular clouds and shapes their sub-structure into a log-normal density PDF (Vazquez-Semadeni, 1994; Padoan *et al.*, 1997; Padoan & Nordlund, 2002). Since the free-fall time is density dependent, only the densest parts of the clouds would undergo gravitational collapse (see e.g. Krumholz *et al.*, 2005), which lowers the overall SF efficiency (for numerical tests see e.g. Federrath & Klessen,2013).

The 'molecular star formation law'.

More recently, Bigiel *et al.* (2008) and Leroy *et al.* (2008) could show that there is a tight and almost linear correlation of $\Sigma_{\rm SFR}$ with the molecular gas surface density $\Sigma_{\rm Mol.Gas}$, which is inferred from CO observations using a CO-to-H_2 conversion factor, $\alpha_{\rm CO}$. The linear relation is still under debate. Shetty *et al.* (2014a, 2014b) carry out a Bayesian analysis of the Bigiel *et al.* (2008) sample and report sub-linear slopes, while e.g. Momose *et al.* (2013) report super-linear slopes. In Fig. 1, we compile all popular scaling relations of $\Sigma_{\rm SFR} \propto \Sigma_{\rm Gas}$ and $\Sigma_{\rm SFR} \propto \Sigma_{\rm Mol.Gas}$, which have been inferred from observations.

To understand the molecular KS relation, one has to additionally consider the transition from atomic to molecular gas. Krumholz *et al.* (2008) determine the atomic-to-molecular transition for idealised, spherical gas clouds under external radiation. Sternberg *et al.* (2014) present an alternative derivation, which leads to a similar formulation for the molecular-to-atomic gas ratio as a function of metallicity and external radiation field. These models predict a metallicity-dependent transition from atomic to predominantly molecular gas at $\Sigma_{\rm Gas} \approx 10\,{\rm M}_\odot\,{\rm pc}^{-2}$ (for solar metallicity) and a steep decline in molecular gas fraction for lower $\Sigma_{\rm Gas}$. This is consistent with observations of resolved disc galaxies e.g. by Schruba *et al.* (2011).

Here it is worth pointing out that molecules, although their presence is tightly connected with star formation, seem to be a companion (forming at high density and low temperature) rather than the cause of star formation. The contribution of molecular cooling to the total cooling rate is negligible (Glover & Clark, 2012). Dust shielding seems to be the dominant physical process to allow gas cooling and star formation. Recently, Clark & Glover (2014) report that dust shielding naturally explains the apparent star formation threshold at $\Sigma_{\rm Gas} \approx 120\,{\rm M}_\odot\,{\rm pc}^{-2}$ (Evans *et al.*, 2009; Lada *et al.*, 2010; Heiderman *et al.*, 2010; Lada *et al.*, 2012). Alternatively, Burkert & Hartmann (2013) show that this surface density could just coincide with the typical density at which self-gravity becomes dominant.

In any case, the star formation efficiency, i.e. the offset of the molecular KS relation, is still low, $\epsilon_{\rm SF} \sim 1-3\%$ and thus, the formation of dense molecular clouds seems not to be the rate limiting step in the star formation process. Feedback from massive stars on molecular cloud scales is still required to moderate $\epsilon_{\rm SF}$.

Scale dependence.

With increasing spatial resolution, the notion of a constant depletion time scale gradually breaks down. Onodera *et al.* (2010) and Schruba *et al.* (2010) show that this happens at scales below ~ 200 pc. On smaller scales, where one would either center on peaks in the molecular gas distribution to measure $\Sigma_{\rm Mol.Gas}$ or on luminous star forming regions, the spatial correlation between the two seems to diverge and the KS relation, respectively the molecular KS relation, fails on molecular cloud scales (Kruijssen *et al.* 2014).

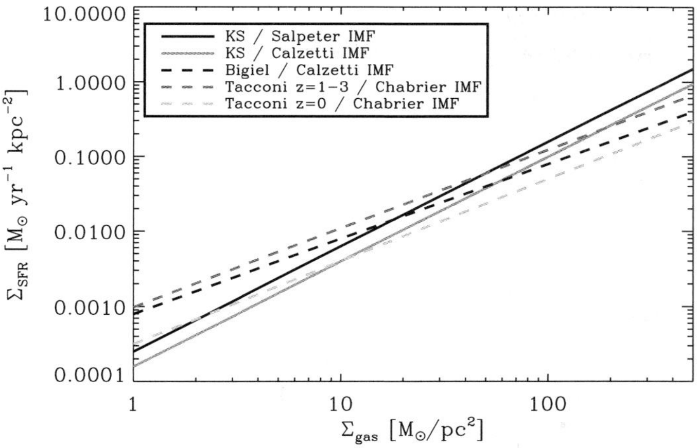

Figure 1. A compilation of proposed power-law scaling relations of the SFR surface density, $\Sigma_{\rm SFR}$, vs. total gas surface density, $\Sigma_{\rm Gas}$ (*full lines*), or molecular mass surface density, $\Sigma_{\rm Mol.Gas}$ (*dashed lines*), respectively. The scaling relations of $\Sigma_{\rm SFR}$ vs. $\Sigma_{\rm Gas}$ follow a power law index of 1.4. The offset depends on the underlying stellar IMF. Here we show two possibilities: (1)*Black line:* the standard Kennicutt-Schmidt (KS) relation (Kennicutt, 1998) as derived for a Salpeter IMF (Salpeter, 1955), and (2) *green line:* the KS relation scaled with the IMF from Calzetti *et al.* 2007, which is the standard IMF in STARBURST99 (Leitherer *et al.*,1999). The scaling relations of $\Sigma_{\rm SFR}$ vs. $\Sigma_{\rm Mol.Gas}$ are approximately linear. We show the results of (3) *blue line:* Bigiel *et al.* (2008), who use the same IMF as (2); (4) *red line:* Tacconi *et al.* (2013), their fit to high redshift galaxies, and (5) *yellow line:* Tacconi *et al.* (2013), their fit to low redshift galaxies, both of which assume a Chabrier IMF (Chabrier, 2001). We note that $\Sigma_{\rm Mol.Gas}$ should approach $\Sigma_{\rm Gas}$ for high gas surface densities, where basically all gas is in molecular form. *Figure taken from Walch et al. (2015). (A color version is available online.)*

The local star formation efficiency per free-fall time seems to be enhanced to $\sim 10\%$ on molecular cloud scales (Lada *et al.*, 2010; Murray, 2011; Evans, 2014), which implies shorter gas depletion times of 10 – 100 Myr (see Agertz *et al.*, 2015 for an overview plot).

2. Is feedback regulating star formation on galactic scales?

What is the cause of the low star formation efficiency in galaxies and in molecular clouds? Since it has been recognised that molecular clouds are probably not long-lived objects but are governed by supersonic turbulence and evolve on a dynamical time scale (see e.g. Pringle *et al.*, 2001), a stabilisation of the clouds against gravitational collapse by e.g. magnetic fields is most likely not a viable explanation for the low observed star formation efficiencies. In addition, turbulence decays on a crossing time (Stone *et al.* , 1998; MacLow *et al.* , 1998) and needs to be continuously replenished. Stellar feedback in its different forms, i.e. ionizing radiation and radiation pressure, stellar winds and Supernova explosions, is a viable mechanism to quench star formation in molecular clouds by unbinding gas, to drive turbulence within the interstellar medium (ISM), as well as to drive galactic fountains and outflows.

In 2010, Ostriker *et al.* developed a model of self-regulated star formation in galactic discs. Assuming that the intensity of ionizing radiation scales linearly with $\Sigma_{\rm SFR}$ and that the star formation rate is proportional to the mass in the self-gravitating component, they derive a model in which the thermal (based on UV heating and cooling) and vertical dynamical equilibria are simultaneously fulfilled by matching the thermal (of the warm, diffuse phase) and mid-plane gas pressures. The model predicts that $\Sigma_{\rm SFR} \propto \Sigma_{\rm Gas}\sqrt{\rho_{\rm stars+DM}}$, where $\rho_{\rm stars+DM}$ is the mid-plane density of stars plus dark matter. Star formation then

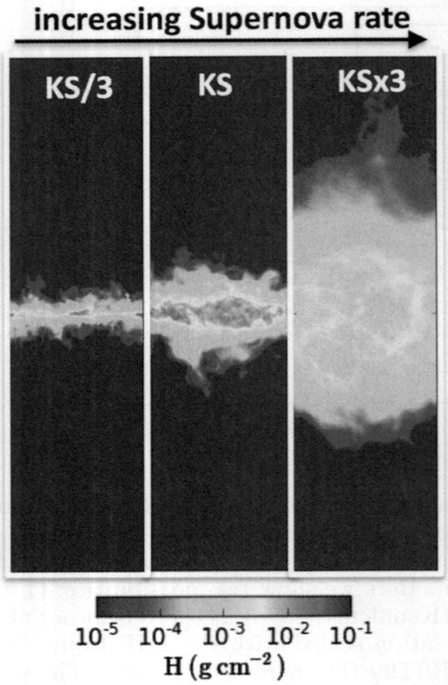

Figure 2. Column density distribution of atomic hydrogen as seen edge on for SILCC simulations including Supernova feedback which is correlated with dense gas (peak driving). *Left to right:* with increasing the Supernova rate from $\frac{1}{3}\times$ the typical KS value to $3\times$ the typical KS value the disc scale height increases. See Walch *et al.* (2015) and Girichidis *et al.* (2016) for further information. *(A color version of this figure is available online.)*

self-regulates, since high SFRs lead to an increase in mid-plane pressure, which causes a decrease in mid-plane density and thus, a decrease in SFR, leading to a decrease in mid-plane pressure etc. The model works well for many different galaxies. In particular, it is able to explain the observed star formation rates in HI dominated outer discs (Roychowdhury *et al.*, 2015) as it predicts – due to the decrease of thermal equilibrium pressure in the outer disc – that star formation can still occur in low density gas in the outer discs rather than having a sharp cut-off radius as predicted by models which tightly couple the SFR to the molecular gas fraction.

Supernova feedback.

Energetically, it seems that Supernovae are the most important feedback mechanism (e.g. Agertz *et al.*, 2013). However, considering that they only explode after a few Myr, when the first massive stars reach the end of their life times, their role for quenching or triggering star formation within individual MCs is still unclear and needs to be investigated in numerical studies. Furthermore, Supernovae cause the formation of the observed hot, volume-filling gas phase (McKee & Ostriker, 1977).

In a series of papers, Kim & Ostriker (e.g. 2011, 2013, 2015) investigate the self-regulation by feedback from massive stars, here in the form of type II Supernovae, that are correlated with the dense, star-forming gas and explode instantaneously to quench local collapse. These simulations quickly establish a dynamical equilibrium in the ISM, where gravitational collapse is balanced by turbulent and thermal pressure. They recover the scaling relation of Ostriker *et al.* (2010).

In the recent work by Gatto *et al.* (2015) the evolution of the Supernova-driven multi-phase ISM is studied in 3D simulations with the AMR code FLASH (Fryxell *et al.*, 2000). We use a periodic box of size $(256\text{ pc})^3$ with different initial densities, and compare three different driving schemes for the Supernovae – peak driving (all Supernovae are associated with density peaks), random driving (all Supernovae explode in random positions), and mixed driving (mix of peak and random driving). The results clearly show that peak driving leads to an ISM which is too cold (very low hot gas fractions) and therefore not in agreement with observations. Random driving on the other hand, leads to an ISM that is too hot: Supernovae primarily explode in low density gas which becomes hotter and hotter and compresses most of the mass into dense, molecular clumps. Here we are limited by the numerical setup, which doesn't allow the gas to escape in the vertical direction as would happen in a stratified disc that would develop a galactic outflow.

The SILCC project: Supernova-driven ISM in stratified discs.
In the SILCC project (SImulating the Life-Cycle of molecular Clouds; Walch *et al.*, 2015; Girichidis *et al.*, 2016), we study the Supernova-driven, multi-phase ISM in a stratified galactic disc including chemistry, cooling, and self-gravity. We choose solar neighbourhood conditions (gas surface density of $\Sigma_{\rm gas} = 10\ {\rm M}_\odot\ {\rm yr}^{-1}$) and solar metallicity and explore the effect of different Supernova rates ($\frac{1}{3}$, 1, and $3\times$ the rate expected from a star formation rate according to the KS relation and a massive star fraction in agreement with the Chabrier IMF; Chabrier, 2001). We find that peak driving leads to a dynamical equilibrium in the ISM, but at the same time, the ISM is lacking a hot component (since SNe in dense gas cool quickly) as well as a molecular component (since molecular gas is heated and dispersed by the feedback). In Fig. 2 we show the column density of atomic hydrogen for runs with peak driving at different Supernova rates at $t = 50$ Myr. The disc scale height increases with increasing Supernova rate, but no galactic fountains or outflows are driven with peak driving. Instead, a random Supernova component is required in order to establish a hot, volume-filling phase and drive outflows. The downside is that – unlike in simulations without self-gravity (e.g. Joung & MacLow, 2006; Hill *et al.* 2012; Gent *et al.* 2013) – there is no dynamical equilibrium established in simulations with self-gravity and a significant fraction of random Supernovae, as the dense molecular clouds cannot be disrupted. In addition, the delay time of Supernovae in dense regions (5-40 Myr, depending on the mass of the progenitor) renders them inefficient in dispersing dense gas and regulating star formation. Therefore the inclusion of other feedback processes (ionizing radiation, stellar winds, radiation pressure) is required to regulate star formation.

Other feedback processes.
Ionizing radiation can disperse typical, small molecular clouds with $M \approx 10^4\ {\rm M}_\odot$ in a few Myr (Walch *et al.*, 2012). In Walch *et al.* (2012), we used a new approach to describe the initial MC structure with a clumps-within-clumps algorithm, which is parameterized in terms of a fractal dimension, D. The fractal dimension D describes the clumpiness of the underlying MC structure. We use the method of Bisbas *et al.* (2009) to model the expansion of HII regions in these clouds and we were able to show that the cloud dispersal is independent of D. This finding is supported by other authors (e.g. Dale *et al.*, 2012). In particular, Dale *et al.* (2012) show that the cloud escape velocity, $v_{\rm esc}$, is the critical parameter which determines the susceptibility of a cloud to ionizing radiation, where clouds with $v_{\rm esc} \lesssim 10$ km/s can be heavily affected. This implies that more massive clouds with $M \approx 10^{5...6}\ M_\odot$ are affected by ionzing radiation, but the expansion of the HII region would stall before it reaches the edge of the cloud and drives outflows. Therefore, additional feedback like e.g. Supernovae is required to disperse more massive clouds. Analytically, the impact of different feedback mechanisms on molecular clouds (MCs) has been estimated by Matzner *et al.* (2002), Fall *et al.* (2011), and Murray *et al.* (2010). The latter two works stress that radiation pressure could be important in massive MCs (see also Krumholz & Matzner, 2009). Due to space limitations I refrain from further discussing radiation pressure here.

Even before the Wolf-Rayet phase, massive stars have typical mass loss rates of $10^{-6}\ {\rm M}_\odot\,{\rm yr}^{-1}$ and drive a wind with a velocity of a few 1000 km s^{-1}. Therefore they

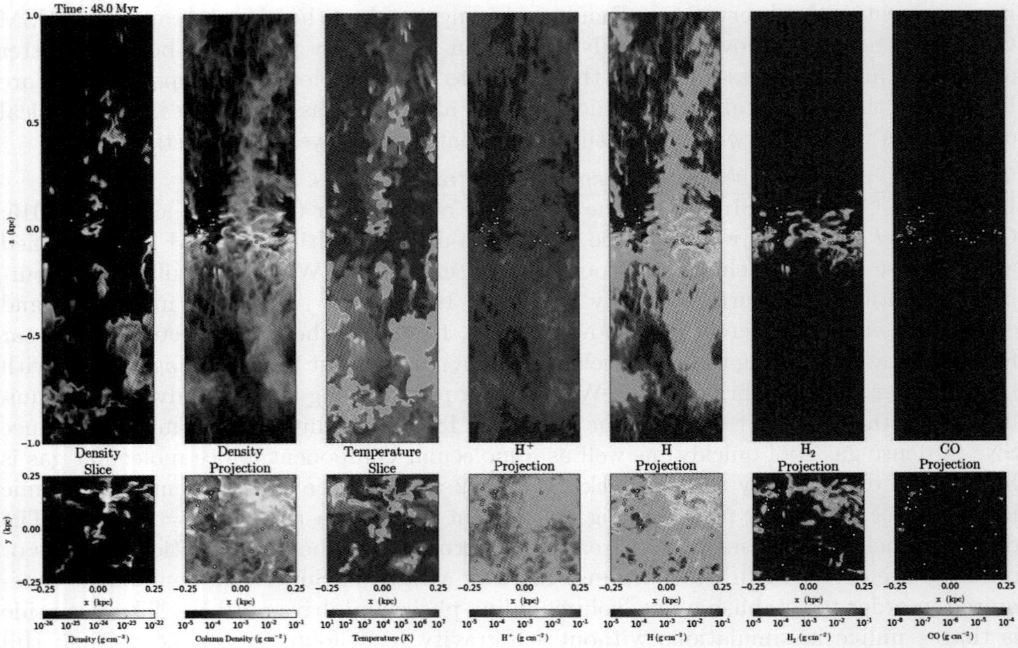

Figure 3. *Top:* Snapshot of a SILCC run with cluster sink particles and magnetic fields at $t = 48$ Myr. Only a section of the simulation box, which measures 0.5 kpc $\times$ 0.5 kpc $\times$ 2 kpc, is shown. *From left to right:* Gas density (slice), column density (projection), gas temperature (slice), and column densities of H^+, H and H_2, and CO. All slices are taken at $y = 0$ (elongated panels), or at $z = 0$ (square panels), respectively. Accordingly, the column density is projected along the y- or z-axis. The winds and Supernovae from the massive stars inside the clusters are followed. Figure from Gatto *et al.* (in prep.). *(A color version is available online.)*

could be an important source of energy and momentum. However, it is unclear how quickly the hot wind would cool as this depends on the stability of the wind-blown shell (Dwarkadas *et al.*, 2007). It seems that the indirect effect of pressure-driven winds do not have a large dynamical impact on the surrounding MCs, as shown in recent numerical studies by Dale *et al.* (2014). Nevertheless, the momentum input of winds could still be significant (Rogers & Pittard, 2013; van Marle *et al.*, 2015). In models of uniform ambient media with ionization and winds, the wind is generally trapped within the HII region and thus has a limited effect (Toalá & Arthur, 2011; Geen *et al.*, 2015). Overall, wind feedback might be negligible compared to the effects of ionizing radiation (Matzner, 2002).

The SILCC project II: Winds and Supernovae.

In a follow-up study (Gatto *et al.*, in prep.), we now include self-consistent star formation, which is modelled using sink particles that represent star clusters. We follow the evolution of each individual massive star that is formed within a cluster by tracking its stellar wind feedback and its Supernova explosion. In these new models (an example is shown in Fig. 3), we find that stellar winds are important regulators of the star formation efficiency in these discs, as they limit the accretion onto the star-forming MCs and therefore quench the formation of gravitationally bound molecular clouds and star formation before the first Supernovae explode.

3. Conclusions

There are several viable models that aim to explain the observed KS and molecular KS relations on galactic scales. New theoretical models work to incorporate changes of $n_{\rm KS}$ with Galactocentric radius and/or galaxy type. However, almost all of the models require stellar feedback to explain the low global star formation efficiency of just $\sim$1%. Although Supernova feedback adds the largest momentum to the ISM, it is most likely not the dominant mechanism to limit $\epsilon_{\rm SF}$ because Supernovae explode late during the evolution of a stellar cluster and are not well correlated with the dense gaseous component. Also, a Supernova fraction in low density gas seems to be required in order to push gas into a galactic fountain or outflow. Other feedback processes are necessary to prevent star formation in the first place. Here we could demonstrate that stellar winds can limit the accretion onto young clusters and significantly reduce $\epsilon_{\rm SF}$. Future high-resolution models that include different forms of stellar feedback will most likely be able to solve the quest of what is regulating star formation on small scales.

References

Agertz, O., Kravtsov, A. V., Leitner, S. N., & Gnedin, N. Y. 2013, *ApJ*, 770, 25

Agertz, O. & Kravtsov, A. V., 2015, *ApJ*, 804, 18

Barnes, K. L., van Zee, L., Côté, S., & Schade, D., 2012, *ApJ*, 757, 64

Bisbas, T. G., Wünsch, R., Whitworth, A. P., & Hubber, D. A. 2009, *A&A*, 497, 649

Bigiel, F., Leroy, A., Walter, F., Brinks, E., de Blok, W. J. G., Madore, B., & Thornley, M. D., 2008, *AJ*, 136, 2846

Burkert, A. & Hartmann, L., 2013, *ApJ*, 773, 48

Calzetti, D., Kennicutt, R. C., Engelbracht, C. W., Leitherer, C., Draine, B. T., Kewley, L., Moustakas, J., *et al.* , 2007, 666, 870

Chabrier, G., 2001, *ApJ*, 554, 1274

Clark, P. C. & Glover, S. C. O., 2014, *MNRAS*, 444, 2396

Dale, J. E., Ercolano, B., & Bonnell, I. A. 2012, *MNRAS*, 424, 377.

Dale, J. E., Nogoumou, J., Ercolano, B., & Bonnell, I. A. 2014, *MNRAS*, 442, 694.

Dwarkadas, V. V. 2007, *ApJ*, 667, 226

Elmegreen, B., 1997, *RMxAC*, 6, 165

Elmegreen, B. G. & Hunter, D. A., 2015, *ApJ*, 805, 145

Elmegreen, B. G., 2015, *ApJ*, 814, 30

Evans, I. I., N. J. *et al.* , 2009, *ApJS*, 181, 321

Evans, I. I., N. J., Heiderman, A., & Vutisalchavakul, N., 2014, *ApJ*, 782, 114

Fall, S. M., Krumholz, M. R., & Matzner, C. D. 2011, *ApJ*, 710, 142

Federrath, C. & Klessen, R. S., 2013, *ApJ*, 763, 51

Fryxell, B., Olson, K., Ricker, P., & Timmes, *et al.* 2000, *ApJS*, 131, 273

Gatto, A., Walch, S., MacLow, M.-M., & Naab, T., *et al.* 2015, *MNRAS*, 449, 1057

Gatto, A., Walch, S., Girichidis, P., *et al.* in preparation

Geen, S., Rosdahl, J., Blaizot, J., Devriendt, J., & Slyz, A. 2015, *MNRAS*, 448, 3248

Gent, F. A., Shukurov, A., Fletcher, A., Sarson, G. R., & Mantere, M. J., 2013, *MNRAS*, 432, 1396

Genzel, R., Tacconi, L. J., Gracia-Carpio, J., Sternberg, A., Cooper, M. C., Shapiro, K., Bolatto, A., Bouché, N., *et al.* , 2010, *MNRAS*, 407, 2091

Genzel, R., Tacconi, L. J., Lutz, D., Saintonge, A., Berta, S., Magnelli, B., Combes, F., García-Burillo, S., *et al.* 2015, *ApJ*, 800, 20

Girichidis, P., Walch, S., Naab, *et al.* 2016, *MNRAS*, 436, 3432

Glover, S. C. O. & Clark, P. C., 2012, *MNRAS*, 421, 9

Harfst, S., Theis, Ch., & Hensler, G., 2006, *A&A*, 449, 509

Heiderman, A., Evans, N. J., II, Allen, L. E., Huard, T., & Heyer, M., 2010, *ApJ*, 723, 1019

Hill, A. S., Joung, M. R., Mac Low, M.-M., Benjamin, R. A., Haffner, L. M., Klingenberg, C., & Waagan, K., 2012, *ApJ*, 750, 104
Joung, M. K. R. & Mac Low, M.-M., 2006, *ApJ*, 653, 1266
Kennicutt, Jr., R. C., 1989, *ApJ*, 344, 685
Kennicutt, Jr., R. C., 1998, *ApJ*, 498, 541
Kim, C.-G., Kim, W.-T., Ostriker, E. C, *ApJ*, 743, 25
Kim, C.-G., Ostriker, E. C, Kim, W.-T., *ApJ*, 776, 1
Kim, C. & Ostriker, E. C. 2015, *ApJ*, 802, 99
Kruijssen, J. M. D. & Longmore, S. N., 2014, *MNRAS*, 439, 3239
Krumholz, M. R. & McKee, C. F., 2005, *ApJ*, 630, 250
Krumholz, M. R. & Tan, J. C., 2007, *ApJ*, 654, 304
Krumholz, M. R., McKee, C. F., & Tumlinson, J., 2008, *ApJ*, 689, 865
Krumholz, M. R. & Matzner, C. D. 2009, *ApJ*, 703, 1352
Lada, C. J., Lombardi, M., & Alves, J. F., 2010, *ApJ*, 724, 687
Lada, C. J., Forbrich, J., Lombardi, M., & Alves, J. F., 2012, *ApJ*, 745, 190
Leitherer, C., Schaerer, D., Goldader, J. D., González Delgado, R. M., Robert, C., Kune, D. F., de Mello, D. F., Devost, D., *et al.* , 1999, *ApJS*, 123, 3
Leroy, A. K., Walter, F., Brinks, E., Bigiel, F., de Blok, W. J. G., Madore, B., & Thornley, M. D., 2008, *AJ*, 136, 2782
Matzner, Christopher D. 2002, *ApJ*, 566, 302
Mac Low, M.-M., Smith, M. D., Klessen, R. S., & Burkert, A., 1998, *ApSS*, 261, 195
Momose, R., Koda, J., Kennicutt, R. C., Jr., Egusa, F. Calzetti, D., Liu, G., Donovan Meyer, J., & Okumura, S. K., 2013, *ApJ*, 772, 13
Murray, N., 2011, *ApJ*, 729, 133
Onodera, S., Kuno, N., Tosaki, T., Kohno, K., Nakanishi, K., Sawada, T., Muraoka, K., Komugi, S., *et al.* , 2010, *ApJ*, 722, 127
McKee, C. F. & Ostriker, J. P. 1977, *ApJ*, 218, 148
Ostriker, E. C., McKee, C. F., & Leroy, A. K., 2010, *ApJ*, 721, 975
Padoan, P., Nordlund, A., & Jones, B. J. T., 1997, *MNRAS*, 288, 145
Padoan, P., Nordlund, A., *ApJ*, 576, 870
Pringle, J. E., Allen, R. J., & Lubow, S. H., 2001, *MNRAS*, 327, 663
Murray, N., Quataert, E., & Thompson, T. A. 2010, *ApJ*, 709, 191
Rogers, H., & Pittard, J. M., *MNRAS*, 431, 1337
Roychowdhury, S., Huang, M.-L., Kauffmann, G., Wang, J., Chengalur, J. N., it MNRAS, 2015, 449, 3700
Schmidt, M. 1959, *ApJ*, 129, 243
Schruba, A., Leroy, A. K., Walter, F., Sandstrom, K., & Rosolowsky, E., 2010, 722, 1699
Schruba, A., Leroy, A. K., Walter, F., Bigiel, F., Brinks, E., de Blok, W. J. G., Dumas, G., Kramer, C., Rosolowsky, E., *et al.* , 2011, 142, 37
Salpeter, E. E., 1955, *ApJ*, 121, 161
Shetty, R., Kelly, B. C., Rahman, N., Bigiel, F., Bolatto, A. D., Clark, P. C., Klessen, R. S., & Konstandin, L. K., 2014, 437, 61
Shetty, R., Clark, P. C., & Klessen, R. S., 2014, 442, 2208
Silk, J., 1997, *ApJ*, 481, 703
Sternberg, A., Le Petit, F., Roueff, E., & Le Bourlot, J., 2014, *ApJ*, 790, 10
Stone, J. M., Ostriker, E. C., & Gammie, C. F., 1998, it ApJL, 508, 99
Tacconi, L. J., Neri, R., Genzel, R., Combes, F., Bolatto, A., Cooper, M. C., Wuyts, S., Bournaud, F., *et al.* ,2013, *ApJ*, 768, 74
&Tasker, E. J.; Bryan, G. L., 2006, *ApJ*, 641, 878
Toalá, J. A. & Arthur, S. J. 2011, *ApJ*, 737, 100
van Marle, A. J., Meliani, Z., & Marcowith, A. 2015, *A&A*, 584, 49
Vazquez-Semadeni, E., 1994, *ApJ*, 423, 681
Wada, K. & Norman, C. A., 2007, *ApJ*, 660, 276
Walch, S.,Whitworth, A. P., Bisbas, T., Wünsch, R., Hubber, D. 2012, *MNRAS*, 427, 625
Walch, S., Girichidis, P., Naab, T., *et al.* 2015, *MNRAS*, 454, 238

From Interstellar Clouds to Star-Forming Galaxies: Universal Processes?
Proceedings IAU Symposium No. 315, 2015
P. Jablonka, Ph. André & F. van der Tak, eds.

doi:10.1017/S174392131600747X

Local Physics and Star Formation in Galaxies

Adam K. Leroy, Frank Bigiel, Annie Hughes, Eva Schinnerer, Antonio Usero and the PAWS, EMPIRE, and HERACLES Dense Gas Collaborations

Abstract. A next main step in understanding star formation is to link the sharp but narrow view of Galactic molecular cloud studies to the wider context accessed by less detail by extragalactic work. In this proceeding, we discuss how new technology and large programs at millimeter wavelengths are improving our ability to access physical conditions in the interstellar medium (ISM) of other galaxies. We highlight results from the multi-line survey of Usero *et al.* (2015), which measured density sensitive lines across nearby galaxy disks, and two new mapping studies of M51: the high resolution Plateau de Bure Arcsecond Whirlpool Survey (PAWS) and the EMPIRE multi-line mapping survey. These results argue for a context-dependent role for gas density in star formation; that is, gas at a particular density does not appear to form stars in a universal way. They also demonstrate the influence of cloud-scale conditions, especially surface density and the velocity dispersion, in setting the small-scale density distribution and highlight gravitational boundedness as a main driver of the ability of gas to form stars. Beyond these specific results, we argue that ability to gauge detailed physical conditions in the star-forming gas of other galaxies promises major advances that will help unify the fields of Galactic and extragalactic star formation in the next few years.

---

1. Introduction

The last decade has seen major strides in how we understand star formation on the scale of galaxies and on the scale of individual molecular clouds. A key next challenge is to connect these two views and better unify the fields of Galactic and extra-galactic star formation. The main obstacle to this has been the dramatically different scales accessed by studies targeting nearby Milky Way clouds and those focused on other galaxies. The former heavily resolve individual clouds, so that cloud substructure and local conditions like gas volume density and the sonic Mach number can be directly compared to the ability of gas to form stars (e.g., Kainulainen *et al.* 2009; Lada *et al.* 2012; André *et al.* 2014). Extragalactic observations have, so far, struggled to access these conditions in a systematic way. In exchange for this more "blurry" view, however, they provide an unparalleled accounting of gas and star formation across the universe and the galaxy population (e.g., Young *et al.* 1995; Saintonge *et al.* 2011, Utomo *et al.*, in prep. and in this volume). Observations that resolve galaxies but not clouds also provide our best tool, so far, to link the process of star formation to galactic structure (e.g., Leroy *et al.* 2008; Bigiel *et al.* 2008; Schruba *et al.* 2011; Leroy *et al.* 2013). Thus, extragalactic observations yield the best answer to date for "How does the process of star formation depend on the larger galactic context?"

This situation is improving rapidly. Over the last few years, Galactic studies have improved in their perspective, with wide field, high resolution maps of interstellar medium (ISM) structure from *Herschel* (e.g., Schneider *et al.* 2015) and the contrast between the Solar Neighborhood and the Central Molecular Zone offering prime examples (Longmore

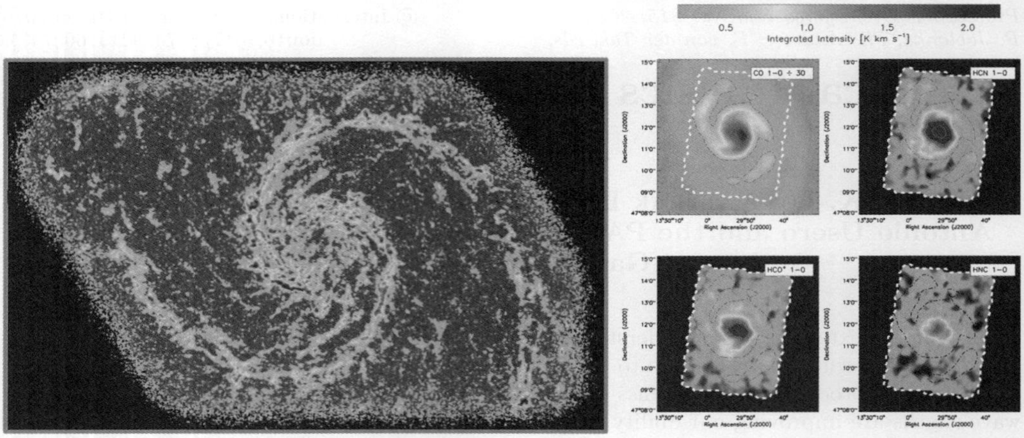

Figure 1. Resolution and spectroscopic mapping. (*left*) The PAWS view of the inner region of M51, from which we know the surface density, velocity dispersion, and boundedness of gas at 40 pc resolution. (*right*) The EMPIRE spectral mapping survey's view of M51 in the density-sensitive line HCN, HCO^+, and HNC. Combining these views give new access to the physical state of the gas in galaxies. *Color versions of the figures are available online.)*

et al. 2013; Kauffmann *et al.* 2013). Many of these efforts are discussed elsewhere in these proceedings. In this contribution, we emphasize improvements from the other direction: thanks to a mixture of new technology and major commitments from existing facilities, extragalactic observations increasingly access some of the small-scale physics previously restricted to studies of the Milky Way and the nearest galaxies in the Local Group.

We demonstrate the bright future for this approach using results from new spectroscopic surveys (Usero *et al.* 2015, and "EMPIRE" — Bigiel *et al.* in prep.) and the PdBI Arcsecond Whirlpool Survey (Schinnerer *et al.* 2013; Pety *et al.* 2013). These demonstrate a changing ratio of star formation rate and high critical density line emission, which we argue to be best explained by a context-dependent role for density in star formation. In M51, they also show a close link between the "cloud scale" ($\sim$ 40 pc) ISM structure traced by PAWS and the density distribution traced by the EMPIRE spectroscopy. We also show that in PAWS, gravitational boundedness appears to be the structural property most closely associated with star formation. The new results comparing these approaches for M51 are presented in Leroy *et al.* (in prep.).

2. Two Old Tools That Are New Again: Resolution and Spectroscopy

Based on both theoretical and observational work, local conditions in the ISM are believed to affect its ability to form stars. For purposes this proceeding, there are two main families of ISM properties expected to drive star formation in gas. First, the *gas volume density* is considered pivotal by almost all theories of star formation. Turbulent models (e.g., Krumholz & McKee 2005) consider the free fall time, $\tau_{\rm ff} \propto \rho^{-0.5}$, the governing timescale for star formation. As a result, in these models, the normalized rate of star formation, SFR-per-unit-gas, tracks the gas volume density closely. Another family of models, mostly motivated by observations, posit a threshold volume density above which star formation proceeds in a more or less universal way (e.g., Gao & Solomon 2004; Heiderman *et al.* 2010; Lada *et al.* 2012; Evans *et al.* 2014). In this case, the main determinant of the ability of gas to form stars is the fraction of gas mass above this threshold, $f_{\rm dense} = M_{\rm dense}/M_{\rm gas}$.

The local gas volume density is primarily accessible to extragalactic observers via multi-line mm-wave spectroscopy. A main result mapping Galactic molecular clouds is that the dense, star-forming gas is concentrated into very small structures, ~ 0.1 pc-wide filaments and $\sim$ pc scale dense clumps. Even with the Atacama Large Millimeter/submillimeter Array, such structures are accessible to direct imaging in other galaxies only under special circumstances and with great effort. Fortunately, the mm-wave band contains transitions from molecules with a wide range of dipole moments and so a wide range of density sensitivities. The distribution of gas density within a beam can be accessed by observing both molecules like ^{12}CO, which — after accounting for optical depth — trace almost all molecular with molecules like HCN, which is effectively excited at H_2 densities $\gtrsim 10^6$ cm^{-3} (before accounting for optical depth).

The power of combining lines with different density sensitivities has long been recognized and has been exploited in classic work by Gao & Solomon (2004). Until recently, the faintness of high critical density lines (often $\sim 1/30$ the intensity of CO or less; see Usero *et al.* 2015) has limited their widespread deployment to understand the density distribution in normal galaxies. With the improved sensitivity and bandwidth of new receivers (like EMIR on the IRAM 30-m) and new facilities (especially ALMA) this situation is changing. Usero *et al.* (2015) used EMIR to make the first systematic survey of density-sensitive lines spanning the disks of normal star-forming galaxies, a complement to the active galaxies and center-dominated measurements of Gao & Solomon (2004). The IRAM large program EMPIRE (P.I. F. Bigiel) is now mapping lines with a suite of critical densities across the entire disk of 9 galaxies, beginning with a prototype program mapping several lines across the whole disk of M51 (Bigiel *et al.*, in prep. and this volume; see also Chen *et al.* 2015). This is the first full-galaxy data set covering lines with a wide range of critical densities and has the advantage of overlapping the high resolution view of PAWS.

In turbulent models, the *cloud-scale structure of the ISM* affects the ability of gas at a given density to form stars (i.e., the star formation efficiency per free-fall time; e.g., Federrath & Klessen 2012). The mean density of a cloud, related to the cloud-scale surface density, Σ, relates to the free-fall time on the scale of the whole cloud. For a fixed temperature, the cloud-scale velocity dispersion, σ, is a close cognate of the turbulent Mach number, which affects the width of the density distribution. The combination of these two parameters tracks the gravitational boundedness of the cloud; this is often expressed as the virial parameter, $\alpha_{\rm vir} = 0.5\ KE/UE$. In Leroy *et al.* (in prep.) we argue that in full-fidelity, high resolution imaging like PAWS the beamwise quantity Σ/σ^2, which we call pixelwise "boundedness," captures the physics of the virial parameter.

Unlike the very small-scale distribution of densities, these cloud-scale quantities are potentially accessible from direct imaging using ALMA or NOEMA for a large set of galaxies. Arcsecond resolution CO imaging can readily resolve the structure of the ISM at ~ 50 pc scales out to the Virgo cluster. The PAWS survey serves a clear demonstration of the power of this approach, showing clear variations of cloud properties with environment in the galaxy (Hughes *et al.* 2013; Colombo *et al.* 2014). While a few times 10 pc resolution only marginally resolves a Galactic cloud, the physical link between gas structure at this scale and star formation is much larger than for ISM properties measured over beams $\sim$ kpc in area or more. To some approximation, one can think of this high resolution, high sensitivity, full-fidelity CO mapping as providing galaxy-wide maps of the initial conditions for star formation in molecular clouds.

Connecting these approaches, it is possible to relate cloud-scale properties to the small scale density distribution and to link both to the ability of gas to form stars. As the prototype for all of these types of observations, M51 is a natural first target for such

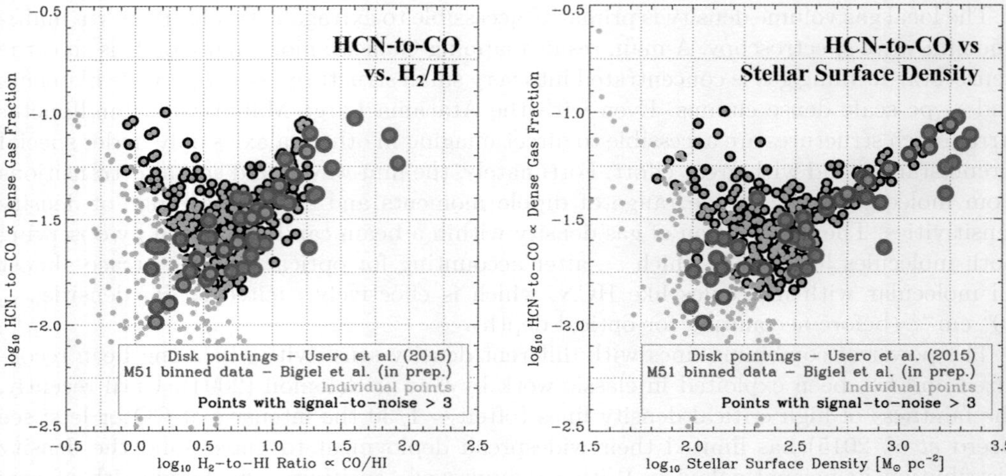

Figure 2. HCN-to-CO ratio, indicative of the fraction of gas mass as a function of the molecular-to-atomic gas ratio (*left*) and stellar surface density (*right*), which both track average ISM pressure or density. Blue points show individual disk pointings from Usero *et al.* (2015), gray points show individual lines of sight in M51 from Bigiel *et al.* (in prep.), with black outlines indicating where signal-to-noise is greater than 3. Red points show the binned trend for M51. The fraction of mass in a dense phase increases with increasing surface density, molecular fraction, and ISM pressure.

studies. Results for individual approaches targeting M51 have appeared in Meidt *et al.* (2013); Usero *et al.* (2015, Bigiel *et al.* in prep.); In Leroy *et al.* (in prep.), we combine these views.

3. A Context-Dependent Role for Gas Density

In Usero *et al.* (2015) and Bigiel *et al.* (in prep.), we measure the distribution of lines with varying critical densities across the disks of spiral galaxies - the whole disk of M51 in Bigiel *et al.* (in prep.) and pointings at a range of radii in 30 disk galaxies in Usero *et al.* (2015). Here we highlight results for the HCN (1-0) transition but both surveys measure several transitions. Following Gao & Solomon (2004), HCN (1-0) has been commonly identified as tracing specifically star-forming, high density gas. A nearly fixed relationship between HCN (1-0) and infrared luminosity has been linked to results from the Solar Neighborhood to argue for an approximately universal role for gas above this threshold density in star formation. Matched measurements of CO, HCN, and recent star formation (mostly traced here by IR luminosity) across galaxy disks thus allow us to ask two key questions: "What environmental factors regulate the fraction of dense gas mass?" and "Does high density gas always form stars at approximately the same rate?" Perhaps surprisingly, the whole-disk map of M51 by Bigiel *et al.* (in prep.) and the disk pointing survey of Usero *et al.* (2015) yield consistent results for both questions.

In Usero *et al.* (2015), we show that the fraction of dense gas, traced by the HCN-to-CO ratio (thanks to their very different effective densities), varies strongly across galaxies. The sense of the variation is that a higher fraction of gas is dense in the inner, high surface density parts of galaxies where most of the gas is molecular. Though neither relationship is necessarily causal, this behavior is well-encapsulated by a correlation of the ratio HCN-to-CO, tracing the dense gas mass fraction, with stellar surface density, $\Sigma_\star$, and the molecular-to-atomic ratio, $\Sigma_{\rm H2}/\Sigma_{\rm Hi}$. In Bigiel *et al.* (in prep.), we show that

M51 obeys the same behavior, with individual ≈ 1 kpc resolution lines of sight exhibiting a correlation that overlaps the pointings from ≈ 30 disk galaxies in Usero *et al.* (2015). Thus, in the high surface density, highly molecular parts of galaxies, where the ISM pressure and mean ISM density is also high, a larger fraction of the molecular gas tends to be in a dense phase. In Leroy *et al.* (in prep.), we show that in M51, the full pattern of line ratios from ^{12}CO (1-0) through HCN (1-0) are consistent with an interpretation in which this changing HCN-to-CO ratio indeed reflects changes in the mean ISM density.

Finding more dense gas in the denser parts of galaxies is expected, though not a given (molecular gas might simply have a higher filling factor in these locales). The more surprising result of these surveys of density-sensitive lines is that the link between dense gas and star formation seems to vary with environment. Specifically, in Usero *et al.* (2015) we show that the ratio of SFR-to-HCN, a close cognate of the rate at which dense gas forms stars, anti-correlates with the stellar surface density and molecular fraction. That is, the deeper the stellar potential well and the more molecular the ISM, the less efficient dense gas is at forming stars. This result is reinforced by results from full-disk mapping in M51. This agrees with the sense of results contrasting the Solar Neighborhood with the Galactic Center (e.g., Longmore *et al.* 2013), but strongly disagrees with the family of results that posit a universal process of star formation in the dense gas traced by HCN (e.g., Gao & Solomon 2004; Lada *et al.* 2012).

The natural explanation for this varying IR-to-HCN ratio is a context-dependent role for density in star formation. As a simple example, if star formation is driven by a density contrast with a marginally bound background cloud, then the density of the background cloud will define what qualifies as "dense" gas for purposes of star formation. Put another way, HCN may trace immediately star-forming gas in a region where the mean density of cold clouds is $\langle n \rangle \sim 10^3$ cm^{-3}. But in a region with higher average density, say $\langle n \rangle \sim 10^4$ cm^{-3}, HCN will trace a larger fraction of each cloud, including potentially non self-gravitating, non star-forming gas. Versions of this scenario have been put forward by Krumholz & Thompson (2007), Narayanan *et al.* (2008), and are nicely synthesized in Federrath & Klessen (2012).

Our observations of an IR-to-HCN ratio that decreases when the ISM appears to become more dense on average provide strong observational support to this theoretical scenario. They argue that the role of gas at a particular volume density depends on context, especially the mean density and gravitational boundedness of the interstellar medium on larger (cloud) scales.

4. A Link Between Cloud-Scale ISM Structure and Star Formation

Results from multi-line spectroscopy thus suggest an important role for cloud-scale ISM properties in the ability of gas to form stars. Thanks to PAWS, we know the local surface density (Σ), velocity dispersion (σ), and boundedness ($\Sigma/\sigma^2 \propto \alpha_{\rm vir}^{-1}$) of the molecular gas at 40 pc resolution. In Leroy *et al.* (in prep.), we convolve these local measurements to the ≈ 1 kpc resolution of the infrared data and multi-line spectroscopy. During this convolution, we weight the small-scale properties by the local CO luminosity so that we measure the luminosity-weighted property within the beam; this is closely related to a mass weighting and is only possible starting from the high resolution PAWS data. Thus Σ measured in this way over a kpc is effectively the average surface density at 40-pc from which a CO photon arises in the PAWS map; σ and Σ/σ^2 are analogous.

Following the discussion in the previous two parts, PAWS offers us the opportunity to ask how cloud-scale ISM structure affects the ability of gas to form stars, here traced by the ratio IR-to-CO, and the fraction of gas in a dense phase. The figures here (from

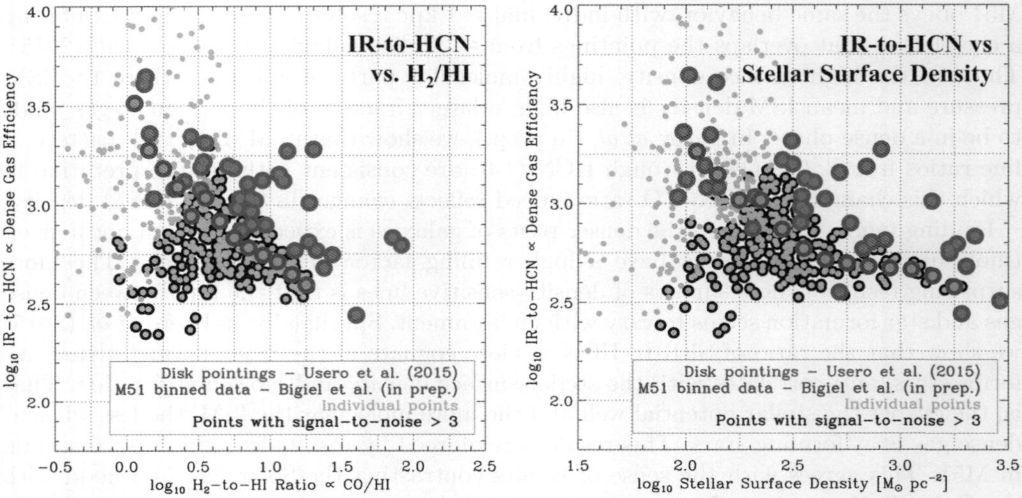

Figure 3. As Figure 1 but now showing the ratio of star formation traced by IR emission to dense gas traced by HCN emission as a function (*left*) and stellar surface density (*right*), which both track average ISM pressure or density. Blue points show individual disk pointings from Usero *et al.* (2015), gray points show individual lines of sight in M51 from Bigiel *et al.* (in prep.), with black outlines indicating where signal-to-noise is greater than 3. Red points show the binned trend for M51. Contrary to hypotheses based on the Solar Neighborhood, the star formation per unit dense gas varies systematically in galaxies. The sense is that in the denser parts of galaxies, the star formation per unit dense gas is lower. This is consistent with a context-dependent role for gas density in star formation, in which contrast with the background matters as much as the absolute value of the density.

Leroy *et al.* in prep) show that a high cloud scale surface density and a high velocity dispersion do correspond to a larger fraction of gas in a dense, star-forming phase. But the relationship of these quantities to the ability of all of the gas to form stars, tracked by IR-to-CO, is less clear, with both high and low star formation efficiencies at high values of Σ and σ. Instead, the apparent boundedness of the gas, traced by Σ/σ^2 offers our best cloud-scale predictor for how well gas forms stars.

These results agree well with the picture outlines in the previous section: gas volume density exhibits a context dependent role in star formation. There will be a higher fraction of dense gas in regions with high average (cloud-scale) densities. But this gas also tends to show a *lower* ratio of recent star formation rate to dense gas (IR-to-HCN). A natural explanation for this is that the HCN emitting gas in these regions traces not only the highly self-gravitating tail but also a larger part of the non-self-gravitating density distribution. Other considerations, especially a variable optical depth of the dense gas tracers (Jesus-Jimenez in prep.) may also help explain this effect, but we argue that the pattern of line ratios and the visible change in cloud-scale surface density is consistent with this interpretation.

Regardless of interpretation these results undermine a main set of arguments for models in which star formation proceeds in a universal way about some volume density. These models have invoked a universal scaling between HCN and tracers of recent star formation as a demonstration as the main extragalactic evidence that dense gas forms stars in a universal way (e.g., Gao & Solomon 2004; Lada *et al.* 2012). Although there remains an broad correlation between between IR luminosity and HCN luminosity across many

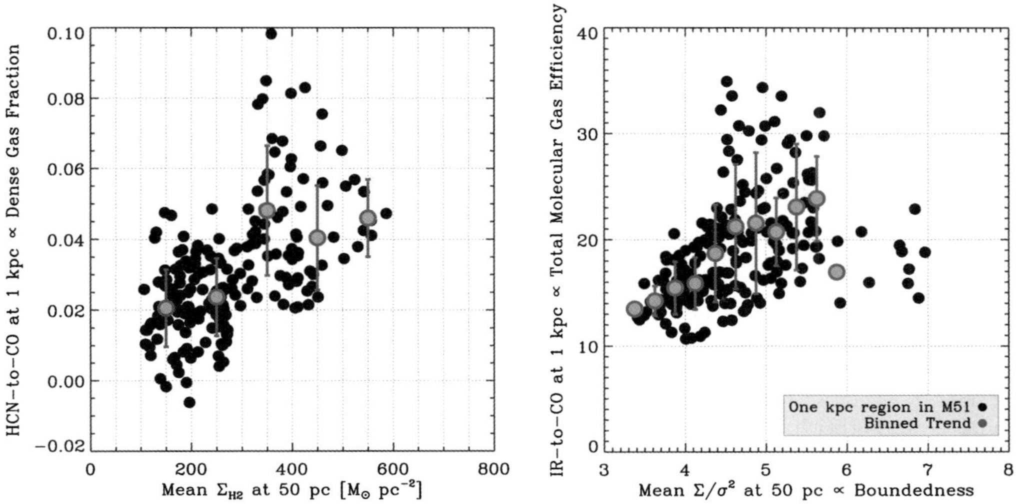

Figure 4. Link between cloud-scale (40 pc resolution) ISM structural properties measured from PAWS and the ability of the ISM to form stars at larger scales. (*left*) The fraction of dense gas mass correlates with the mass weighted cloud-scale surface density in a beam. (*right*) The overall ability of gas to form stars correlates with the mass-weighted gravitational boundedness in a beam. The link between the overall ability of the molecular gas to form stars, the distribution of small-scale densities traced by spectroscopy, and the cloud-scale structural properties show how galaxy-scale processes connect to the small-scale physics of star formation, a link that promises to be productively explored with ALMA, NOEMA, the GBT, and similar facilities in coming years.

decades (Gao & Solomon 2004; Wu *et al.* 2005), in detail the star formation rate per unit HCN shows real physical variations.

5. Outlook

Beyond the specific results highlighted above, a main goal of this proceeding is to sound an optimistic note regarding studies of star formation in the next few years. The ability of telescopes including ALMA and NOEMA to resolve cloud-scale ISM structure opens a new window to link galaxy-scale processes to the immediate sites of star formation. The sensitivity of facilities including ALMA, NOEMA, the Green Bank Telescope (now a mm-wave spectroscopic facility), and the IRAM 30-m telescope also at last allows multi-line mm-wave spectroscopy to be deployed as a survey tool, not restricted to individual case studies or the most extreme systems. This promises a window into small scale conditions including gas volume density, which is commonly viewed as a governing parameter for the ability of gas to form stars. Here we show that the large progress may be expected from combining these two views, and thus measuring how galactic structure sets cloud scale conditions, how these drive small-scale physics, and how they all relate to star formation in clouds.

References

André, P., Di Francesco, J., Ward-Thompson, D., Inutsuka, S.-I., Pudritz, R. E., & Pineda, J. E. 2014, *Protostars and Planets VI*, 27

Bigiel, F., Leroy, A., Walter, F., Brinks, E., de Blok, W. J. G., Madore, B., & Thornley, M. D. 2008, *AJ*, 136, 2846

Chen, H., Gao, Y., Braine, J., & Gu, Q. 2015, *ApJ*, 810, 140

Colombo, D., Hughes, A., Schinnerer, E., Meidt, S. E., Leroy, A. K., Pety, J., Dobbs, C. L., Garcia-Burillo, S., Dumas, G., Thompson, T. A., Schuster, K. F., & Kramer, C. 2014, ArXiv e-prints

Evans, II, N. J., Heiderman, A., & Vutisalchavakul, N. 2014, *ApJ*, 782, 114

Federrath, C. & Klessen, R. S. 2012, *ApJ*, 761, 156

Gao, Y. & Solomon, P. M. 2004, *ApJ*, 606, 271

Heiderman, A., Evans, II, N. J., Allen, L. E., Huard, T., & Heyer, M. 2010, *ApJ*, 723, 1019

Hughes, A., Meidt, S. E., Colombo, D., Schinnerer, E., Pety, J., Leroy, A. K., Dobbs, C. L., García-Burillo, S., Thompson, T. A., Dumas, G., Schuster, K. F., & Kramer, C. 2013, *ApJ*, 779, 46

Kainulainen, J., Beuther, H., Henning, T., & Plume, R. 2009, *AAP*, 508, L35

Kauffmann, J., Pillai, T., & Zhang, Q. 2013, it ApJL, 765, L35

Krumholz, M. R. & McKee, C. F. 2005, *ApJ*, 630, 250

Krumholz, M. R. & Thompson, T. A. 2007, *ApJ*, 669, 289

Lada, C. J., Forbrich, J., Lombardi, M., & Alves, J. F. 2012, *ApJ*, 745, 190

Leroy, A. K., Walter, F., Brinks, E., Bigiel, F., de Blok, W. J. G., Madore, B., & Thornley, M. D. 2008, *AJ*, 136, 2782

Leroy, A. K., Walter, F., Sandstrom, K., Schruba, A., Munoz-Mateos, J.-C., Bigiel, F., Bolatto, A., Brinks, E., de Blok, W. J. G., Meidt, S., Rix, H.-W., Rosolowsky, E., Schinnerer, E., Schuster, K.-F., & Usero, A. 2013, ArXiv e-prints

Longmore, S. N., Bally, J., Testi, L., Purcell, C. R., Walsh, A. J., Bressert, E., Pestalozzi, M., Molinari, S., Ott, J., Cortese, L., Battersby, C., Murray, N., Lee, E., Kruijssen, J. M. D., Schisano, E., & Elia, D. 2013, *MNRAS*, 429, 987

Meidt, S. E., Schinnerer, E., García-Burillo, S., Hughes, A., Colombo, D., Pety, J., Dobbs, C. L., Schuster, K. F., Kramer, C., Leroy, A. K., Dumas, G., & Thompson, T. A. 2013, *ApJ*, 779, 45

Narayanan, D., Cox, T. J., Shirley, Y., Davé, R., Hernquist, L., & Walker, C. K. 2008, *ApJ*, 684, 996

Pety, J., Schinnerer, E., Leroy, A. K., Hughes, A., Meidt, S. E., Colombo, D., Dumas, G., García-Burillo, S., Schuster, K. F., Kramer, C., Dobbs, C. L., & Thompson, T. A. 2013, *ApJ*, 779, 43

Saintonge, A., Kauffmann, G., Wang, J., Kramer, C., Tacconi, L. J., Buchbender, C., Catinella, B., Graciá-Carpio, J., Cortese, L., Fabello, S., Fu, J., Genzel, R., Giovanelli, R., Guo, Q., Haynes, M. P., Heckman, T. M., Krumholz, M. R., Lemonias, J., Li, C., Moran, S., Rodriguez-Fernandez, N., Schiminovich, D., Schuster, K., & Sievers, A. 2011, *MNRAS*, 415, 61

Schinnerer, E., Meidt, S. E., Pety, J., Hughes, A., Colombo, D., García-Burillo, S., Schuster, K. F., Dumas, G., Dobbs, C. L., Leroy, A. K., Kramer, C., Thompson, T. A., & Regan, M. W. 2013, *ApJ*, 779, 42

Schneider, N., Bontemps, S., Girichidis, P., Rayner, T., Motte, F., André, P., Russeil, D., Abergel, A., Anderson, L., Arzoumanian, D., Benedettini, M., Csengeri, T., Didelon, P., Di Francesco, J., Griffin, M., Hill, T., Klessen, R. S., Ossenkopf, V., Pezzuto, S., Rivera-Ingraham, A., Spinoglio, L., Tremblin, P., & Zavagno, A. 2015, *MNRAS*, 453, L41

Schruba, A., Leroy, A. K., Walter, F., Bigiel, F., Brinks, E., de Blok, W. J. G., Dumas, G., Kramer, C., Rosolowsky, E., Sandstrom, K., Schuster, K., Usero, A., Weiss, A., & Wiesemeyer, H. 2011, *AJ*, 142, 37

Usero, A., Leroy, A. K., Walter, F., Schruba, A., García-Burillo, S., Sandstrom, K., Bigiel, F., Brinks, E., Kramer, C., Rosolowsky, E., Schuster, K.-F., & de Blok, W. J. G. 2015, *AJ*, 150, 115

Wu, J., Evans, II, N. J., Gao, Y., Solomon, P. M., Shirley, Y. L., & Vanden Bout, P. A. 2005, it ApJL, 635, L173

Young, J. S., Xie, S., Tacconi, L., Knezek, P., Viscuso, P., Tacconi-Garman, L., Scoville, N., Schneider, S., Schloerb, F. P., Lord, S., Lesser, A., Kenney, J., Huang, Y.-L., Devereux, N., Claussen, M., Case, J., Carpenter, J., Berry, M., & Allen, L. 1995, *ApJS*, 98, 219

From Interstellar Clouds to Star-Forming Galaxies: Universal Processes?
Proceedings IAU Symposium No. 315, 2015
P. Jablonka, Ph. André & F. van der Tak, eds.

doi:10.1017/S1743921316007481

Star Formation Thresholds: The View from Inside the Galaxy

James Di Francesco

National Research Council of Canada,
Herzberg Programs in Astronomy & Astrophysics,
5071 West Saanich Road, Victoria, BC, Canada, V9E 2E7
email: james.difrancesco@canada.ca

Abstract. We explore the relationship between the total gas surface density and star formation rate surface density, a.k.a., the "Kennicutt-Schmidt relation," in a Galactic context. Specifically, we probe the origins of thresholds in the behaviour of the K-S relation at 10 $M_\odot$ pc^{-2} and 100-200 $M_\odot$ pc^{-2} using images from the Herschel Hi-GAL and Gould Belt surveys. In both cases, pervasive filamentary structures are seen, possibly due to turbulent motions. The Hi-GAL image supports the view that at ~10 $M_\odot$ pc^{-2} gas becomes molecular, leading to the formation of clouds that harbour star formation. The GBS images suggest the 100-200 $M_\odot$ pc^{-2} threshold originates from the nature of filaments being stable until a critical column density of ~160 $M_\odot$ pc^{-2} is reached. Therefore, the transition between non-star-forming and star-forming gas in clouds (and galaxies) may be set universally by the dynamical properties of filaments.

Keywords. stars: formation, ISM: clouds, ISM: structure, galaxies: ISM, submillimeter

1. Introduction

"Stars form out of dense cores in molecular clouds." This statement or variations thereof has been for decades a recurring first sentence in numerous publications about Galactic star formation. In general, "dense cores" have been defined as pockets of gas typically within larger molecular clouds with sizes of ~0.1 pc, masses of ~0.5 $M_\odot$, and densities of >10^4 cm^{-3} (see Di Francesco *et al.* 2007). Indeed, the strong connection between young stellar objects (YSOs) and dense gas goes back to the pioneering work of Myers & Benson (1983), who found associations between NH_3 (2,2) emission and the locations of "emission line stars" in Taurus-Auriga. Today, NH_3 emission is used extensively to trace dense gas associated with cores in other Galactic star-forming regions (e.g., Friesen *et al.* 2009, Friesen *et al.* 2015, in prep). In addition, very strong correlations have been seen between the numbers of young stars produced and the amounts of gas seen at densities greater than 10^4 cm^{-3} (i.e., extinctions of $A_V > 8$) for nearby and Galactic clouds (see Johnstone, Di Francesco & Kirk 2004; Wu *et al.* 2005; Lada, Lombardi & Alves 2010; Sadavoy *et al.* 2014).

The clear association of dense gas with star formation has made the well-known "Kennicutt-Schmidt" (K-S) scaling relation hard for some Galactic astronomers (at least this one) to understand at face value. Indeed, the K-S relation is a tight power-law correlation between the surface density of the star formation rate (Σ_{SFR}) and the surface density of the *total* gas in galaxies (Kennicutt 1989; Kennicutt 1998). The total gas surface density is defined as the sum of the atomic and molecular hydrogen surface densities, Σ_{HI+H_2}, with no overt connection to line-of-sight amounts of dense gas. For reference, the density of a cold neutral HI cloud is ~10 cm^{-3} and that of a molecular cloud is ~$10^{1.5-2}$ cm^{-3} , not the ~10^4 cm^{-3} densities (or higher) of dense cores. Nevertheless, the origin

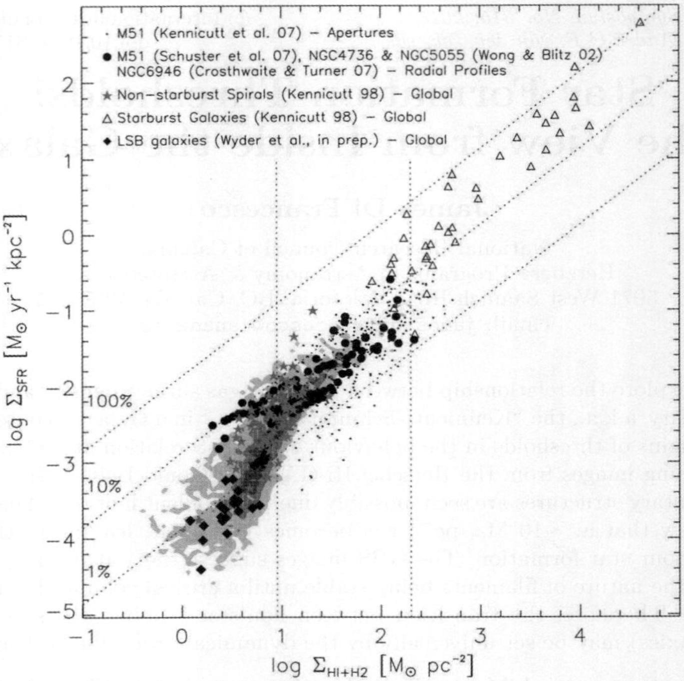

Figure 1. The relationship between total gas surface density (Σ_{HI+H_2}) and star formation rate surface density (Σ_{SFR}) (a.k.a. the "Kennicutt-Schmidt relation") over a variety of galaxy types from Bigiel *et al.* (2008). The colour contours are from galaxies from their paper. Small and large black dots, stars, triangles, and black diamonds are other galaxy populations. Notably, the K-S relation is tested here on sub-kpc scales using resolved observations of several galaxies. The horizontal dashed lines indicate the two "thresholds," one at ~10 $M_\odot$ pc^{-2} and another at 100-200 $M_\odot$ pc^{-2} where the behaviour of the relation changes. The diagonal dashed lines show lines of constant star formation efficiency, indicating the level needed to consume 1% (lower) to 10% (middle) to 100% (upper) of available gas in 10^8 years. See Bigiel *et al.* (2008) for assumptions regarding IMF, CO line ratio, X-factors and galaxy inclinations.

and interpretation of the K-S law have been matters of great debate among extragalactic observers and theorists for almost two decades (e.g., see Kennicutt & Evans 2012).

2. The Kennicutt-Schmidt Relation and its Thresholds

Figure 1 shows a determination of the K-S relation by Bigiel *et al.* (2008) that includes observational data from low surface brightness (LSB) galaxies, non-starburst spirals, starburst galaxies, and resolved observations of the nearby galaxies M51, NGC 4736, NGC 5055, and NGC 6946. The achievement here was that the K-S relation was explored on sub-kpc scales. Though a degree of overall correlation is still seen, the power-law behaviour of the K-S relation is seen to vary somewhat over total gas surface density. At lower values, i.e., $\Sigma_{HI+H_2} < 10$ $M_\odot$ pc^{-2}, the relation is broad and relatively steep, but at higher values, the relation is shallower. Furthermore, somewhere around Σ_{HI+H_2} ~ 100-200 $M_\odot$ pc^{-2}, the relation becomes steeper. The two values of total gas surface density appear to correspond to *thresholds* for star formation production in galaxies, at least when averaged on sub-kpc scales along the line of sight.

The spatial averaging that is inescapable when observing external galaxies indeed makes interpreting the K-S relation tricky. Examining where star-forming Galactic clouds

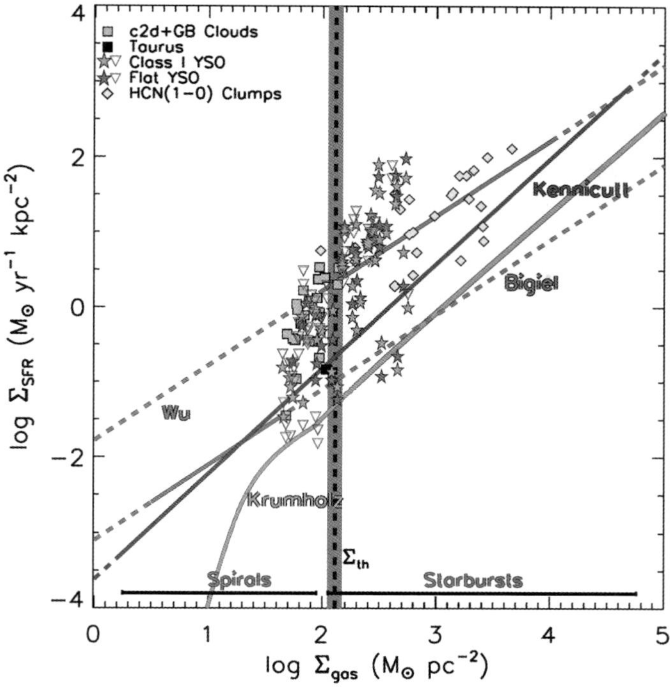

Figure 2. The relationship between gas surface density (Σ_{gas}) and SFR surface density (Σ_{SFR}) for Galactic objects from Heiderman *et al.* (2010). Squares indicate nearby clouds whose YSO populations were enumerated by Spitzer, stars indicate protostellar cores and diamonds indicate "massive" clumps. The extragalactic K-S scaling relations found by Kennicutt (1998) and Bigiel *et al.* are indicated by blue and red lines. The threshold in gas surface density where the relationship changes is marked by a vertical grey bar and black dashed line. Note how the Galactic objects have Σ_{SFR} values much larger than those expected from the extragalactic scaling relations.

lie on a K-S relation diagram is instructive. For example, Figure 2 shows a determination of the K-S relation for Galactic clouds by Heiderman *et al.* (2010). Using Spitzer data to determine the actual YSO content of nearby clouds, these authors found that such clouds have $\Sigma_{\rm SFR}$ values about an order of magnitude above those expected from Figure 1 for given total (molecular) gas surface densities. For individual protostellar cores, the discrepancy was found to be even larger. If such clouds and cores were placed in external galaxies, however, the observed total gas surface densities would be much lower, as they would be included in a spatial element with other non-star-forming HI and molecular gas. In addition, their YSOs would be hard to detect (if at all). Hence, a different conclusion about the relation between $\Sigma_{\rm SFR}$ and $\Sigma_{\rm HI+H_2}$ would result. As an aside, it is worth noting that when Heiderman *et al.* added dense Galactic clumps traced by HCN (1–0) into the mix, they found the trend in $\Sigma_{\rm SFR}$ with $\Sigma_{\rm HI+H_2}$ changing at 129 ± 14 $M_\odot$ pc^{-2}, a value similar to the second threshold seen by Bigiel *et al.*

The transitions seen in Figure 1 can be understood as depending on what clouds at which phase are dominant within a spatial element of a galaxy. A single spatial element may encompass multiple phases of the ISM, including several neutral HI and molecular clouds. At lower $\Sigma_{\rm HI+H_2}$ values, the gas *on average* is <10 $M_\odot$ pc^{-2}, a value corresponding (within a factor of 2) to a line-of-sight extinction $A_V \sim 1$. Below that extinction, self-shielding of the ISRF by H_2 becomes ineffective, and molecular gas cannot be sustained

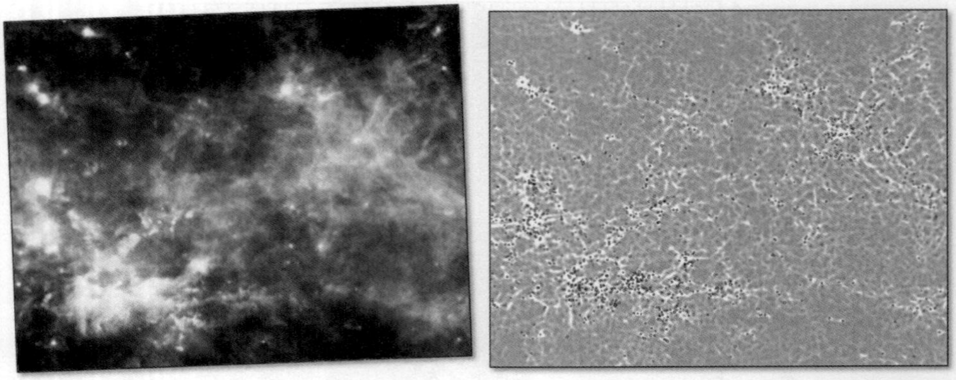

Figure 3. (*left*): Three-colour image composed of 70 μm (blue), 160 μm (green) and 350 μm (red) of the Hi-GAL field around $l = 59°$ from Molinari *et al.* (2010). The left panel is tilted slightly to correspond spatially with features in the right panel. (*right*): A multidirectional second-derivative image of the same field as in the left from 250 μm data, also from Molinari *et al.* (2010). White denotes locations of maximal emission, which to first order trace locations of peak column density. Blue circles represent compact sources detected in the 250 μm image. This representation clearly reveals a web-like structure in the ISM due to filaments. The compact structures, likely clouds and dense cores within them, are found preferentially where $A_V > 1$.

(see Krumholz *et al.* 2009). Hence, the low Σ_{HI+H_2} regions of galaxies have *on average* low amounts of molecular gas. Star formation, however, is still found in such elements because there can be within them at least some molecular gas with pockets of dense gas that form stars. Such gas is relatively rare, however, and so star formation may be effectively "stochastic" with no true relation to total gas density (Kennicutt & Evans 2012). Above $\Sigma_{HI+H_2} \approx 10$ M$_\odot$ pc^{-2}, however, spatial elements include *on average* gas that exceeds the threshold for molecular persistence. In such clouds, a more regular trend is seen between Σ_{HI+H_2} and Σ_{SFR}, indicating some percentage of gas dense enough to form stars is commonly present. The dashed lines in Figure 1 suggest star formation efficiencies in this regime as being 1-10%, and the large scatter suggests molecular clouds do not have a common percentage of gas within them that is dense enough to form stars. Around $\Sigma_{HI+H_2} \approx 100$-200 M$_\odot$ pc^{-2}, a second transition is seen where star formation efficiencies jump to 10-100%. At these values, the spatial elements (here, entire starburst galaxies) now encompass *on average* dense gas with extinctions $A_V > 8$ that forms stars, and the SFR surface density rises accordingly.

3. Herschel Continuum Observations of Galactic Clouds

Though the nature behind the first transition is fairly well understood, the same cannot be said about the second. What makes gas where extinctions exceed $A_V > 8$ so efficient at forming stars? Recently, continuum observations of Galactic clouds by ESA's Herschel Space Observatory have provided important clues about this threshold. Herschel was a 3.5-m diameter submillimeter space observatory situated at the Sun-Earth L2 point that operated from 2009 to 2013. Two of its instruments, PACS (Poglitsch *et al.* 2010 and SPIRE (Griffin *et al.* 2010) had photometric modes that allowed simultaneous mapping of emission from dust at 70 μm or 100 μm and 160 μm (PACS) and 250 μm, 350 μm, and 500 μm (SPIRE). These wavelengths are unobservable from the ground due to atmospheric absorption, and Herschel allowed wide-field mapping of cold Galactic clouds (among other targets) at unprecedented sensitivity and dynamic range. In particular, Herschel detected from clouds faint large-scale emission that was previously unseen by ground-based observatories utilizing atmospheric windows (e.g., at 850 μm) due to the necessary removal from such data of large-scale emission from the atmosphere itself. In the following, we discuss some results from a few Herschel projects related to star formation thresholds.

Figure 3 (left) shows a multi-color image obtained with Herschel of the $l = 59°$ field for the large Hi-GAL survey of the Galactic Plane by Molinari *et al.* (2010). The ~3.5 deg^2 field exhibits considerable complexity with emission seen at all scales and the brighter knots denote star-forming cloud complexes. Notably, colour in this image broadly reflects temperature, with redder locations being dominated by 500 μm emission and thus cold. Figure 3 (right) is the same field as Figure 3 (left) but shows the second derivatives of the 250 μm emission, done to reveal spatial relationships between locations of locally bright emission. Here, white shows locations of maximal emission, and these appear connected in a web-like arrangement of filamentary structures across the entire field. Blue circles indicate the locations of compact sources at 250 μm, and a very close correspondence is seen between the locations of such objects and the filamentary structure. Indeed, the compact sources, here molecular clouds or dense structures within them, are seen only at column densities equivalent to extinctions $A_V > 1$. This correspondence demonstrates the close connection between the persistence of molecular gas and extinctions of $A_V > 1$, i.e., the first threshold in Figure 1. Crucially, however, Figure 3 (right) suggests a filamentary origin of the molecular clouds, possibly one shaped by dynamical (e.g., turbulent) flows where atomic gas is shaped into filaments within which only the densest locations exceed $A_V \sim 1$ and become molecular.

Closer to home, Herschel data show nearby molecular clouds have ubiquitous *internal* filamentary structures. With a more-detailed look at such clouds than generally possible with Hi-GAL, the conditions for star formation can be better seen. For example, Figure 4 shows mid-scale spatial components of column densities derived from SPIRE and PACS data of the nearby Aquila Rift (left) and Polaris Flare (right) molecular clouds by the Herschel Gould Belt Survey (André *et al.* 2010; Ward-Thompson *et al.* 2010; Könyves *et al.* 2010). Here, the column densities from compact and diffuse structures have been suppressed via a curvelet decomposition. The Aquila Rift is an active star-forming cloud containing the Serpens South and W40 clusters and the Polaris Flare is a relatively low column density cloud with no known star formation. Regardless of star-forming activity, however, Figure 4 shows that filamentary structures pervade both clouds. Indeed, the presence of filaments in both suggests an origin for the structure that is independent of star formation itself, and, as with the larger-scale filaments seen in Figure 3, one perhaps related to turbulence. Figure 5 shows a histogram of deconvolved filament widths determined from non-decomposed column density images of eight nearby clouds, including Aquila and Polaris (see Arzoumanian *et al.* 2011; André *et al.* 2014). Interestingly, the filamentary structures in these clouds have a narrow range of widths around ~0.1 pc, despite a factor of 4 range in distance. The 0.1 pc scale is where thermal and turbulent velocities are equivalent (see Larson 1981), again suggesting a turbulent origin.

Figure 4 (left) also shows a close association of gravitationally bound prestellar cores and known Class 0 objects with the densest filaments, i.e., those exceeding an extinction $A_V \sim 8$, in Aquila. Quantitatively, ~75% of prestellar cores are found above the equivalent column density of 130 M$_\odot$ pc^{-2}. Clearly, core formation is more efficient within sufficiently dense filaments. Indeed, the filamentary morphology so prevalent within clouds may provide the key to understanding the initial conditions for star formation, and perhaps also the key to the second threshold of Figure 1. Representing filaments

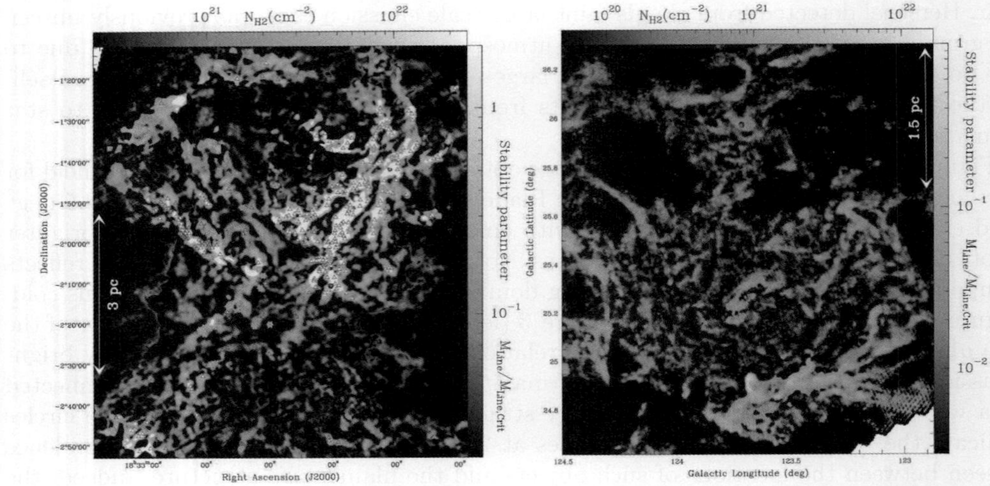

Figure 4. Column density maps of subfields of the Aquila Rift (*left* and the Polaris Flare (*right*) derived from SPIRE and PACS data obtained by the Herschel Gould Belt Survey from André *et al.* (2010). The maps have been spatially decomposed to highlight mid-scale spatial structures in these clouds. Large-scale, diffuse structure and small-scale, compact structures have been effectively suppressed. Linear filamentary features are seen to pervade both clouds. The colour scale is different for each image. On the left, the column densities for Aquila range between $\sim 2 \times 10^{20}$ cm^{-2} (black) and $\sim 2 \times 10^{22}$ cm^{-2} (white). On the right, the range for Polaris is $\sim 7 \times 10^{19}$ cm^{-2} to $\sim 10^{22}$ cm^{-2}. Assuming a common filamentary width of 0.1 pc, the colour scale also reflects the local lines per unit length, with the ratio of $M_{\rm line}/M_{\rm line,crit}$ extending from $\sim$0.02 to $\sim$2 on the left and from $\sim$0.002 to 1 on the right. Blue triangles denote prestellar cores and green stars denote Class 0 young stellar objects. Note the clear spatial correspondence between such objects and filaments of line per unit length above the critical value. Of the two clouds, only in Aquila are such high column densities reached, and consequently star formation is ongoing.

theoretically as isothermal cylinders, the stability of such structures to fragmentation can be understood as depending simply on a critical mass per unit length defined as $M_{\rm line} = 2c_s^2/{\rm G}$ (Inutsuka & Miyama 1997; see also Ostriker 1964). For typical temperatures of 10 K in a molecular cloud, this critical value is 16 $M_\odot$ pc^{-1}. Above this value, perturbed filaments will segment, i.e., fragment locally along their long axes, into cores rather than globally collapse (Pon *et al.* 2011). The colour scale of Figure 4 has been fixed so that filaments whose $M_{\rm line}$ values exceed the critical value are white, illustrating the close connection between star formation activity and such filaments.

Given the seemingly common filament width of $\sim$0.1 pc (see Figure 5), the critical $M_{\rm line}$ translates to a critical column density of $\sim$160 $M_\odot$ pc^{-2} which is approximately equivalent to an extinction $A_V \sim 8$. Hence, we can understand the higher efficiency of star formation in filaments as resulting from the ability of such structures to fragment effectively into cores. Real filaments, however, are not strictly isothermal cylinders and exhibit small (<5 K) temperature dips in their interiors (see Palmeirim *et al.* 2013) and they likely have internal turbulent velocities on par with their thermal velocities. Moreover, they exhibit a small but finite distribution of widths around 0.1 pc (see Figure 5), and may be inclined relative to the plane of sky. For these reasons, the observed "threshold" for core/star formation is not necessarily sharp at 160 $M_\odot$ pc^{-2} (see André *et al.* 2014).

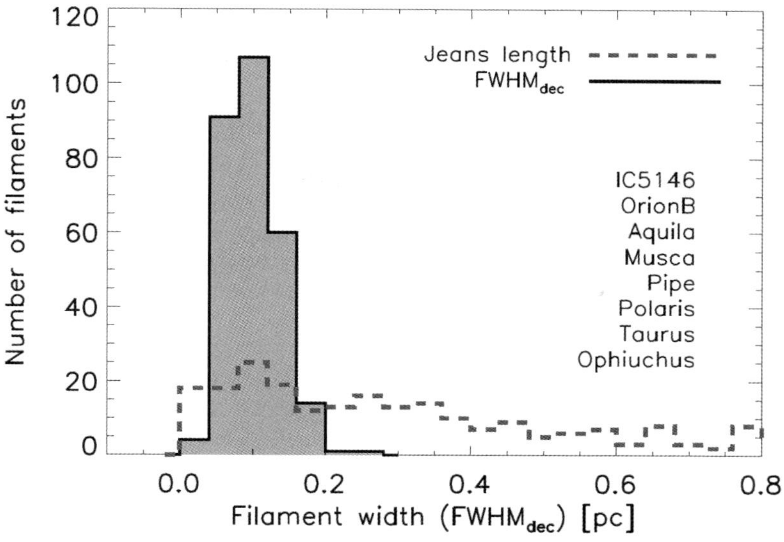

Figure 5. Histogram of deconvolved FWHMs of 278 filaments recovered in eight molecular clouds by Herschel GBS observations from André *et al.* (2014). The widths were determined from Gaussian fits to the filament profiles obtained from non-decomposed (all-scale) column density maps. The distribution is peaked at ~0.1 pc with a width of ~0.04 pc. For comparison, the much wider distribution of Jeans lengths corresponding to the central column densities of the filaments is also shown as the blue-dashed histogram.

4. A Universal Threshold for Star Formation?

The similarity between the critical column densities seen for filaments in nearby clouds and the second threshold of Figure 1 for galaxies strongly suggests that the filamentary morphology and notably its critical requirement for instability are responsible physically and universally for the transition between diffuse, non-star-forming molecular gas and dense, star-forming molecular gas. Indeed, filaments may provide an effective means for gas to condense within clouds to produce through fragmentation a mass distribution of cores that is similar in shape to the stellar IMF, as seen directly in Aquila by Könyves *et al.* (2010) (see also Könyves *et al.* 2015). In Aquila, the peak of the core mass function is 0.6 $M_\odot$ which is roughly equal to the locally critical Bonnor-Ebert mass of 0.5 $M_\odot$. Moreover, a segment of 0.1 pc length in a critical isothermal filament has a typical mass of 1.6 $M_\odot \approx 3 \times M_{BE}$, indicating that the typical critical filament is locally unstable. Hence, we may understand the peak of the clump mass function, and perhaps the IMF, as indeed due to thermal fragmentation (Larson 1985). Furthermore, an recent analysis of 1-D fluctuations along 80 filaments by Roy *et al.* (2015) revealed a mean power spectrum slope of -1.6, similar to the slope of -1.5 from Kolmogorov turbulence. Inutsuka (2001) previously showed such fluctuations could evolve naturally towards a core mass function with a observed Salpeter-like slope ($\alpha \approx$ -2.35) at higher masses.

Though highly suggestive, further work is needed to test the robustness of the connection between the critical column density of filamentary fragmentation and the observed threshold to higher star forming efficiencies in Figure 1. Notably, the Herschel GBS has concentrated primarily on local clouds that do not produce the kinds of high-mass stars used to determine SFRs in distant galaxies. It would be highly instructive to explore the relationship between filamentary structures in giant molecular clouds and high-mass star formation. Preliminary work of this kind has begun using data of giant molecular clouds within 3 kpc from the Herschel OB Young Star (HOBYS) survey (Hill *et al.* 2011;

Schneider *et al.* 2012). Furthermore, studies of the dynamics of filaments using robust kinematic tracers are also needed, to determine how such structures are evolving. Preliminary work also has begun in this direction. (Schneider *et al.* 2010; Arzoumanian *et al.* 2014; Friesen *et al.* 2015, in prep).

Acknowledgement

The author thanks Philippe André, Derek Ward-Thompson, Shu-Ichiro Inutsuka, Ralph Pudritz, and Jaime Pineda for the stimulating collaboration on which this paper is based. The author also thanks the Hi-GAL, GBS, and HOBYS Key Programme teams.

References

André, Ph., Di Francesco, J., Ward-Thompson, D., Inutsuka, S.-I., Pudritz, R., & Pineda, J. 2014, in: H. Beuther, R. Klessen, P. Dullemond, & Th. Henning (eds.), *Protostars and Planets VI* (Arizona: Tucson), p. 27

André, Ph., *et al.* 2010, *A&A*, 518, L102

Arzoumanian, D., *et al.* 2013, *A&A*, 553, A119

Arzoumanian, D., *et al.* 2011, *A&A*, 529, L6

Bigiel, F., Leroy, A., Walter, F., Brinks, E., de Blok, W. J. G., Madore, B., & Thornley, M. 2008, *AJ*, 136, 2846

Di Francesco, J., Evans, N. J., II, Caselli, P., Myers, P. C., Shirley, Y., Aikawa, Y., & Tafalla, M. 2007, in B. Reipurth, D. Jewitt, & K. Keil (eds.), *Protostars and Planets V* (Arizona: Tucson), p. 17

Friesen, R. K., Di Francesco, J., Shirley, Y., & Myers, P. C. 2009, *ApJ*, 697, 1457

Griffin, M., *et al.* 2010, *A&A*, 518, L3

Heiderman, A., Evans, N. J., II, Allen, L. E., Huard, T., & Heyer, M. 2010, *ApJ*, 723, 1019

Hill, T., *et al.* 2011, *A&A*, 533, A94

Inutsuka, S.-I. 2001, *ApJ*, 559, L149

Inutsuka, S.-I. & Miyama, S. 1997, *ApJ*, 480, 681

Johnstone, D., Di Francesco, J, & Kirk, H. M. 2004 *ApJ*, 611, L45

Kennicutt, R. 1998, *ApJ*, 498, 541

Kennicutt, R. 1989, *ApJ*, 344, 685

Kennicutt, R. & Evans, N. J., II 2012, *ARAA*, 50, 531

Krumholz, M. R., McKee, C. F., & Tumlinson, J. 2009 *ApJ*, 693, 216

Könyves, V., *et al.* 2015, *A&A*, in press

Könyves, V., *et al.* 2010, *A&A*, 518, L106

Lada, C. J., Lombardi, M., & Alves, J. 2010, *ApJ*, 724, 687

Larson, R. B. 1985 *MNRAS*, 214, 379

Larson, R. B. 1981, *MNRAS*, 194, 809

Molinari, S., *et al.* 2010, *A&A*, 518, L100

Myers, P. C. & Benson, P. 1983, *ApJ*, 266, 309

Ostriker, J. 1964, *ApJ*, 140, 1056

Palmeirim, P., *et al.* 2013, *A&A*, 550, A38

Poglitsch, A., *et al.* 2010, *A&A*, 518, L2

Pon, A., Johnstone, D., & Heitsch, F. 2011, *ApJ*, 740, 88

Roy, A., *et al.* 2015, *A&A*, in press

Sadavoy, S., *et al.* 2014, *ApJ*, 787, L18

Schneider, N., *et al.* 2012, *A&A*, 540, L11

Schneider, N., *et al.* 2010, *A&A*, 520, A49

Ward-Thompson, D., *et al.* 2010, *A&A*, 518, L92

Wu, J., Evans, N. J., II, Gao, Yu., Solomon, P. M., Shirley, Y. L., & Vanden Bout, P. A. 2005, *ApJ*, 635, L173

From Interstellar Clouds to Star-Forming Galaxies: Universal Processes?
Proceedings IAU Symposium No. 315, 2015
P. Jablonka, Ph. André & F. van der Tak, eds.

doi:10.1017/S1743921316007493

How does metallicity affect the gas and dust properties of galaxies?

Suzanne C. Madden[1], **Diane Cormier**[2] **and Aurélie Rémy-Ruyer**[3,1]

[1]CEA/DSM/IRFU/Service d'Astrophysique, AIM (UMR 7158), 91191, Gif-sur-Yvette, France
email: suzanne.madden@cea.fr

[2]Institute for Theoretical Astrophysics, Center for Astronomy, University of Heidelberg, 69120, Heidelberg, Germany
email: diane.cormier@zah.uni-heidelberg.de

[3]Institut d'Astrophysique Spatiale, CNRS, 91405, Orsay, France
email: ruyeraurelie.astro@gmail.com

Abstract. Comparison of the ISM properties of a wide range of metal poor galaxies with normal metal-rich galaxies reveals striking differences. We find that the combination of the low dust abundance and the active star formation results in a very porous ISM filled with hard photons, heating the dust in dwarf galaxies to overall higher temperatures than their metal-rich counterparts. This results in photodissociation of molecular clouds to greater depths, leaving relatively large PDR envelopes and difficult-to-detect CO cores. From detailed modeling of the low-metallicity ISM, we find significant fractions of CO-dark H_2 - a reservoir of molecular gas not traced by CO, but present in the [CII] and [CI]-emitting envelopes. Self-consistent analyses of the neutral and ionized gas diagnostics along with the dust SED is the necessary way forward in uncovering the multiphase structure of galaxies.

Keywords. galaxies: dwarf, galaxies: ISM, galaxies: starburst, infrared: galaxies

1. Introduction

Dwarf galaxies, often carrying signatures of chemical youth, are unique laboratories to obtain insight on star formation and interstellar medium (ISM) properties in low-metallicity environments that may be relevant to understanding early universe conditions, when metal abundances were very low. The decrease of metals reduces shielding effects which have important consequences, for example, on the molecular gas reservoir, which, in turn, should affect how we observe the ISM properties. Our knowledge in this subject has been growing at a fast pace due to the increasingly sensitive infrared (IR) space missions *IRAS*, *ISO*, *Spitzer*, *AKARI* and *Herschel*, as well as ever-growing powerful ground-based millimeter (mm) and submillimeter (submm) telescopes. The Dwarf Galaxy Survey (DGS; Madden *et al.* 2013), has compiled a large observational database of 48 low-metallicity galaxies, motivated by *Herschel* PACS (Poglitsch *et al.* 2010) and SPIRE (Griffin *et al.* 2010) photometric and spectroscopic observations covering 55 to 500 μm. These data, together with the ancillary database, create a rich legacy for dust and gas analyses in unprecedented detail in low-metallicity galaxies from ultraviolet (UV) to mm wavelengths in environments covering the largest metallicity range achievable in the local Universe, as low as $\sim 1/40 Z_{\odot}$ (Rémy-Ruyer *et al.* 2013; 2015). Here we describe enigmatic issues surrounding the nature of the dust and gas properties in low-metallicity galaxies.

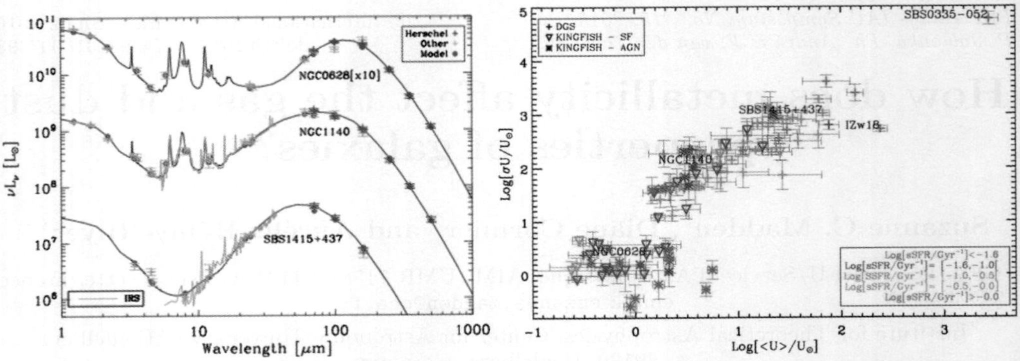

Figure 1. *(left)* Examples of SEDs, demonstrating the increasing broadening of the SEDs, or the wider distribution of the starlight intensities (σU). Metallicity is decreasing from the metal-rich galaxy, NGC0628, to the metal-poor galaxy, SBS1415+437. *(right)* Broadening of the SED (σU) as a function of average starlight intensity ($< U >$) for the DGS + KINGFISH galaxies. The color-code is the specific star formation rate (sSFR $\equiv$ SFR/$M_\star$) of the sources. Notice that the low-metallicity dwarf galaxies tend to have higher $< U >$ and higher σU (broader dust SEDs) while the more metal-rich KINGFISH galaxies tend towards lower sSFR values, lower $< U >$ (lower T$_d$), and narrower SEDs (see Rémy-Ruyer *et al.* 2015). *(Color version available online.)*

2. Metallicity effects on dust properties and gas-to-dust mass ratios

2.1. *Dust spectral energy distributions and dust temperatures*

The dust spectral energy distributions (SEDs) of the DGS galaxies and the *Herschel* KINGFISH survey, consisting of mostly metal-rich galaxies (Kennicutt *et al.* 2011; Dale *et al.* 2012) have been modeled by Rémy-Ruyer *et al.* (2015) using the model of Galliano *et al.* (2011). This complete, consistently modeled sample consists of 109 galaxies combined, covering almost 2 dex in metallicity, making it a valuable sample to investigate the effects of metallicity on the SED shape and the dust properties.

Low-metallicity galaxies tend to have higher average starlight intensity ($< U >$) and broader SED peaks (σU) shifted to shorter wavelengths compared to the more metal-rich KINGFISH galaxies (Fig. 1, *left*), as a result of their high star formation activity. This is reflected in the overall dust temperatures (T$_d$): the low-metallicity galaxies span a broader range in temperature and a higher mean T$_d$ of 26 K, in contrast to the higher-metallicity KINGFISH galaxies which span a narrower temperature distribution and a lower mean T$_d$ of 20 K (Rémy-Ruyer *et al.* 2013; 2015). A trend of increasing T$_d$ with decreasing metallicity is likewise observed. Also seen in Fig. 1 *(right)* is the trend of increasing $< U >$ and σU with increasing specific star formation rate sSFR showing that the sSFR is the parameter controlling the dust SED shape, while we observe a weaker correlation of $< U >$ and σU with metallicity, placing metallicity in a secondary role in shaping the observed SED. The DGS galaxies contain a wider range of warmer dust, primarily due to their intensive star formation activity, not necessarily due to the fact that they are low metallicity. The secondary effect of the low metallicity is reflected in the higher T$_d$. The consequence of the lower dust attenuation is the heating of the dust deeper into the molecular clouds.

2.2. *Gas-to-dust mass ratio versus metallicity*

With dust masses accurately and consistently measured for 109 DGS plus KINGFISH galaxies over the full MIR to submm wavelength range and covering a broad metallicity range, it is now possible to characterize the impact of metallicity on the gas-to-dust mass ratio (G/D) over almost 2 dex range of metallicity (Fig. 2, *left*; Rémy-Ruyer *et al.*

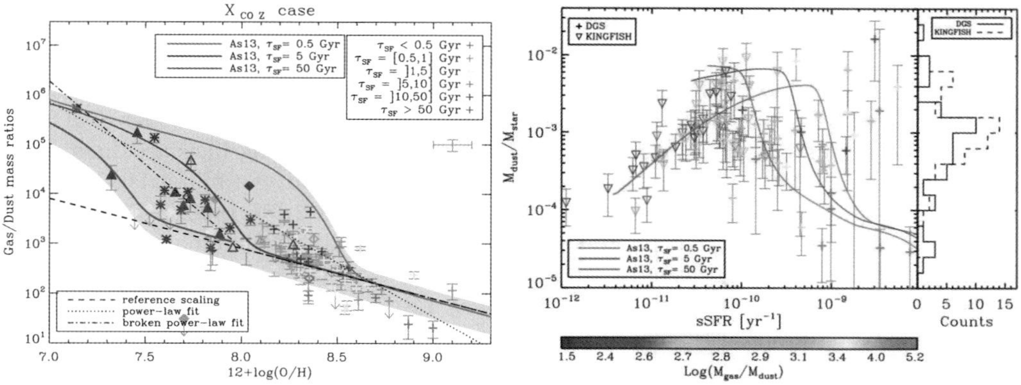

Figure 2. *(left)* Observed gas-to-dust mass ratio (G/D) for the DGS + KINGFISH galaxies compared to models of Asano *et al.* (2013) for various star formation time scales (τ) and star formation histories. From the study of Rémy-Ruyer *et al.* (2014). *(right)* Dust-to-stellar mass ratios for the DGS (crosses) and KINGFISH (downward triangles) galaxies as a function of sSFR, color-coded by M_{gas}/M_{dust} (from Rémy-Ruyer *et al.* 2015). The distribution of M_{dust}/M_{star} is indicated on the side: solid line for DGS and dashed line for KINGFISH. *(A color version of this figure is available online.)*

2014). We observe a broad degree of scatter over all metallicities and a steep rise in the G/D for the lower-metallicity galaxies for which 12 + log (O/H) $\lesssim$ 8. This behavior is found to be consistent with chemical evolution models that require grain growth in the ISM as a source of dust production as well as episodic star formation history (e.g. Zhukovska *et al.* 2014; Asano *et al.* 2013). Dust production becomes more efficient when enough metals can accumulate, and this occurs in the observations and models at around 12 +log (O/H) $\sim$ 8. Normalizing the dust mass by the stellar mass gives us a view of how the dust mass builds itself up as a function of the stellar mass. Previous studies of the dust-to-stellar mass ratio of mostly rather metal-rich galaxies (e.g. da Cunha *et al.* 2010; Skibba *et al.* 2011; Cortese *et al.* 2012) have shown that the dust-to-stellar mass ratio decreases for increasing stellar mass (or metallicity) and decreasing sSFR. We extend the dust-to-stellar mass ratio studies to lower-metallicity, low-mass galaxies with our DGS galaxies and we do not find any correlation for the dust-to-stellar mass ratio with metallicity or sSFR over the full sample. The results for only the more metal-rich KINGFISH galaxies are in agreement with the findings of previous studies. However, the low-metallicity sample does not extend the same behavior of the metal-rich galaxies to higher sSFR. Instead we see a wide scatter in dust-to-stellar mass ratio for the DGS galaxies and a very low dust-to-stellar mass ratio for some of the lowest metallicity (high G/D) galaxies (Fig. 2, *right*). The different behavior of the dust-to-stellar mass ratios for the low-metallicity DGS galaxies can be explained by their chemical evolutionary stage, also traced by the G/D (Rémy-Ruyer *et al.* 2014). The evolutionary tracks of the chemical evolution model of Asano *et al.* 2013 (Fig. 2) show a turnover with the scatter related to various star formation time scales (τ) and star formation histories. Dust growth processes saturate when all the available metals are locked up in the grains. In the meantime, star formation continues, consuming the gas reservoir and increasing the stellar mass. This results in increasing the sSFR, decreasing the dust-to-stellar mass ratio and decreasing the G/D. The prominent non-linearity of the G/D vs metallicity at low metallicity, as well as the wide degree of scatter, even for the metal-rich galaxies, call for caution when attempting to determine the total gas mass from the dust mass with an assumed G/D simply scaled linearly with metallicity.

3. The FIR fine-structure lines reveal a porous ISM structure

The DGS galaxies were surveyed in the most important FIR fine-structure lines, such as 158 μm [CII], 63 and 145 μm [OI], 88 μm [OIII] and more rarely, due to sensitivity limits, 57 μm [NIII], 122 and 205 μm [NII] (see Madden *et al.* 2013 and Cormier *et al.* 2015 for more details of the DGS observations). These important cooling lines are diagnostics probing the FUV flux, the gas density and temperature, and the filling factor of the ionized gas and photodissociation regions (PDRs; e.g. Wolfire *et al.* 1990; Kaufman *et al.* 2006; Le Petit *et al.* 2006). The photoelectric effect is normally the dominant source of gas heating in PDRs and [CII] usually ranks foremost in the PDR cooling lines in galaxies, followed by the 63 μm [OI]. Thus the [CII]/L_{FIR} ratio, or ([CII]+ [OI]/L_{FIR}, indicates the efficiency of photoelectric heating. Fig. 3 *(left)* shows the ([CII]+ [OI])/L_{FIR} ratio as a function of L_{FIR}. The ratios of dwarf galaxies are higher than those of galaxies in other surveys of mostly metal-rich galaxies. These relatively larger efficiency factors, from 1% to 2%, can be a consequence of relatively normal PDR gas densities (often on the order of 10^3 to 10^4 cm^{-3}) and low average ambient radiation fields over full galaxy scales (Cormier *et al.* 2015). Note, in contrast, the FIR line deficit seen in the most luminous sources (e.g. Luhman *et al.* 2003; Gracia-Carpio *et al.* 2011; Diaz Santos *et al.* 2013; Farrah *et al.* 2013) where dustier HII regions are present leading to a high ionization parameter on global scales. On the other hand, the 88 μm [OIII] line, requiring an ionization energy of 35 eV, is the brightest FIR cooling line in almost all of the dwarf galaxies (Cormier *et al.* 2015) - not the [CII] line, as in the normal-metallicity galaxies. While higher stellar effective temperatures in the low metallicity star-bursting dwarf galaxies already can account for elevated [OIII] line emission compared to more metal-rich stellar atmospheres (e.g. Hunter *et al.* 2001), the predominance of the [OIII] line demonstrates the ease at which such hard photons can leak out of the medium surrounding stars and traverse the ISM on full galaxy scales (see also Cormier *et al.* 2012; Lebouteiller *et al.* 2012).

The dwarf galaxies also show extreme [CII]/CO ratios (Fig. 3, *right*): while most normal star-forming galaxies are observed to have [CII]/CO $\sim$ 1 000 to 4 000, the low-metallicity galaxies can reach much higher ratios - up to an order of magnitude higher, or more, in [CII]/CO over galaxy-wide scales (e.g. Madden 2000; Cormier *et al.* 2014). Elevated $L_{[C\,II]}/L_{FIR}$ and $L_{[C\,II]}/L_{CO(1-0)}$ values as seen above, seem to be attributed to the effects that are unique to the low-metallicity and clumpy ISM: a consequence of the lower dust abundance and a longer mean-free path of the FUV photons which penetrate molecular clouds, photodissociating the CO and leaving a relatively larger [CII]-emitting envelope surrounding a small CO core. This is consistent with the fact that we see such luminous 88 μm [OIII] line emission over full galaxy-wide scales, and rarely can detect significant CO line emission. The decreased attenuation and the presence of intense, hard radiation fields, together facilitate the photodissociation of the molecular clouds, highlighting the different nature and structure of the ISM of low-metallicity galaxies.

4. H_2 mass in dwarf galaxies and the CO-dark gas

While CO is routinely used to trace the H_2 reservoir in local as well as high-z galaxies, its utility can be called into question for dwarf galaxies, which are notoriously deficient in CO (e.g. Schruba *et al.* 2012; Cormier *et al.* 2014 and references within), leaving us with large uncertainties in quantifying the molecular gas mass, and compounding the difficulty in understanding the process of both local and global star formation in low-metallicity environments. This road block is necessary to overcome in order to refine recipes for converting gas into stars under early-universe conditions and thus further our

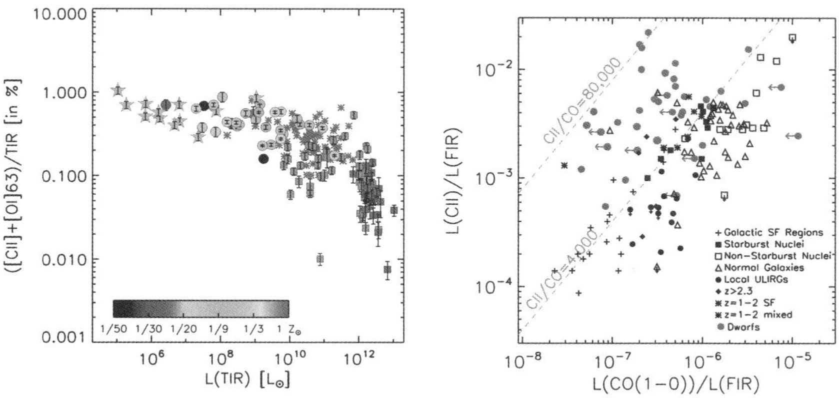

Figure 3. *(left)* ([CII]+[OI])/L_{TIR} vs L_{TIR}, from Cormier *et al.* (2015). *(right)* [CII]/L_{FIR} vs CO(1-0)/L_{FIR} for the dwarf galaxies (red symbols) and a wide range of metal-rich galaxies (Madden *et al.* in prep.). The metal-rich galaxies, black closed and open symbols, are from Stacey *et al.* (1991). *(A color version of this figure is available online.)*

understanding of how galaxies evolve. We are left with the question: where is the bulk of the fuel for the obvious star formation taking place in many of the dwarf galaxies?

While extreme observed [CII]/CO values of the dwarf galaxies suggest very active star formation, many of the dwarf galaxies are thought to be HI-rich rather than H_2-rich. The fact that their molecular gas reservoirs remain uncertain renders their location on the Schmidt-Kennicutt diagram ambiguous. Depending on how the total gas surface densities are determined (from HI, HI+H_2, CO, dust, or HI+H_2-modeled) several star-forming dwarfs deviate from the Schmidt-Kennicutt correlation with elevated star formation rates compared to their gas masses (Fig. 4; Cormier *et al.* 2014).

The lower dust abundance in dwarf galaxies allows the CO molecular clouds to be subjected more easily to photodissociation by UV photons, making the detections of the smaller CO cores with single-dish telescopes difficult. H_2, on the other hand, is photodissociated by absorption of Lyman-Werner band photons. These bands become optically thick at modest A_V values and the H_2 becomes self-shielded. This self-shielding effect can leave a potentially significant reservoir of H_2 in the [CI] and [CII]-emitting regions, outside of the CO core (e.g. Roellig *et al.* 2006; Wolfire *et al.* 2010; Glover *et al.* 2012). This is the CO-dark molecular gas and the [CII], [CI] and/or other FIR fine-structure lines, may have the capability to trace this "missing" molecular gas that is not probed by CO. This dark gas reservoir was first uncovered in a few low-metallicity galaxies using [CII] (Poglitsch *et al.* 1995; Israel *et al.* 1996; Madden *et al.* 1997), but *Herschel* has now increased the sample with deeper sensitivity. More recently [CII] observations of the Milky Way have found that on average the fraction of CO-dark gas, compared to the total H_2 can be as high as $\sim 30\%$ (Pineda *et al.* 2013).

Determination of their true source of the molecular gas reservoir requires more than CO observations, which have proven to not be a reliable tracer of the total H_2 gas reservoir. The MIR and FIR fine-structure lines have recently been modeled for each galaxy of the DGS survey to quantify the predicted total H_2 reservoir determined from the best-fitted model (Cormier *et al.* in prep). This is then compared to the H_2-determined from the CO observations, the difference being the CO-dark H_2 (Fig. 4, *right*). The mass fraction of CO-dark gas / H_2 traced by CO, spans at least 4 orders of magnitude. The contribution of the CO-dark gas is greatest under conditions of the lowest A_V environments as well low values of density and radiation field which correlate with higher [CII]/L_{FIR} rates

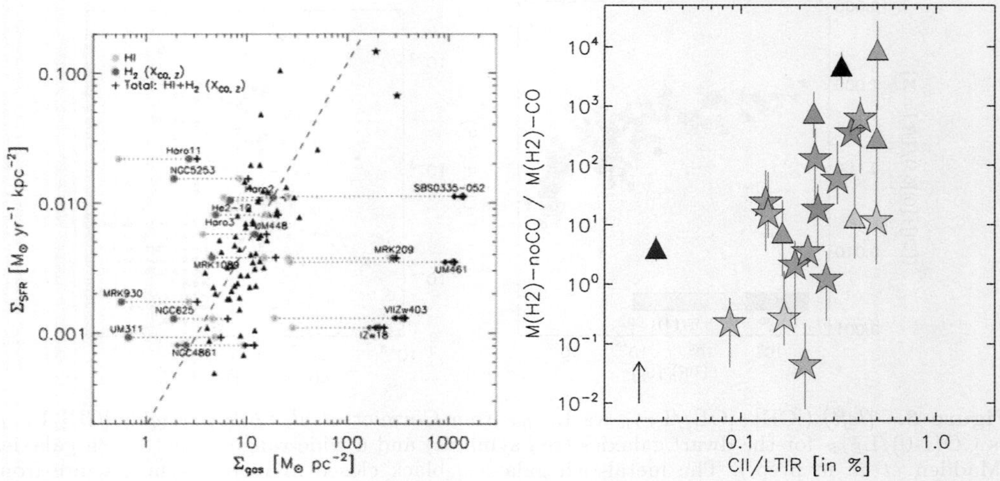

Figure 4. *(left)* Star formation rate surface density (Σ_{SFR}) versus gas surface density (Σ_{gas}) for metal-rich starbursts and spirals from Kennicutt 1998 (black triangles) and the low metallicity DGS galaxies. Taking into account the large reservoir of HI brings the dwarf galaxies closer to the Schmidt-Kennicutt relationship (dotted line with power index of 1.4), while some remain as outliers to the right and some to the left of the relationship (see Cormier *et al.* 2014 for more details). *(right)* The CO-dark H_2 gas mass /H_2 determined from CO alone versus the observed [CII]/L_{TIR} from modeling of the DGS (Madden *et al.* in prep). The triangles are upper limits due to lack of CO detections. The color indicates metallicity values as in the color scale in Fig. 3. *(A color version of this figure is available online.)*

(with large scatter). Calibrating this empirically is work in progress. While we find that both [CII] and [CI] do indeed trace the CO-dark gas under the conditions we find for the dwarf galaxies, we also find that [CI] traces more of the CO-dark mass than [CII] (Madden *et al.* in prep) making observations of the submm [CI] lines potentially the most valuable tracer of the CO-dark gas (see also Glover *et al.* 2015).

To obtain greater detailed comprehension of the photodissociation of molecular clouds in the vicinity of a massive star cluster within a low-metallicity galaxy, Chevance *et al.* (2015, submitted) have studied the spatial variation of the ISM properties surrounding the R136 cluster in our nearest low metallicity galaxy, the Large Magellanic Cloud. Modeling the suite of MIR and FIR Herschel and Spitzer fine-structure emission lines (Fig. 5, *left*) has uncovered a significant dark-gas reservoir, where more than 80% of the total H_2 is not traced by CO in the highly porous environment of 30 Doradus (Fig. 5, *right*), demonstrating the value of the [CII] and [CI] lines.

5. Way forward: modeling complexities of the low-metallicity ISM

With the wealth of tracers of the galactic phases present today, it is a challenge to determine the local physical conditions within galaxies and to relate them to the global properties. With the wide range of MIR and FIR cooling lines available from *Spitzer* and *Herschel*, for example, we are now able to model the physical conditions of the various phases of the gas and dust over full galaxy scales *as well as* spatially within galaxies. In this way we can obtain a more accurate prescription of the role of metallicity and star formation activity in governing the evolution of galaxies.

Using limited MIR and FIR gas diagnostics to deduce the conditions for star formation and the structure of the galaxy is effectively averaging ensembles of PDRs and physically

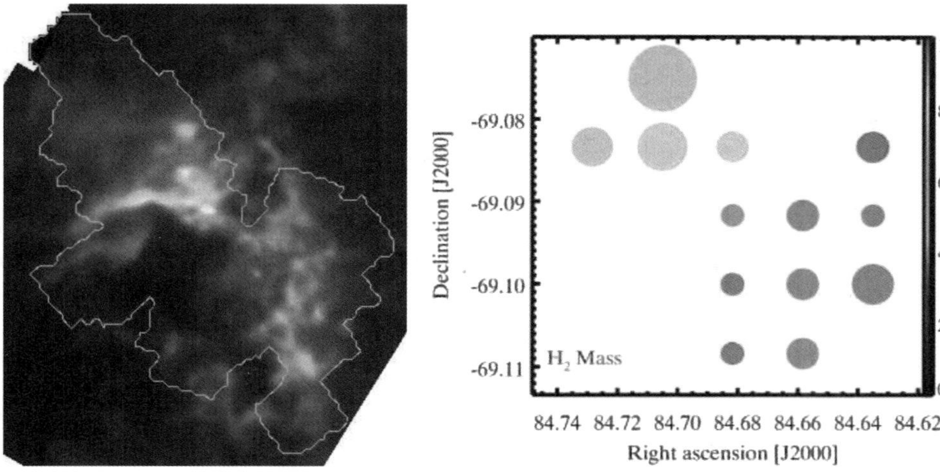

Figure 5. *(left)* 3-color image of 30 Doradus in the Large Magellanic Cloud. Red: [SIV]10.5 μm; Green: [NeII]12.8 μm; Blue: PACS [CII]158 μm. The blue contours outline the limit of the [CII] map. The different layers of the gas from the highly ionized medium near the star cluster R136 (image center with dark hole), to the PDR regions (blue), are evident (Chevance *et al.* 2015, submitted). *(right)* Fraction of the CO-dark gas calculated by comparing the predicted H_2 column density from PDR modeling to the H_2 derived from CO(1-0) observations. The color code is the CO-dark fraction which is $> 80\%$ throughout, with red the largest fraction of CO-dark gas. The size of the circles indicates the total H_2 mass, between 50 and 500 $M_\odot$ pc^{-2}. *(A color version of this figure is available online.)*

different gas phases mixed within a telescope beam. The challenge in moving forward with interpretation is to cover a range of diagnostics having varying critical densities and ionization potentials, to be able to characterize the dense and diffuse ionized gas and neutral atomic and molecular gas which have very different filling factors even for the range of dwarf galaxies *Herschel* has surveyed. Analysis of these various phases comprising galaxies cannot be done without considering the gas and dust tracers together. An example of one of the most thorough analysis that has been approached, given the wide selection of diagnostics for any one galaxy, is demonstrated in Cormier *et al.* (2012) with ~20 FIR fine-structure lines and CO observations as well as the photometry constraints to model the full SED of Haro 11. While this galaxy was not resolved with the observations, the self-consistent photoionization and photodissociation modeling was able to determine the mass fraction and filling factor of the different ISM phases. This is the approach that is necessary now with the existence of many observational constraints. To go beyond this relatively simple, yet novel multiphase model though, including more realistic geometries, will really make that significant leap forward.

6. Conclusion

From studies of the MIR and FIR gas properties and MIR to submm dust properties, a picture of the structure of low-metallicity galaxies is emerging. The decreased attenuation and the presence of intense, hard radiation fields conspire together to highlight the unique place the low-metallicity star-forming galaxies take compared to their more metal-rich counterparts. The porosity of the ISM linked with the star formation activity in dwarf galaxies, accounts for the extreme [OIII]/[CII] and [CII]/CO. Much of the ISM is filled with hard photons which, due to the decrease in global dust attenuation, allows the permeating hard photons to heat the dust to higher temperatures, and fill a relatively large volume of the galaxy and successfully photodissociate CO clouds effectively, making single-dish CO observations challenging. We find that a significant reservoir of H_2, not traced by CO (the CO-dark gas), can exist and is traceable by [CII] and [CI].

7. Acknowledgements

This research was made possible through the financial support of the French "Agence Nationale de la Recherche" (ANR) through the programme SYMPATICO (Program Blanc Projet NR-11-BS56-0023).

References

Asano,R. S., Takeuchi,T. T., Hirashita,H., & Inoue, A. K. 2013, *Earth, Planets,, Space*, 65, 213
Cormier, D., Lebouteiller, V., Madden, S. C., *et al.* 2012, *A&A*, 548, 20
Cormier, D., Madden, S. C., Lebouteiller, V., *et al.* 2014, *A&A*, 564, 121
Cormier, D., Madden, S. C., Lebouteiller, V., *et al.* 2015, *A&A*, 578, 53
Cortese, L., Ciesla, L., Boselli, A., *et al.* 2012, *A&A*, 540, 52
da Cunha, E., Eminian, C., Charlot, S., & Blaizot, J. 2010, *MNRAS*, 403, 1894
Dale, D. A., Aniano, G., Engelbracht, C. W., *et al.* 2012, *ApJ*, 745, 95
Díaz-Santos, T., Armus, L., Charmandaris, V., *et al.* 2013, *ApJ*, 774, 68
Farrah, D., Lebouteiller, V., Spoon, H. W. W., *et al.* 2013, *ApJ*, 776, 38
Galliano, F., Hony, S., Bernard, J.-P., *et al.* 2011, *A&A*, 536, 88
Glover, S. C. O. & Clark, P. C. 2012, *MNRAS*, 421, 9
Glover, S. C. O. & Clark, P. C. 2015, *2015arXiv150901939G*
Graciá-Carpio, J., Sturm, E., Hailey-Dunsheath, S., *et al.* 2011, *ApJ*, 728, 7
Griffin, M. J., Abergel, A., Abreu, A., *et al.* 2010, *A&A* (Letter), 518, 3
Hunter, D. A., Kaufman, M., Hollenbach, D. J., *et al.* 2001, *ApJ*, 553, 121
Israel, F. P., Maloney, P. R., Geis, N., *et al.* 1996, *ApJ*, 465, 738
Kaufman, M. J., Wolfire, M. G., & Hollenbach, D. J. 2006, *ApJ*, 644, 283
Kennicutt, Jr., R. C. 1998, *ApJ*, 498, 541
Lebouteiller, V., Cormier, D., Madden, S. C., *et al.* 2012, *A&A*, 548, 91
Le Petit, F., Nehmé, C., Le Bourlot, J., & Roueff, E. 2006, *ApJS*, 164, 506
Luhman, M. L., Satyapal, S., Fischer, J., *et al.* 2003, *ApJ*, 594, 758
Madden, S. C., Poglitsch, A., Geis, N., Stacey, G. J., & Townes, C. H. 1997, *ApJ*, 483, 200
Madden, S. C. 2000, *NewAR*, 44, 249
Madden, S. C., Rémy-Ruyer, A., Galametz, M., *et al.*, 2013, *PASP*, 125, 600
Pineda, J. L., Langer, W. D., Velusamy, T., & Goldsmith, P. F. 2013, *A&A*, 554, 103
Poglitsch, A., Krabbe, A., Madden, S. C., *et al.* 1995, *ApJ*, 454, 293
Poglitsch, A., Waelkens, C., Geis, N., *et al.* 2010, *A&A* (Letter), 518, 2
Rémy-Ruyer, A., Madden, S. C., Galliano, F., *et al.* 2013, *A&A*, 557, 95
Rémy-Ruyer, A., Madden, S. C., Galliano, F., *et al.* 2014, *A&A*, 563, 31
Rémy-Ruyer, A., Madden, S. C., Galliano, F., *et al.* 2015, *A&A*, 582, 121
Röllig, M., Ossenkopf, V., Jeyakumar, S., Stutzki, J., & Sternberg, A. 2006, *A&A*, 451, 917
Schruba, A., Leroy, A. K., Walter, F., *et al.* 2012, *AJ*, 143, 138
Skibba, R. A., Engelbracht, C. W., Dale, D., *et al.* 2011, *ApJ*, 738, 89
Stacey, G. J., Geis, N., & Genzel, R., *et al.* 1991, *ApJ*, 373, 423
Wolfire, M. G., Tielens, A. G. G. M., & Hollenbach, D. 1990, *ApJ*, 358, 116
Wolfire, M. G., Hollenbach, D., & McKee, C. F. 2010, *ApJ*, 716, 1191
Zhukovska, S. 2014, *A&A*, 562, 76

From Interstellar Clouds to Star-Forming Galaxies: Universal Processes?
Proceedings IAU Symposium No. 315, 2015
P. Jablonka, Ph. André & F. van der Tak, eds.

doi:10.1017/S174392131600750X

Molecular gas in galaxies: changing conditions from disks to starbursts

Christine D. Wilson

Department of Physics and Astronomy, McMaster University,
Hamilton, Ontario, Canada L8S 4M1
email: `wilsoncd@mcmaster.ca`

Abstract. In understanding galaxy evolution, one central question is how star formation is regulated in galaxies. Changes in star formation rates are likely tied to changes in the interstellar medium, particularly the molecular gas which is the fuel for star formation. I will discuss our recent results which use data from the Herschel Space Observatory, the Atacama Large Millimeter/submillimeter Array, and other telescopes to determine the typical density, temperature, and surface density of the molecular gas in various nearby galaxies. Comparing the properties of molecular gas in starburst and other active galaxies with more quiescent spiral disks provides some clues as to how changes in the physical state of the gas, such as mean density, can lead to enhanced star formation rates.

Keywords. ISM: molecules, galaxies: spiral, galaxies: starburst

1. Introduction

The interstellar medium plays an important role in galaxies, providing both the immediate fuel for star formation (the molecular gas) and a larger reservoir for future star formation (the atomic gas). In the local Universe the interstellar medium accounts for a relatively small fraction of the total baryonic mass in galaxies. However, it is in nearby galaxies that radio and far-infrared telescopes have sufficient resolution to measure the gas and star-formation properties on the kiloparsec scales that are necessary for testing many models of star formation. The hope is that the physical processes uncovered in the local universe can then be applied to galaxies at high redshift, where the interstellar medium accounts for a much larger fraction of the mass in galaxies (e.g. Tacconi *et al.* 2013).

One interesting question concerns whether and how the molecular gas content of galaxies is affected by the broader environment in which both the gas and the galaxies live. Particularly in galaxy clusters, the effects of processes such as ram-pressure stripping and galaxy harassment can be seen in the distribution of the atomic gas; which processes affect the molecular gas and its rate of star formation is less well understood. In addition, the effects of local environment (e.g. arm versus interarm regions) inside spiral galaxies remains an area of interest.

A second interesting question concerns how the molecular gas in disk galaxies, where the star formation is proceeding at a steady rate in a kind of equilibrium, differs from that in starburst galaxies and mergers, where the huge rate of star formation limits the duration of this phase to perhaps a few hundred Myr. Given that many spectacular starburst galaxies are formed from the interaction of two disk galaxies, it is particularly interesting now to try to track the change in the molecular gas properties as the merger progresses, by studying a number of galaxies spread along the merger sequence.

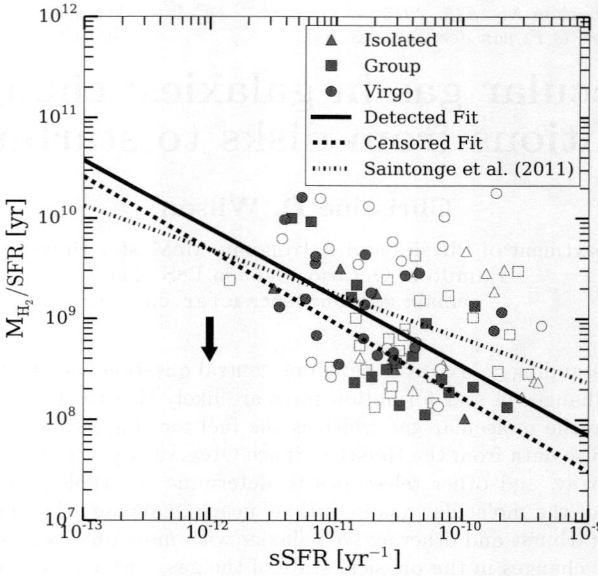

Figure 1. Molecular gas depletion time as a function of the specific star formation rate for spiral galaxies from the NGLS. The different symbols identify Virgo, group, and isolated spiral galaxies. Filled symbols are detections while open symbols are upper limits (the direction of the limits is indicated by the large arrow). The solid line is a linear fit to the detections while the dashed line is a fit including the upper limits (see Mok et al. 2016 for details). A similar fit from Saintonge *et al.* (2011) is shown as the dot-dashed line. Figure from Mok et al. 2016. *(A color version of this figure is available online.)*

In this contribution, I give an overview of some of the recent work that focuses on these two questions. Section 2 looks at the properties of nearby spiral galaxies, while Section 3 looks in detail at the particularly well-studied grand-design spiral, M51. Section 4 presents some new results for the Antennae, the closest merger between two large spirals, while Section 5 gives an overview of efforts to trace molecular gas properties along the merger sequence.

2. How uniform are molecular gas properties in spiral galaxies?

To first order, we may be tempted to view spiral galaxies in the local universe as a relatively uniform population in terms of their gas and star formation properties . For example, work by Bigiel *et al.* (2011) and others has shown that the star formation surface density correlates nearly linearly with the H_2 surface density when averaged over kiloparsec scales in nearby spiral galaxies. However, evidence for changes in the CO-to-H_2 conversion factor in galaxy centers as well as variations in the dust-to-gas mass ratio with metallicity suggest environmental dependencies lurk even inside this relatively uniform sample (Sandstrom *et al.* 2013).

One of the interesting questions that we can try to answer is whether and how the large-scale environment affects the molecular gas content of spiral galaxies. It has been clear for many years that the amount and distribution of atomic gas in galaxies can be strongly affected by environment, with many spirals showing a deficiency of atomic gas in cluster environments (e.g. Haynes, Giovanelli, & Chincarini 1984, Solanes et al. 2001). However, Koopmann *et al.* (2006) showed that the star formation properties of galaxies are not as strongly affected by the cluster environment as is the atomic gas. Attempts to determine the effect of molecular gas on environment (Kenney & Young 1989; Fumagalli & Gavazzi 2008) have been hampered by a lack of a non-cluster sample of galaxies to use as a control sample. More recent studies of Virgo spirals include work by Cortese et al. (2012) on the dust scaling relations and Boselli *et al.* (2014) on the cold gas properties.

Mok *et al.* (2016 and this volume) have used the JCMT Nearby Galaxies Legacy Survey (NGLS, Wilson *et al.* 2012) to examine the effect of environment on the molecular gas content of spiral galaxies. The 98 spiral galaxies are classified as either Virgo Cluster members, members of other smaller groups of galaxies, or isolated galaxies. Galaxies in the Virgo cluster are found to have larger masses of molecular gas but relatively longer molecular gas depletion times relative to galaxies that reside in smaller groups. Interestingly, this suggests that the cluster environment is aiding in the conversion of atomic to molecular gas, but suppressing the rate at which that molecular gas turns into stars. Enhanced pressure from the intracluster medium might play a role in both these properties (see Usero *et al.* 2015 and Leroy, this volume).

The NGLS is unusual in being an HI-selected sample, in contrast to other surveys which are more commonly selected based on optical or near-infrared luminosity or on stellar mass. It is thus interesting to examine the correlations between various global properties to see if they are recovered in our sample. Fig. 1 shows a negative correlation between molecular gas depletion time and specific star formation efficiency (SFR/M_*) that is consistent with results from other teams (Saintonge *et al.* 2011, Huang & Kauffmann 2014, Boselli et al. 2014). This correlation implies that molecular gas is being converted more quickly into stars in those galaxies where the current star formation rate is high relative to the integrated past star formation rate. In effect, there is a link between the number of new stars being formed, the amount of gas available to form those stars, and the total number of stars that have been formed over the galaxies lifetime (see also Kannappan et al. 2013).

3. M51: a case study of a disk galaxy

As a gas-rich spiral seen relatively face-on, M51 (NGC 5194) has been a prime target for studying molecular gas in galaxies. Recent work includes Vlahakis *et al.* (2013), Schinnerer et al. (2013), and Pety *et al.* (2013). M51 was also one of 13 galaxies that were the target of the Very Nearby Galaxies Survey, a Guaranteed Time Key Program with the Herschel Space Observatory (PI: C. Wilson). Resolved modelling of the far-infrared spectral energy distribution using the Draine and Li (2007) models suggested somewhat elevated ultraviolet radiation fields compared to the solar neighborhood and a very low fraction of the dust being heated by young stars (Mentuch Cooper *et al.* 2012).

We have used data from the *Herschel Space Observatory* to approach the problem of determining the density and temperature of the molecular gas in M51 in two different ways. Parkin et al. (2013) used maps of far-infrared fine structure lines such as [CII], [OI], and [NII] combined with photon-dominated region (PDR) models from Kaufman et al. (1999) to constrain the molecular hydrogen density and the strength of the ultraviolet radiation field, G_o. Our analysis showed that the density and UV field strength were similar in the arm and interarm regions (Fig. 2). This result suggests that the average cloud properties are similar in the arm and interarm regions, despite having different gas and star formation rate surface densities on average. Both the average density and UV field strength were found to be higher towards the center of the galaxy.

We have also used observations of CO and [CI] at submillimeter wavelengths with the imaging Fourier Transform Spectrometer on Herschel to constrain these same physical conditions (Schirm 2015). We fit the CO lines with two components using the RADEX

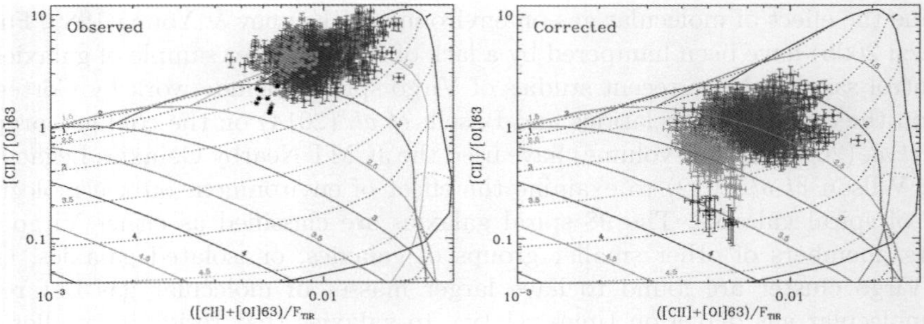

Figure 2. Fine-structure line data for M51 are shown overlaid on the PDR model grid from Kaufman *et al.* (1999). Dotted lines show $\log n_{H_2}$ and solid lines show $\log G_o$. Black, green, red, and blue pixels show data from the nucleus, center, arm, and interarm regions, respectively. The left panel shows observed quantities while the right panel shows the data after removal of the ionized gas contribution to [CII] and correction of [OI] for optical depth effects. Figure from Parkin *et al.* (2013). *(A color version of this figure is available online.)*

radiative transfer model (van der Tak *et al.* 2007) combined with a Bayesian likelihood code (Kamenetzky *et al.* 2011) to determine the density and temperature of the gas. These results are then compared to the PDR models from Hollenbach *et al.* (2012) to constrain the density and UV field strength of the molecular gas. For the cooler, lower density component in M51 which dominates the total mass in the models, we find similar average densities as in Parkin *et al.* (2013) but significantly lower UV field strength ($\log G_o \leqslant 2$ compared to $\log G_o \sim 3$). Possible reasons for the discrepancy include the use of different PDR models in the two analyses as well as the addition of a constraint due to the CO line to far-infrared luminosity ratio in the analysis of Schirm (2015).

4. The Antennae: a case study of a galaxy merger

The merger system NGC 4038/9 known as "the Antennae" represents the closest example of a major merger between two gas-rich spiral galaxies. As such, it has been extensively studied at many wavelengths. It contains perhaps the largest collection of young massive star clusters of any single nearby galaxy (Whitmore et al. 1999) as well as a collection of super-giant molecular clouds (Wilson *et al.* 2003). However, its total gas content and global star formation rate are each within a factor of a few of similar properties of M51. Thus, a comparison of the Antennae with M51 may serve to highlight the similarities and differences between these two systems.

Schirm *et al.* (2014) carried out an analysis of the submillimeter CO and [CI] lines of the Antennae to determine the gas density and temperature. The method used was similar to that described above for M51. The two-component analysis of the molecular gas revealed that only about 0.3% of the molecular gas was in a warm, high-pressure phase (Fig. 3). This mass fraction is similar to what Schirm (2015) found for M51, but is substantially lower than the warm mass fraction of 10% seen in M82 (Kamenetzky *et al.* 2012) and Arp 220 (Rangwala *et al.* 2011).

Recently, we have used the Atacama Large Millimeter/submillimeter Array (ALMA) to make high angular resolution maps of the Antennae in the dense gas tracers HCN, HCO^+, and HCN (Schirm *et al.* 2016). We see significant variations in the strengths of these lines relative to each other and also relative to CO. In particular, HCN appears significantly enhanced relative to CO in the nuclear regions of the two galaxies that make up the

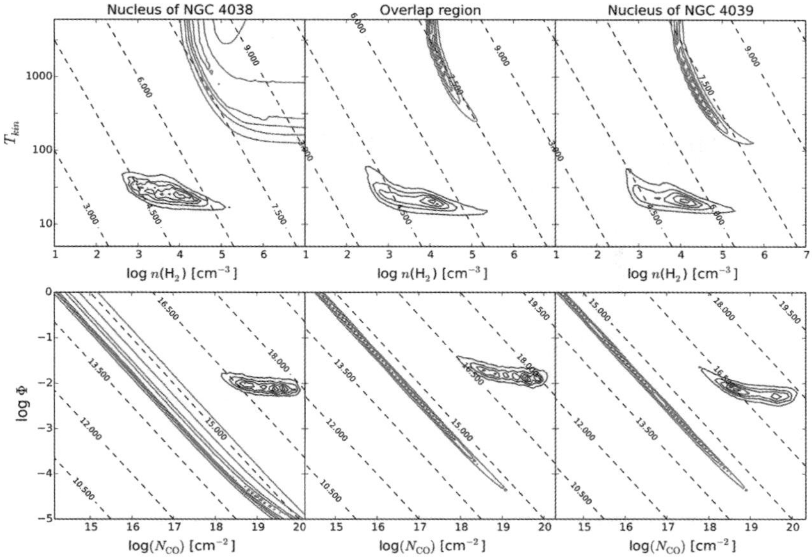

Figure 3. Contour probability plots from the Bayesian analysis of the CO and [CI] lines for three positions in the Antennae. The warm solution is shown as red contours and the cold solution as the blue contours. The top row plots kinetic temperature and density while the bottom row plots the filling factor and column density of the H_2 gas. The dashed diagonal lines in the top row indicate log(Presssure) units of Kcm^2; the dashed lines in the bottom row indicate the log of the average CO column density in units of cm^{-2}. Figure from Schirm *et al.* (2014). *(A color version of this figure is available online.)*

Antennae system. These enhanced line ratios may trace increase gas pressure driven by the stronger potential of the stellar bulge in these regions (cf. Helfer & Blitz 1993).

5. Gas properties along the merger sequence

As the closest ultraluminous infrared galaxy, Arp 220 presents probably the most extreme environment in terms of its star formation and gas properties in the local universe. We observed Arp 220 with the *Herschel Space Observatory* (Rangwala *et al.* 2011), revealing a submillimeter spectrum rich with emission and absorption lines of many molecules. A two-component analysis of the CO emission lines showed that 10% of the molecular gas mass was contained in an extremely warm (T = 1350 K) component, while both the warm and the cooler components showed moderate H_2 densities of $\log n \sim 3$. High resolution follow-up of the CO J=6-5 emission with ALMA to trace the morphology of the hot component revealed a very complex structure, with both red and blue-shifted absorption line and high optical depth towards each of the two nuclei (Rangwala *et al.* 2015 and this volume).

A central question in understanding the changes that take place in the star formation rates during galaxy interactions in mergers is determining how the physical properties of the molecular gas evolve as the merger progresses. Sliwa *et al.* (2012, 2013, 2014, 2015) have used high-resolution CO and ^{13}CO data for a small sample of merging galaxies from the sample of Wilson *et al.* (2008) to measure the physical conditions (density, temperature) and other properties such as ^{13}CO abundance and CO-to-H_2 conversion factor in five ystems ranging from early to late phase mergers (Arp 55 (Fig. 4), Arp 299, NGC 1614, NGC 2623, and VV 114). This work has revealed interesting changes in the

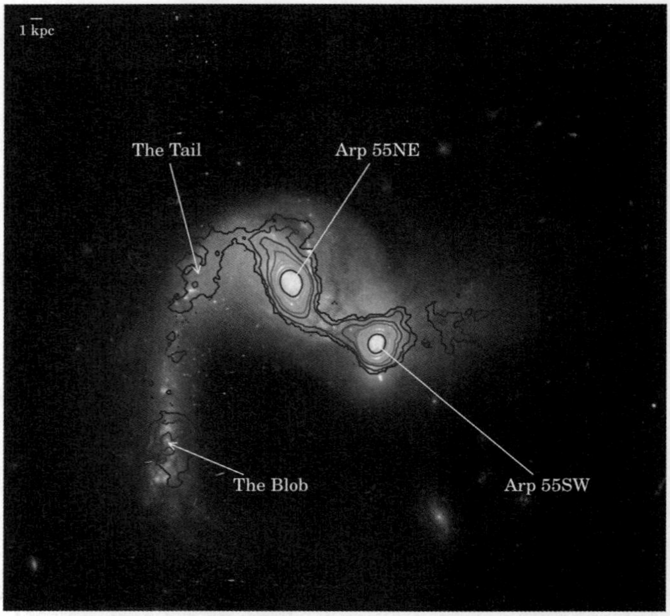

Figure 4. CO J=1-0 map from CARMA of the early-phase merger Arp 55 overlaid on an HST image. Figure from Sliwa (2015). *(A color version of this figure is available online.)*

mean molecular gas properties (see also Sliwa *et al.*, this volume). Our analysis reveals that early and intermediate stage mergers typically have molecular gas that is colder and denser than that found in the late-type mergers (VV114, NGC2623). The data for Arp 220 (late) and the Antennae (early-intermediate) from Rangwala *et al.* (2011) and Schirm et al. (2014) conform to this trend as well.

High density gas *is* seen in Arp 220 (Wilson *et al.* 2014, Scoville *et al.* 2015) and in the Antennae (Schirm et al. 2016) in high resolution images from ALMA. However, our analysis suggests that the gas which is producing most of the CO emission has a substantially lower density than the gas found near the two nuclei in Arp 220 and in the Antennae. Given the complicated optical depth effects traced in the CO J=6-5 images of Arp 220 (Rangwala *et al.* 2015), this high density component may be nearly hidden from view in global analyses.

This result of lower density gas in advanced mergers is rather unexpected given the canonical picture of gas falling into the central regions as the merger progresses; in this picture, we would have expected higher density gas in the late-type mergers. One possible explanation is that the increasing feedback from stellar winds and supernova is beginning to dissipate the molecular gas in the most advanced systems, which could result in a lower average volume density. Recent observations showing evidence for massive molecular winds (e.g. Sturm *et al.* 2011) provide some support for this interpretation.

6. Conclusions

Data from new telescopes such as Herschel and ALMA combined with large surveys from single dish telescopes such as IRAM and the JCMT are opening up new possibilities for studying the link between star formation and the molecular interstellar medium in galaxies in the nearby universe. We are beginning to be able to answer the question of how environment affects the star formation process, be it inside a single galaxy comparing arm and inter-arm regions or radial trends, or larger external factors such as group or cluster environment, or the extreme cases of galaxy mergers. Initial surveys on the effect of cluster environment on the molecular gas in spirals (Mok et al. 2016) should soon be extended to other nearby clusters such as Fornax and Coma. Similarly, expanding the sample of mergers with well determined molecular gas properties beyond the seven systems studied so far (Sliwa 2015) will allow us to map out the evolution of the interstellar medium during galaxy mergers. Both directions are important for providing insights into evolutionary processes at work in the early universe.

References

Bigiel, F., Leroy, A. K., Walter, F., *et al.* 2011, *AJ*, 730, L13
Boselli, A., Cortese, L., Boquien, M., Boissier, S., Catinella, B., Lagos, C., & Saintonge, A. 2014, *A&A*, 564, A66
Cortese, L., Ciesla, L., Boselli, A., *et al.* 2012, *A&A*, 540, A52
Draine, B. T. & Li, A. 2007, *ApJ*, 657, 810
Fumagalli, M. & Gavazzi, G. 2008, *A&A*, 490, 571
Haynes, M. P., Giovanelli, R., & Chincarini, G. L. 1984, *ARAA*, 22, 445
Helfer, T. T. & Blitz, L. 1993 *ApJ*, 419, 86
Hollenbach, D., Kaufman, M. J., Neufeld, D., Wolvire, M., & Goicoechea, J. R. 2012, *ApJ*, 754, 105
Huang, M.-L. & Kauffmann, G. 2014, *MNRAS*, 443, 1329
Kamenetzky, J., Glenn, J., Maloney, P. R.. *et al.* 2011, *ApJ*, 731, 83
Kamenetzky, J., Glenn, J., Rangwala, N. *et al.* 2012, *ApJ*, 753, 70
Kannappan, S. J., Stark, D. V., Eckert, K. D., *et al.* 2013, *ApJ*, 777, 42
Kaufman, M. J., Wolfire, M. G., Hollenback, D. J., & Luhmn, M. L. 1999, *ApJ*, 527, 795
Kenney, J. D. P. & Young, J. S. 1989, *ApJ*, 334, 171
Koopmann, R. A., Haynes, M. P., & Catinella, B. 2006, *AJ*, 131, 716
Mentuch Cooper, E., Wilson, C. D., Foyle, K. *et al.* 2012, *ApJ*, 755, 165
Mok, A., Wilson, C. D., Golding, J. *et al.* 2016, *MNRAS*, submitted
Parkin, T. J., Wilson, C. D., Schirm, M. R. P., *et al.* 2013, *ApJ*, 776, 65
Pety, J., Schinnerer, E., Leroy, A. K. *et al.* 2013, *ApJ*, 779, 43
Rangwala, N., Maloney, P. R., Glenn, J. *et al.* 2011, *ApJ*, 743, 94
Rangwala, N., Maloney, P., Wilson, C. D., Glenn, J., Kamenetzky, J., & Spinoglio, L. 2015, *ApJ*, 806, 17
Saintonge, A., Kauffmann, G., Wang, J., *et al.* 2011, *MNRAS*, 415, 61
Sandstrom, K. M., Leroy, A. K., Walter, F., *et al.* 2013, *ApJ*, 777, 5
Schinnerer, E., Meidt, S. E., Pety, J. *et al.* 2013, *ApJ*, 779, 42
Schirm, M. R. P., Parkin, T. J., Wilson, C. D., *et al.* 2014, *ApJ*, 781, 101
Schirm, M. R. P. 2015, Ph.D. thesis, McMaster University
Schirm, M. R. P., Wilson, C. D., Madden, S. C., & Clemens, D. L. 2016, *ApJ*, submitted
Scoville, N., Sheth, K., Walter, F., *et al.* 2015, *ApJ*, 800, 70
Sliwa, K., Wilson, C. D., Petitpas, G. R. *et al.* 2012, *ApJ*, 753, 46
Sliwa, K., Wilson, C. D., Krips, M. *et al.* 2013, *ApJ*, 796, L15
Sliwa, K., Wilson, C. D., Iono, D., Peck, A., & Matsushita, S. 2014, *ApJ*, 753, 46
Sliwa, K. 2015, Ph.D. thesis, McMaster University
Solanes, J. M., Manrique, A., García-Gómez, C., González-Casado, G., Giovanelli, R., & Haynes, M. P. 2001, *ApJ*, 548, 97
Sturm, E., González-Alfonso, E., Veilleux, S., *et al.* 2011, *ApJ*, 733, L16
Tacconi, L., Neri, R., Genzel, R., *et al.* 2013, *ApJ*, 768, 74
Usero, A., Leroy, A. K., Walter, F., *et al.* 2015, *AJ*, 150, 115
van der Tak, F. F. S., Black, J. H., Schier, F. L., Jansen, D. J., & van Dishoeck, E. F. 2007, *A&A*, 468, 627
Vlahakis, C., van der Werf, P., Israel, F. P., & Tilanus, R. P. J. 2013, *MNRAS*, 433, 1837

Whitmore, B. C., Zhang, Q., Leitherer, C., Fall, S. M., Schweizer, F., & Miller, B. W. 1999, *AJ*, 118, 1551
Wilson, C. D., Scoville, N., Madden, S. C., & Charmandaris, V. 2003, *ApJ*, 599, 1049
Wilson, C. D., Petitpas, G. R., Iono, D. *et al.* 2008, *ApJS*, 178, 189
Wilson, C. D., Warren, B. E., Israel, F., *et al.* 2012, *MNRAS*, 424, 3050
Wilson, C. D., Rangwala, N., Glenn, J., Maloney, P., Spinoglio, L., & Pereira-Ssantaella, M. 2014, *ApJ*, 789, L36

From Interstellar Clouds to Star-Forming Galaxies: Universal Processes?
Proceedings IAU Symposium No. 315, 2015
P. Jablonka, Ph. André & F. van der Tak, eds.

doi:10.1017/S1743921316007511

Gas flows in galactic nuclei: observational constraints on BH-galaxy coevolution

Santiago García-Burillo

Observatorio de Madrid (OAN-IGN),
Alfonso XII, 3, Madrid-28014, Spain
email: s.gburillo@oan.es

Abstract. Galaxy nuclei are a unique laboratory to study gas flows. Their high-resolution imaging in galactic nuclei are instrumental in the study of the fueling and feedback of star formation and nuclear activity in nearby galaxies. Several fueling mechanisms can now be confronted in detail with observations done with state-of-the-art interferometers. Furthermore, the study of gas flows in galactic nuclei can probe the feedback of activity on the interstellar medium of galaxies. Feedback action from star formation and AGN activity is invoked to prevent galaxies from becoming overly massive, but also to explain scaling laws like black hole (BH)-bulge mass correlations and the bimodal color distribution of galaxies. This close relationship between galaxies and their central supermassive BH can be described as *co-evolution*. There is mounting observational evidence for the existence of gas outflows in different populations of starbursts and active galaxies, a manifestation of the feedback of activity. We summarize the main results recently obtained from the observation of galactic inflows and outflows in a variety of active galaxies with current millimeter interferometers such as ALMA or the IRAM array.

Keywords. galaxies:nuclei, galaxies:kinematics and dynamics, galaxies:ISM, galaxies: evolution

1. Introduction

Galaxy scaling relations indicate that the growth of supermassive BH and the assembly of galaxy spheroids are closely connected. In order to understand the physical framework explaining the nature of this connection we must study the complex interplay between gas inflows and outflows in galaxies. The gas component can form stars on its way to the BH. Quantifying the efficiency of angular momentum transport in galaxy disks help address how star formation (SF) and nuclear (AGN) activity can be fed and on what timescales this can occur. Furthermore the gas component can also experience significant feedback from SF and AGN activity on its way to the BH. The study of gas outflows, considered to be a smoking gun evidence of ongoing mechanical feedback, is crucial to assess the efficiency of the feedback loop.

Finding a casual link between the presence of gas inflows/outflows and the existence of ongoing star formation or nuclear activity in galaxies is observationally challenging because both time- and spatial scales associated with these phenomena can be very different. In particular, although timescale associated with the total BH-growth phase can be comparable to the SF one (10^{7-9} years), the BH growth phase is suspected to be a collection of thousands of burst events, each of them lasting only for about 10^5 years. The emerging picture sees AGN accretion as a chaotic and largely time-dependent process. AGN can *flicker* and may oscillate between high-accretion modes, where feedback is dominated by radiation and winds, and low accretion modes, where feedback is mostly kinetic and driven by jets (e.g., Schawinski *et al.* 2015; King & Nixon 2015).

In this paper we discuss how molecular line observations can be used to study fueling and feedback of star formation nuclear activity. In particular we will illustrate the importance of high-resolution observations in the millimeter range done with interferometers to study inflows and outflows in nearby galaxies, where the relevant processes responsible for secular evolution can be studied in great detail.

2. Tracing molecular inflows

The search for a mechanism responsible of feeding nuclear activity is arduous in low-luminosity AGN (LLAGN). Several mechanisms that show unequal success to induce gas inflow have been discussed in the literature. Gravitational barriers can sometimes stop the gas on its way to the nucleus. The case of LLAGN is special as the total BH growth phase in this family of objects is expected to consist in short-lived episodes, each of them likely lasting for less than the lifetime of the agent responsible for feeding the nucleus. Different timescales of potentially more than one mechanism can mask the correlation further (e.g., Hopkins *et al.* 2011). While various model predictions are still debated in the literature, there is still ample room for improvement in the picture drawn from observations. Observations that probe the critical scales for AGN feeding ($< 10-100$ pc) can help to make a significant progress in the field.

The Nuclei of Galaxies (NUGA) survey (see García-Burillo & Combes 2012 and references therein) is an interferometer CO survey of 25 low-L nearby AGNs conducted with the IRAM array. With these observations we were able to probe the distribution and kinematics of molecular gas in the central kpc-scale disks of these galaxies with high spatial resolution ($\sim 10-100$ pc) to search for evidences of ongoing AGN feeding. Together with the CO NUGA maps, we have access to high-resolution NIR maps obtained with HST, Spitzer and/or ground-based telescopes. The NIR imaging is used to derive the stellar potentials, which are combined with the CO maps, to derive the gravity torque budget. Two-dimensional torque maps are used to estimate the torques averaged over the azimuth $t(r)$ using the gas column density $N(x, y)$ derived from CO as weighting function (García-Burillo *et al.* 2005; Haan et al. 2009). With this information at hand we quantitatively evaluated if there were signs of ongoing AGN fueling in the galaxy disks, down to the spatial resolution of our observations.

About one third of the LLAGNs analyzed in NUGA show negative torques $t(r)<0$, indicative of inflow, down to typical radial distances $r<$25–100 pc. Among these galaxies, which can be considered as those showing smoking gun evidence of ongoing fueling, we distinguish three types of objects:

–*Galaxies showing dynamical decoupling of embedded structures*: these include nuclear stellar bars - within - bars/ovals (e.g.; NGC 2782) and nuclear ovals - within -bars (e.g.; NGC 4579).

–*Galaxies with ILR-free large-scale stellar bars*: these galaxies are characterized by the apparent absence of gravity torque barriers in their nuclei (e.g.; NGC 3627).

–*Galaxies with nuclear gas spirals*: long predicted by theoretical models and seen in dusty structures in HST images, these gas instabilities are started to by seen by ALMA in CO line emission, like in the Seyfert 1 galaxy NGC 1566 (Fig. 1). The trailing spiral seen in the central $r \leqslant 50-300$ pc of NGC 1566 reflects the influence of the BH on the gas response, which facilitates gas inflow down to $r \leqslant 50$ pc.

On the contrary, about two thirds of the LLAGNs analyzed in NUGA show positive torques $t(r)>0$, indicative of outflow, down to typical radial distances $r<$300 pc. This *puzzling* gravity torque budget is found in two categories of objects in our sample:

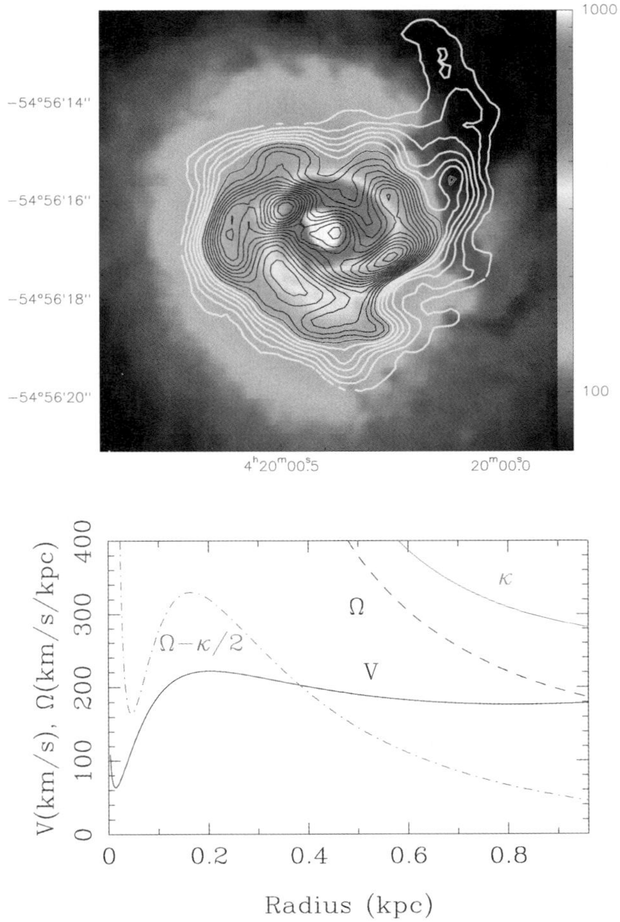

Figure 1. *(Top panel)* The CO(3–2) emission contours obtained by ALMA towards the circum-nuclear disk of the Seyfert 1 NGC 1566 are overlaid on the (F606W) HST image of the galaxy. The CO image shows a trailing spiral structure from $r \leqslant 50$ pc to $r = 300$ pc, which closely follows the dusty spiral seen by the HST. *(Lower panel)* The rotation curve (black), epicyclic frequency κ (red), and $\Omega - \kappa/2$ curve (blue) in NGC 1566. The influence of the super-massive black hole explains reversal of $\Omega - \kappa/2$ curve at $r \leqslant 50$ pc and implies negative torques on the gas spiraling to the nucleus on these spatial scales. Figures adapted from Combes et al. (2014). *(A color version is available online.)*

–*Galaxies showing non-cooperative embedded structures*: in these galaxies, nuclear bars or ovals - within - bars, like those found in NGC 4321 and NGC 6951 are not always conducive to gas inflow at present.

–*Galaxies showing featureless/mostly axisymmetric potentials*: the absence of a clear non-axisymmetric feature in the stellar potential makes the angular momentum transfer in the nucleus a very inefficient process in some galaxies (e.g.; NGC 4826; NGC 7217; NGC 5953).

The overall results of NUGA indicate that high spatial resolution is instrumental in quantifying angular momentum transfer processes at critical scales ($\sim 10 - 100$ pc). Gas flows in the NUGA maps reveal a wide range of large-scale and embedded $m = 2, m = 1$ instabilities in the circumnuclear disks of LLAGNs. The derived gravity torque maps

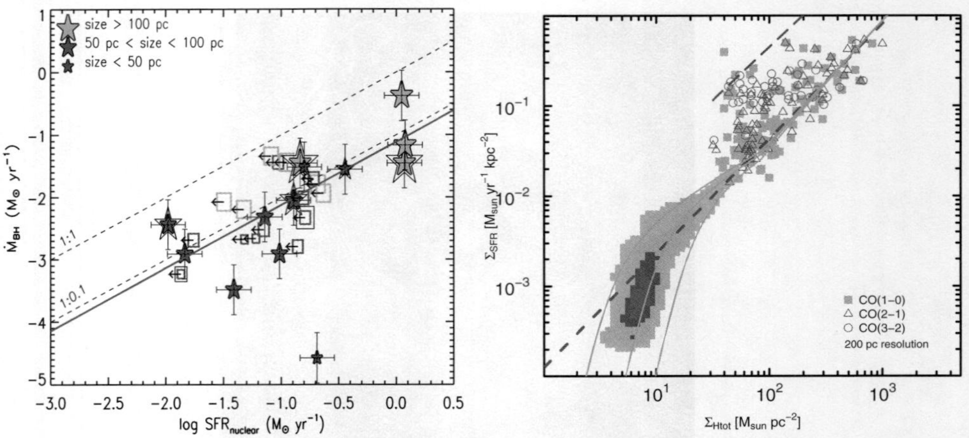

Figure 2. *a)Left panel:* The nuclear ($r \leqslant 100$ pc) $SFR - dM_{\rm BH}/dt$ scaling relation obtained in the sample of 29 Seyfert galaxies analyzed by Esquej *et al.* (2014). The solid line represents the fit to the detections of the nuclear 11.3 μm PAH feature. The sizes/colors of the symbols indicate the different physical sizes probed. *b)Right panel:* The spatially resolved Kennicutt-Schmidt-relation derived for the four active NUGA galaxies studied by Casasola *et al.* (2015) at a spatial resolution of 200 pc, compared to the results from Bigiel *et al.* (2008), and the model predictions from Krumholz *et al.* (2009). The blue line identify the branch of the SF galaxies, while the red line identifies the merger sequence of Genzel *et al.* (2010). *(A color version is available online.)*

indicate that gas is frequently stalled in rings, which are the signposts of gravity torque barriers. Only $\sim 1/3$ of galaxies show negative torques down to ~ 50 pc. These various fueling mechanisms are related to bar cycles.

However, the $r \sim 1-10$ pc-scales are to a large extent uncharted territory for molecular line observers. We need more observational constraints to investigate what are the main fueling mechanisms at work on these scales and confirm the predictions of numerical simulations (e.g., Hopkins & Quataert 2011; Hopkins *et al.* 2015; Wada 2001; Wada *et al.* 2009).

3. The SF-AGN connection

Many works have tried to establish a causal connection between BH and galaxy growth, by studying to what extent the rate of SF is correlated with the BH feeding rate and, most importantly, by finding on what spatial scales (*global*∼kpc or *nuclear*∼100 pc) this correlation may hold. Despite the fact that both SF and BH feeding rely on a common cold gas supply, the fact that the timescales associated to SF and AGN feeding can be very different adds non-trivial complications. Delvecchio *et al.* (2015) have recently studied the relation between AGN accretion and the global SF rate in a sample of 8600 SF galaxies up to $z = 2.5$ from the GOODS and COSMOS surveys, finding a good correlation between these two quantities, in spite of the large variance due to the different spatial/time-scales. However the prediction of the recent numerical simulations of Volonteri *et al.* (2015ab) are in tension with these results, as they anticipate that only on *nuclear* scales the SF-BH growth rates should show any significant correlation.

Despite first claims that the SFR-BHR correlation may be 'blurred' even on nuclear scales, due to an eventual delay in fueling the AGN after a circumnuclear SF episode in Seyfert galaxies, (e.g.; Davies *et al.* 2007), Esquej *et al.* (2014) have recently shown that the SFR-$dM_{\rm BH}/dt$ scaling law obtained using a sample of 29 Seyferts shows evidence of a significant correlation (Fig. 2a), a result that confirms the predictions of numerical

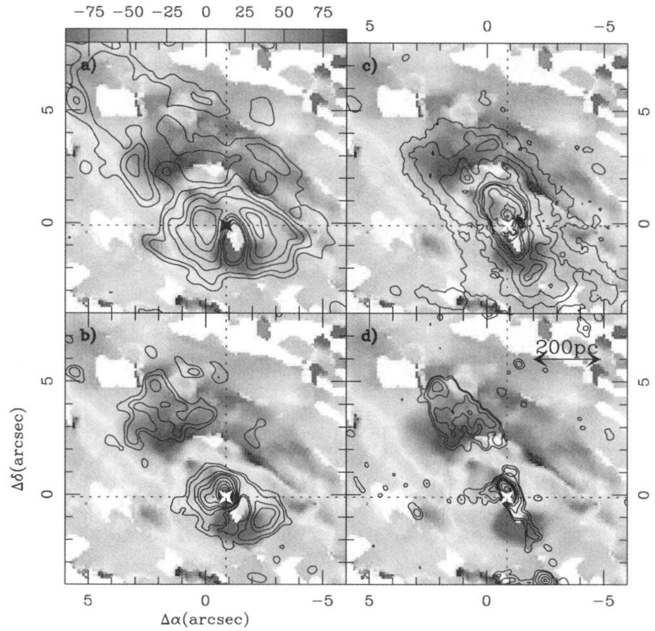

Figure 3. Overlay of the residual velocity field of molecular gas in NGC 1068 observed by ALMA (in color scale) obtained after subtraction of the model rotation curve, with the contours representing: the integrated intensity of CO(3–2) (*a)upper left panel*), the 349 GHz continuum emission (*b) lower left panel*), the HST Paα emission (*c) upper right panel*), and the 22 GHz VLA map of Gallimore *et al.* (1996) (*d) lower right panel*). Figure from García-Burillo *et al.* (2014).

simulations. Taken at face value, the results of Esquej *et al.* (2014) indicate that SF and BH growth may be running in parallel in a symbiotic way in the circumnuclear regions of LLAGN. Furthermore, this picture seems to be confirmed by the work of Casasola et al. (2015), who found that the SF efficiency of molecular gas in the circumnuclear disks of four NUGA targets, analyzed with $\sim$ 200 pc spatial resolution, lies between the values characterising the sequence of SF galaxies and that of mergers (Fig. 2*b*). However, the question of whether AGN activity can significantly quench SF in AGN-hosts is far from being settled. Using larger samples, two different groups (Shimizu *et al.* 2015; Mullaney *et al.* 2015) have found evidence that a significant percentage of AGN hosts seem to be located below the main sequence of SF galaxies, an indication that the brightest AGN may be in the process of quenching.

4. Tracing molecular outflows

Outflows can prevent galaxies from becoming exceedingly massive and help regulate activity. There is observational evidence for the existence of gas outflows in different populations of nearby galaxies, including ultra-luminous infrared galaxies (ULIRGs), radio galaxies, quasars and Seyferts. The outflow phenomenon concerns all the phases of the ISM. In particular, the study of the molecular component, being the most massive phase of the neutral ISM in the central kiloparsec regions, is crucial if we are to quantify the impact of outflows on galaxy evolution.

The quest for molecular outflows using mm-interferometers like ALMA or the IRAM array has yielded an increasing number of detections in ULIRGs (e.g., Feruglio *et al.* 2010, 2013, 2015; Aalto et al. 2012, 2015; Cicone *et al.* 2012, 2014; García-Burillo *et al.* 2015),

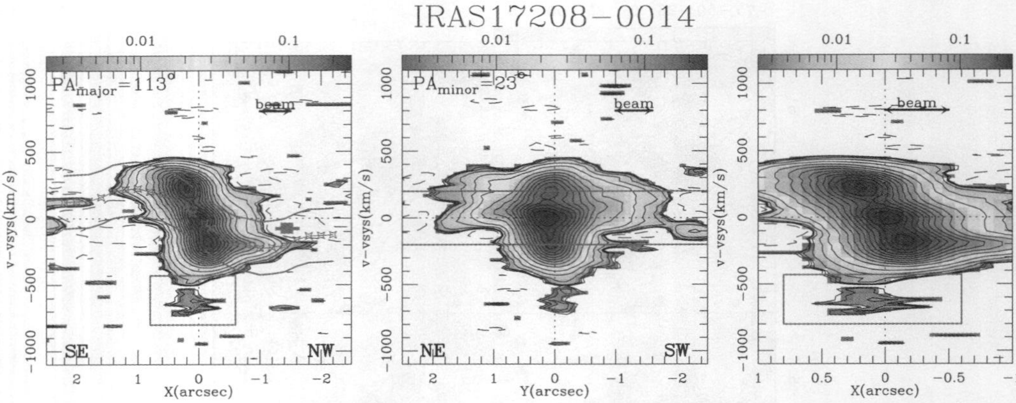

Figure 4. *a)Left panel:* The CO(2–1) p-v plot taken along the kinematic major axis of IRAS 17208-0014 derived with the IRAM array. We delineate the (projected) rotation curve and the allowed virial range around it. Similarly, we delimit the region where emission is detected from the *line wing* betraying the most extreme blue-shifted velocities of the outflow ($v - v_{\rm sys}$ = [-450,-750] km s^{-1}). *b)Middle panel:* The p-v plot taken along the kinematic minor axis. *c)Right panel:* Same as *left panel* but zooming in on the central ±1" region around the center. Figure taken from García-Burillo *et al.* (2015). *(A color version is available online.).*

radio galaxies (e.g., Tremblay 2014) and LLAGN (e.g., Alatalo et al. 2011,2015; Combes *et al.* 2013; Morganti *et al.* 2013, 2015; García-Burillo *et al.* 2014; Sakamoto *et al.* 2014). The high-resolution capabilites of these observations are instrumental in constraining the geometry, as well as the mass, momentum and energy budgets of the molecular outflows ($dM_{\rm out}/dt$, $dP_{\rm out}/dt$, $L^{\rm kin}_{\rm out}$). Having good estimates of these three quantities is a prerequisite in order to study the specific role that star formation versus AGN feedback have in launching and maintaining molecular outflows.

As an illustration of the power of mm-interferometry, the sensitivity and spatial resolution of ALMA have provided an unprecedented view of the distribution and kinematics of the dense molecular gas in the central $r \sim 2$ kpc of the Seyfert 2 galaxy NGC 1068 with spatial resolutions $\sim 0.3" - 0.5"$ ($\sim 20 - 35$ pc) (García-Burillo *et al.* 2014). The gas kinematics from $r \sim 50$ pc out to $r \sim 400$ pc reveal a massive outflow in all molecular tracers mapped by ALMA. The tight correlation between the ionized gas outflow, the radio jet, and the occurrence of outward motions in the disk suggests that the outflow is AGN driven (Fig 3). The outflow rate estimated in the circumnuclear disk (CND), dM/d$t \sim 63^{+21}_{-37}$ $M_\odot$ yr^{-1}, is an order of magnitude higher than the star formation rate at these radii. The molecular outflow could quench star formation in the inner $r \sim 400$ pc of the galaxy on short timescales of $\leqslant 1$ Myr and regulate gas accretion in the CND.

Galaxy evolution scenarios foresee that the feedback of SF and AGN activity can drive the transformation of gas-rich spiral mergers into ULIRGs, and, eventually, lead to the build-up of QSO/elliptical hosts. In an attempt to address the role that star formation and AGN feedback have in launching and maintaining the molecular outflows in starburst-dominated advanced mergers, García-Burillo *et al.* (2015) used the IRAM array to image with high spatial resolution (0.5" − 1.2") the CO(1–0) and CO(2–1) line emissions in NGC 1614 and IRAS 17208-0014, respectively. These observations have detected in both mergers the emission from high-velocity *line wings* that extend up to ±500–700 km s^{-1}, well beyond the estimated virial range associated with rotation and turbulence, a signature of outflows (Fig. 4). In IRAS 17208-0014, gas emission from the blueshifted *line wing* is kinematically decoupled from the disk: its apparent kinematic

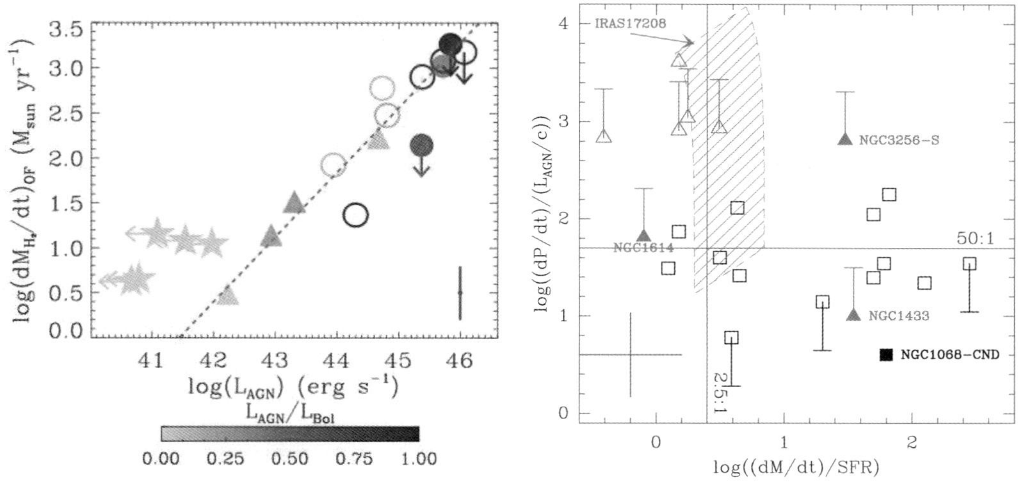

Figure 5. *a)Left panel:* Outflow mass rate (dM/dt) as a function of the AGN bolometric luminosity $L_{\rm AGN}$ for the sample of galaxies analyzed by Cicone *et al.* (2014), who used CO observations to trace molecular gas masses. Filled and open circles represent unobscured and obscured AGNs respectively, LINERs are plotted as triangles and pure starbursts as stars. The red line illustrates the fit to the deteted AGNs. Figure taken from Cicone *et al.* (2014). *b)Right panel:* Momentum boost factor $(dP/dt)/(L_{\rm AGN}/c)$ as a function of mass-loading factor $((dM/dt)/SFR)$ for the galaxies compiled by García-Burillo *et al.* (2015). Open (red) triangles show the data from pure starburst galaxies; open (black) squares identify the data of low and high-luminosity AGNs. Filled symbols identify the data from the new additions. The location of galaxies in this diagram help identify molecular outflows driven by SF (upper left corner), radiation pressure or winds of AGN (lower right corner), or AGN jets (upper right corner) or a combination thereof in mixed/undefined cases (lower left corner). Figure adapted from García-Burillo *et al.* (2015). *(A color version is available online.)*

major axis is reversed. This pattern can be explained by a non-coplanar molecular outflow. This scenario is reinforced after subtraction of the rotation curve, which reveals that the kinematic decoupling of the outflow starts at $\mid v - v_{\rm sys} \mid \geqslant 300$ km s^{-1}. The mass, energy and momentum budget requirements of the molecular outflow in IRAS 17208-0014 can be accounted for by the existence of a so far undetected (*hidden*) AGN of $L_{\rm AGN} \sim 7 \times 10^{11}$ L$_\odot$.

Several diagnostic tools are used in the literature to study the location of different populations of galaxies where molecular outflows have been detected (Cicone *et al.* 2014; García-Burillo *et al.* 2015). These type of diagrams are crucial if we are to identify the driving agent of the outflows: SF versus AGN, but also, more specifically, radiation pressure, winds or jets (Fig 5). Based on their compiled sample of 19 sources, Cicone *et al.* (2014) found that star formation can drive outflows with dM/dt values up to 2–4 times the value of the star formation rate (SFR). However, the presence of an AGN, even if weak ($L_{\rm AGN}/L_{\rm bol} < 0.10$), can strongly boost an outflow with dM/d$t \propto L_{\rm AGN}$. In the most extreme quasars ($L_{\rm AGN}/L_{\rm bol} \geqslant 0.80$), outflow rates can be up to $100 \times SFR$, and therefore have a decisive impact on the fueling of the star formation and nuclear activity of their hosts. This picture is similar to the scenario derived from the numerous observations of outflows detected in the OH and NaID lines (Fischer *et al.* 2010; Surm *et al.* 2011; Rupke & Veilleux 2013; Spoon *et al.* 2013).

There is still a number of key results emerging from observations that remain to be understood. Molecular outflows are energetically relevant, especially, but not exclusively, in luminous starbursts and AGN. In the case of Seyferts, molecular outflows, although less extreme in their properties, play a relevant role in regulating the fueling of activity. As a general feature, the velocities measured in molecular outflows suggest that most of the gas is likely to fountain back to the disk. Furthermore, the radio-mode feedback seems to be at work not only in extreme radio-loud systems, but also in Seyfert galaxies. This may be connected to the changing accretion/feedback modes in *flickering* AGN.

References

Aalto, S., Garcia-Burillo, S., Muller, S., *et al.* 2012,*A&A*, 537, A44
Aalto, S., Garcia-Burillo, S., Muller, S., *et al.* 2015, *A&A*, 574, A85
Alatalo, K., Blitz, L., Young, L. M., *et al.* 2011, *ApJ*, 735, 88
Alatalo, K., Lacy, M., Lanz, L., *et al.* 2015, *ApJ*, 798, 31
Bigiel, F., Leroy, A., Walter, F., *et al.* 2008, *AJ*, 136, 2846
Casasola, V., Hunt, L., Combes, F., & García-Burillo, S. 2015, *A&A*, 577, A135
Cicone, C., Feruglio, C., Maiolino, R., *et al.* 2012, *A&A*, 543, A99
Cicone, C., Maiolino, R., Sturm, E., *et al.* 2014, *A&A*, 562, A21
Combes, F., García-Burillo, S., Casasola, V., *et al.* 2013, *A&A*, 558, A124
Combes, F., García-Burillo, S., Casasola, V., *et al.* 2014, *A&A*, 565, A97
Davies, R. I., Müller Sánchez, F., Genzel, R., *et al.* 2007, *ApJ*, 671, 1388
Delvecchio, I., Lutz, D., Berta, S., *et al.* 2015, *MNRAS*, 449, 373
Esquej, P., Alonso-Herrero, A., González-Martín, O., *et al.* 2014, *ApJ*, 780, 86
Feruglio, C., Maiolino, R., Piconcelli, E., *et al.* 2010, *A&A*, 518, L155
Feruglio, C., Fiore, F., Maiolino, R., *et al.* 2013, *A&A*, 549, A51
Feruglio, C., Fiore, F., Carniani, S., *et al.* 2015, *arXiv:1503.01481*
Fischer, J., Sturm, E., González-Alfonso, E., *et al.* 2010, *A&A*, 518, L41
Gallimore, J. F., Baum, S. A., O'Dea, C. P., & Pedlar, A. 1996, *ApJ*, 458, 136
García-Burillo, S., Combes, F., Schinnerer, E., Boone, F., & Hunt, L. K. 2005, *A&A*, 441, 1011
García-Burillo, S. & Combes, F. 2012, *Journal of Physics Conference Series, 372, 012050*
García-Burillo, S., Combes, F., Usero, A., *et al.* 2014, *AA*, 567, A125
García-Burillo, S., Combes, F., Usero, A., *et al.* 2015, *A&A*, 580, A35
Genzel, R., Tacconi, L. J., Gracia-Carpio, J., *et al.* 2010, *MNRAS*, 407, 2091
Haan, S., Schinnerer, E., Emsellem, E., *et al.* 2009, *ApJ*, 692, 1623
Hopkins, P. F. & Quataert, E. 2011, *MNRAS*, 415, 1027
Hopkins, P. F., *et al.* 2015, *arXiv:1504.05209*
King, A. & Nixon, C. 2015, *MNRAS*, 453, L46
Krumholz, M. R., McKee, C. F., & Tumlinson, J. 2009, *ApJ*, 699, 850
Morganti, R., Frieswijk, W., Oonk, R. J. B., Oosterloo, T., & Tadhunter, C. 2013, *A&A*, 552, L4
Morganti, R., Oosterloo, T., Oonk, R. J. B., Frieswijk, W., & Tadhunter, C. 2015, *A&A*, 580, A1
Mullaney, J. R., Alexander, D. M., Aird, J., *et al.* 2015, *MNRAS*, 453, L83
Rupke, D. S. N. & Veilleux, S. 2013, *ApJ*, 768, 75
Sakamoto, K., Aalto, S., Combes, F., Evans, A., & Peck, A. 2014, *ApJ*, 797, 90
Schawinski, K., Koss, M., Berney, S., & Sartori, L. F. 2015, *MNRAS*, 451, 2517
Sturm, E., González-Alfonso, E., Veilleux, S., *et al.* 2011, *ApJ*, (Letters) 733, L16
Shimizu, T. T., Mushotzky, R. F., Meléndez, M., Koss, M., & Rosario, D. J. 2015, *MNRAS*, 452, 1841
Spoon, H. W. W., Farrah, D., Lebouteiller, V., *et al.* 2013, *ApJ*, 775, 127
Tremblay, G. 2014, *Einstein Fellows Symposium 2014, held 28-29 October, 2014 at the Harvard-Smithsonian Center for Astrophysics, Cambridge, MA*
Volonteri, M., Capelo, P. R., Netzer, H., *et al.* 2015a, *MNRAS*, 449, 1470
Volonteri, M., Capelo, P. R., Netzer, H., *et al.* 2015b, *MNRAS*, 452, L6
Wada, K. 2001, *ApJ* (Letters), 559, L41
Wada, K., Papadopoulos, P. P., & Spaans, M. 2009, *ApJ*, 702, 63

From Interstellar Clouds to Star-Forming Galaxies: Universal Processes?
Proceedings IAU Symposium No. 315, 2015
P. Jablonka, Ph. André & F. van der Tak, eds.

doi:10.1017/S1743921316007523

High resolution ALMA observations of dense molecular medium in the central regions of active galaxies

Kotaro Kohno[1], Ryo Ando[1], Akio Taniguchi[1], Takuma Izumi[1] and Tomoka Tosaki[2]

[1]Institute of Astronomy, The University of Tokyo, Osawa, Mitaka, Tokyo 181-0015, Japan
email: kkohno@ioa.s.u-tokyo.ac.jp
[2]Joetsu University of Education, Yamayashiki, Joetsu, Niigata 943-8512, Japan

Abstract. In the central regions of active galaxies, dense molecular medium are exposed to various types of radiation and energy injections, such as UV, X-ray, cosmic ray, and shock dissipation. With the rapid progress of chemical models and implementation of new-generation mm/submm interferometry, we are now able to use molecules as powerful diagnostics of the physical and chemical processes in galaxies. Here we give a brief overview on the recent ALMA results to demonstrate how molecules can reveal underlying physical and chemical processes in galaxies. First, new detections of Galactic molecular absorption systems with elevated HCO/$H^{13}CO^+$ column density ratios are reported, indicating that these molecular media are irradiated by intense UV fields. Second, we discuss the spatial distributions of various types of shock tracers including HNCO, CH_3OH and SiO in NGC 253 and NGC 1068. Lastly, we provide an overview of proposed diagnostic methods of nuclear energy sources using ALMA, with an emphasis on the synergy with sensitive mid-infrared spectroscopy, which will be implemented by JWST and SPICA to disentangle the complex nature of heavily obscured galaxies across the cosmic time.

Keywords. galaxies: ISM — galaxies: starburst — galaxies: Seyfert — ISM: molecules — submillimeter

1. ALMA detections of new Galactic molecular absorption systems: HCO as an indicator of intense UV fields

Molecular medium in front of bright background continuum sources can often be observed as absorption lines, allowing us to make sensitive studies of physical and chemical properties of the interstellar medium (ISM) in various environments, including quiescent and translucent molecular media in the Milky Way, dense molecular gas in the vicinity of active galactic nuclei (AGNs), and the ISM in the early epoch of the universe. In fact, because the absorption line depths are independent of the distance to the absorption system, they offer a very powerful technique to uncover cold molecular gas in the distant universe, which is often difficult to observe in emission unless it is intensely heated by either star-formation or active nuclei. Since the first cosmological detection of such molecular absorption system toward PKS 1413+135 at $z = 0.247$ (Wiklind & Combes 1994), it has been demonstrated that molecular absorption systems are extremely useful for studying chemical properties of quiescent ISM (e.g., Muller *et al.* 2014), for determining the redshift of a foreground object in a galaxy-galaxy lens system (Wiklind & Combes 1996), and for constraining fundamental physical quantities and constants such as cosmic background temperature (e.g., Henkel *et al.* 2009, Muller *et al.* 2011) and proton-electron mass ratios (e.g., Kanekar *et al.* 2015) across cosmic time. In spite of their usefulness, however, only four molecular absorption systems beyond $z > 0.1$ are currently known at millimeter wavelengths (Combes 2008; Curran *et al.* 2011).

Another important case is a molecular absorption system in which absorption occurs near an AGN, i.e., presumably in the putative dense obscuring tori around AGNs. For instance, Centaurus A is known to exhibit strong absorption line features seen in both molecules (e.g., CO and HCO^+) and neutral hydrogen (e.g., Espada *et al.* 2010; see also Ott *et al.* in this volume). However, the number of such systems, which are believed to originate from such ~pc scale tori, is very small again. NGC 4261 is another example showing nuclear absorption lines in both CO and HI (Jaffe & McNamara 1994) but recent sensitive search for mm-wave absorption features using the Nobeyama Millimeter Array (NMA) and the Plateau de Bure Interferometer (PdBI) yields no significant detections of either CO(1-0) or CO(2-1), giving very small optical depths for these transitions ($\tau_{CO(2-1)} < 0.098$, Okuda *et al.* 2013).

Such molecular absorption lines are also observed in the diffuse medium within the Galaxy against the background extragalactic continuum sources. This also gives an important way to unveil "dark" molecular medium, i.e., invisible in emission because they are often in the equilibrium with the background radiation. Currently, the number of such extragalactic continuum sources with known Galactic molecular absorption lines is limited to ~30. Recently, the first detections of HCO, the formyl radical have been reported toward two extragalactic continuum sources (Liszt *et al.* 2014), opening a new window for studying the impact of UV radiation on these diffuse media because the coexistence of H_2 and C^+, i.e., the presence of photodissociation regions (PDRs), is required for efficient HCO production (Gerin *et al.* 2009). Again we need to expand the sample of Galactic absorption systems with such rare species to obtain a comprehensive picture of the diffuse molecular media within the Galaxy, but a survey of new molecular absorption systems is always very time consuming and entails a significant risk of non-detection.

Ando *et al.* (2015) propose using the ALMA archive to search for new molecular absorption systems in mm/submm wavelengths, because the ALMA archive already offers a vast number of sensitive mm/submm spectra of extragalactic continuum sources, i.e., bandpass and/or visibility calibrators, which are essentially radio-loud quasars including BL-LAC objects. Because of the enormous improvement in sensitivity provided by ALMA, these calibrators in the ALMA archive are potentially ideal *free lunch* in the quest for molecular absorption systems from the Galaxy to the high-z universe without any new investment of observing time!

They analyzed 36 calibrator sources in the ALMA archive to create 3D cubes and search for any absorption features. This resulted in four new detections of absorption line features, and three of which are found to be newly detected Galactic absorption systems. Another one is NRAO530, which is a known absorption system, but strong HCO absorption lines have been detected toward NRAO530 for the first time (Figures 1 and 2). With these detections, this doubles the number of Galactic molecular absorption systems exhibiting HCO absorption features, and it was shown that these Galactic diffuse molecular media are in an extreme condition similar to the Horsehead Nebula, a prototypical PDR (Figure 3).

2. Shock tracers in the central regions of active galaxies: HNCO, CH_3OH, and SiO in NGC 253 and NGC 1068

SiO is considered to be a tracer of shocks, although it could also be related to the presence of X-ray radiation (e.g., García-Burillo *et al.* 2000). SiO abundance is enhanced by ejection of significant Si from sputtered silicate grain cores in high velocity shocks

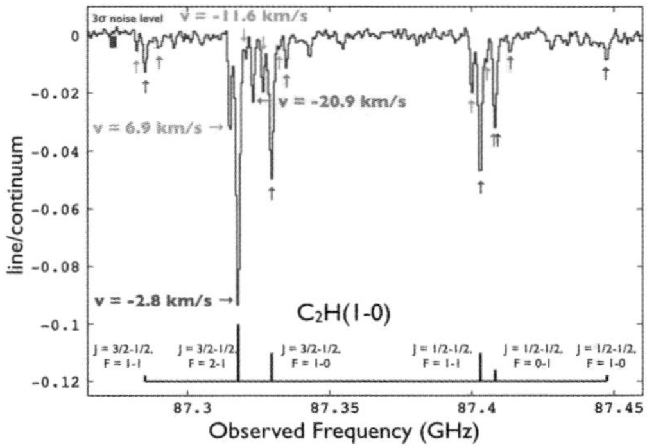

Figure 1. ALMA band-3 spectrum toward the continuum source J1717-337 showing detections of hyperfine components of C_2H in absorption. At least four different velocity components are visible despite of the short integration time (of just a few minutes), demonstrating the superb sensitivity of ALMA. Figure modified from Ando *et al.* (2015). *Color versions of the figures are available online.*

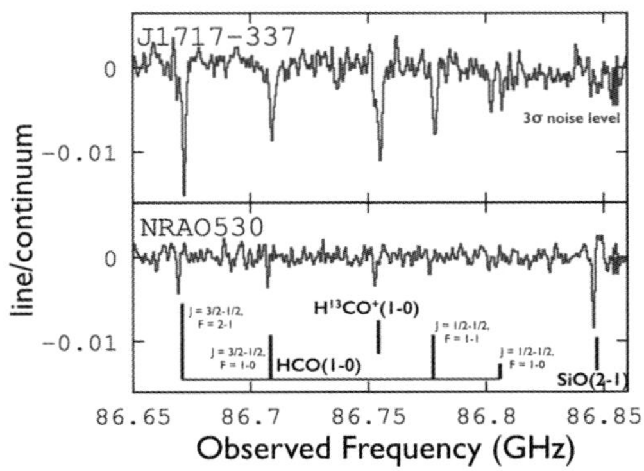

Figure 2. HCO radical and $H^{13}CO^+$ absorption lines toward J1717-337 and NRAO530. Figure modified from Ando *et al.* (2015).

(e.g., Martín-Pintado *et al.* 1992). In contrast, HNCO forms in the grain mantle, and it can be sublimated in weaker shock events; therefore the HNCO/SiO ratio can be an indicator of shock strength (see also Kelly *et al.* in this volume). HNCO is also known to be fragile, being easily photodissociated by UV radiation. It is then proposed that the HNCO/CS ratio can give an indicator of starburst evolutionary stage balancing between shock and photodissociation (Martín *et al.* 2009). Methanol, CH_3OH, is formed through grain-surface reactions. A series of hydrogenations starting with CO in cold ($\sim$10 K) dust (Watanabe & Kouchi 2002) forms CH_3OH efficiently, which then sublimes into the gas phase as the result of shock waves (Viti *et al.* 2011) or warms up dust grains in hot cores (Garrod *et al.* 2008). It is also suggested that methanol will be easily dissociated by cosmic-ray and UV radiation (e.g., Aladro *et al.* 2013).

Recent ALMA observations of the central few-100-pc regions of NGC 253 depict the spatial variation of these shock tracers; SiO(2-1) shows a series of clumps with a significant peak at the very center of NGC 253, where an intense H40α recombination line is observed. Meanwhile, HNCO has no significant peak at the very center of the galaxy, suggesting that HNCO is photodissociated by intense UV radiation (Meier *et al.* 2015), although the overall extent of both emissions is quite similar.

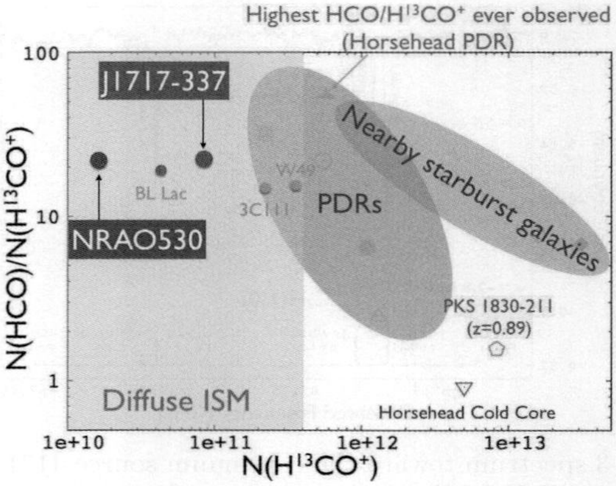

Figure 3. Column density ratios of HCO to $H^{13}CO^+$ as a function of the column density of $H^{13}CO^+$, including two new detections of HCO absorption (J1717-337 and NRAO530; Ando *et al.* 2015). Molecular absorption systems toward J1717-337, NRAO530, 3C111, and BL-LAC show elevated $HCO/H^{13}CO^+$ column density ratios similar to the value in the Horsehead Nebula (Gerin *et al.* 2009), indicating that the diffuse media in the Galaxy seen toward these radio sources are exposed in the intense UV field. Figure modified from Ando *et al.* (2015).

In contrast, in the central ~kpc region of NGC 1068, there are striking differences between tracers of mild and strong shocks. Takano *et al.* (2014) presented the first methanol image of NGC 1068, showing clumps across the starburst ring or arms, as seen in the case of CO and its isotopologs, suggesting that mild shocks are ubiquitous in the arms hosting bursts of massive star formation. Meanwhile, PdBI observations reveal that SiO is concentrated near the circumnuclear disk (CND), without any significant emission across the starbursting spiral arms (García-Burillo *et al.* 2010; see also Kelly, Viti & Garcia-Burillo in this volume for new SiO(3-2) and HNCO(6-5) data taken with PdBI). New ALMA cycle 2 images of CH_3OH and HNCO with improved sensitivity are displayed in Figure 4 (Tosaki *et al.* in prep.). We find a striking similarity between CH_3OH and HNCO, confirming that mild shocks are ubiquitous in the starbursting spiral arms. It is also evident that these two mild shock tracers are clearly detected in the CND, although the modeling of physical and chemical properties of the CND is not straightforward (Viti *et al.* 2014; Nakajima *et al.* 2015).

A new SiO multi-transition study obtaining four transitions of SiO with ALMA in the central region of NGC 1068 is in progress (Taniguchi *et al.* in prep.), and a preliminary $\sim 0''.5$ resolution ALMA image shows that SiO(6-5) is concentrated at the western knot of the CND and no significant emission can be seen in any other regions including the eastern knot, molecular ridges along the inner bar seen in ^{12}CO (e.g., Schinnerer *et al.* 2000; García-Burillo *et al.* 2014) and spiral arms. This would suggest that the shock properties are drastically different between the Western knot and other regions. Further quantitative analysis of SiO fractional abundance distribution employing rotation diagrams of SiO, CS, and ^{13}CO are on-going (Taniguchi *et al.* in prep.) to understand the origin of such sharp and anisotropic variation of SiO properties.

3. Unveiling the nature of obscured galaxies

With the enormous improvement of the sensitivity and spatial resolution in mm/submm interferometry accomplished by ALMA, it is now strongly expected that ALMA will

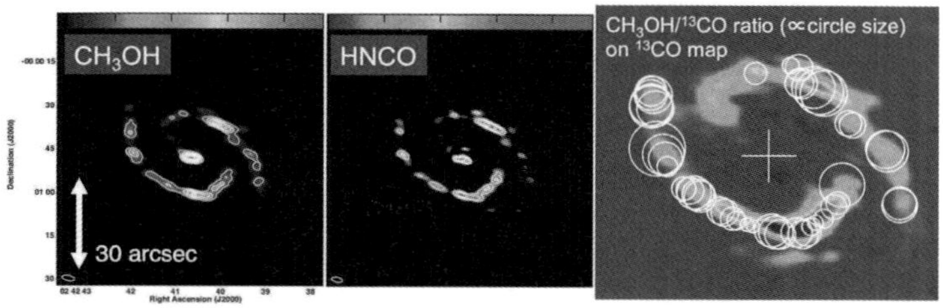

Figure 4. Distribution of CH_3OH (left) and HNCO (middle), tracers of mild shock, in the central ~1 kpc region of NGC 1068. The $CH_3OH/^{13}CO$(1-0) intensity ratios, indicated as the sizes of circles superposed on the ^{13}CO(1-0) integrated intensity image, are displayed in the right panel. Figures adopted from Tosaki *et al.* (in prep.).

provide unique diagnostic probes of energy sources in heavily obscured galaxies by penetrating deep dust condensations via mm/submm imaging and/or spectroscopy.

IR luminosity density and L/M ratio: One of the promising methods is to perform high spatial resolution measurements of the dust continuum to constrain the IR luminosity density and luminosity-to-mass (L/M) ratio in the heart of obscured galaxies. Theoretically, an extreme starburst, which will expel the ISM via radiation pressure onto dust and quench star-formation there, is estimated to have a maximum luminosity surface density of $\sim 10^{13}$ $L_\odot$ kpc^{-2} and L/M ratio of ~1,000 $L_\odot/M_\odot$ (Thompson *et al.* 2005; Sakamoto *et al.* 2008), so spatially resolved measurements of these values would be crucial to test whether such galaxies are powered by a non-stellar source. In fact, PdBI and SMA observations of double nuclei in Arp 220 suggest that the western nucleus of Arp 220 can be heated by an AGN rather than a starburst (Downes & Eckart 2007; Sakamoto *et al.* 2008). Recent ALMA band-9 observations of the 430μm dust continuum at a resolution of $0''.36 \times 0''.20$ suggest that the IR luminosity density and the L/M ratio reach 2×10^{14} $L_\odot$ kpc^{-2} and 540 $L_\odot/M_\odot$, respectively, in the western nucleus of Arp 220, indicating the presence of a Compton-thick, heavily obscured AGN at least in the western nucleus of Arp 220 (Wilson *et al.* 2014).

The HI/HeII recombination line ratio: Another promising, physically clean indicator would be (the ratio of) recombination lines of H, He, and He^+ because they have significantly different ionization potentials (13.6 to 54.4 eV). Specifically, Scoville & Murchikova (2013) demonstrate that the H26α (353.623 GHz) to HeII42α (342.894 GHz) luminosity ratio will show a factor of 50 difference between an AGN and a starburst. These two recombination lines can be simultaneously observed with ALMA band-7 with a reasonable integration time for local IR-luminous galaxies, and they will provide an ambiguous basis to calibrate other proposed diagnostic methods.

The HCN enhancement: A significant enhancement of HCN emission with respect to other molecules such as CO and HCO^+ at the centers of local AGNs was first indicated by Jackson *et al.* (1993) and Tacconi *et al.* (1994) based on the NMA and PdBI observations of HCN(1-0) in NGC 1068. Since then, mounting supporting evidence for such elevated HCN emission at the centers of local Seyfert galaxies has been accumulated (e.g., Kohno *et al.* 2001; Imanishi *et al.* 2006; Krips *et al.* 2008), giving an empirical AGN diagnostic. There are on-going debates on the origin of such elevated HCN, however. X-ray induced chemistry is one of the attractive scenarios because the CND of NGC 1068 is suggested to be a giant X-ray dominated region (XDR; Maloney *et al.* 1996) based on detections of HOC^+ and other molecules that are preferentially enhanced in the XDR (Usero *et al.* 2004), although chemical models often do not reproduce such HCN enhancement in the XDR (e.g., Meijerink*et al.* 2007). More recently, spatially resolved ALMA observations

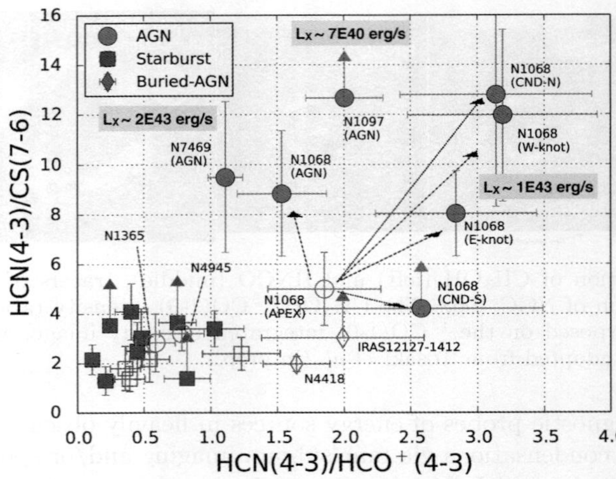

Figure 5. HCN(4-3)/HCO$^+$(4-3) and HCN(4-3)/CS(7-6) line ratios of local AGNs, starburst galaxies, and IR-luminous galaxies with buried AGNs. All these lines can be simultaneously observed using ALMA Band-7. In NGC 1068, these line ratios significantly depend on the size of the observing beam; a measurement with APEX ($\sim$1 kpc scale beam; Zhang *et al.* 2014) is shown as an open red circle, whereas spatially resolved ALMA measurements ($\sim$50 pc; García-Burillo *et al.* 2014) are indicated by red filled circles. It was found that the HCN enhancement at the nucleus of NGC 1068 is not very significant compared with those in the peaks of the CND at $\sim$50 pc resolution, but it is still distinguishable from purely star-forming galaxies, suggesting that these ratios can give an empirical diagnostic of the dominant power sources within the observing beam. Note that NGC 1365 and NGC 4945 (two red open circles), with ratios similar to starbursts, are measured with a single dish telescope (and therefore dominated by circumnuclear starburst). Figure adopted from Izumi *et al.* (2015b).

of HCN(4-3) and HCO$^+$(4-3) at the CND of NGC 1068 at a $\sim$50 pc resolution reveal that the HCN(4-3)/HCO$^+$(4-3) line ratios are indeed highly elevated at the CND but not significantly at the very position of the active nucleus (García-Burillo *et al.* 2014; Viti *et al.* 2014). This indicates that HCN enhancement is not simply caused by X-ray irradiation. In fact, recent HCN(4-3)/HCO$^+$(4-3) measurements of the high-luminosity type-1 AGN NGC 7469 (Izumi *et al.* 2015a) with $L_X = 2 \times 10^{43}$ erg s^{-1} suggest that the degree of HCN enhancement is not very different from that in NGC 1097 (Izumi *et al.* 2013), a much lower luminosity type-1 AGN ($L_X = 7 \times 10^{40}$ erg s^{-1}), also adding further evidence against the XDR origin of enhanced HCN in AGNs (see also Martín *et al.* 2015). Mechanical heating, which is caused by the dissipation of shock-driven energy (Loenen *et al.* 2008) and high-temperature-induced chemistry (Harada *et al.* 2010; Harada *et al.* 2013; Izumi *et al.* 2013) are other possible mechanisms that could explain such HCN enhancement. Elevated HCN/HCO$^+$ ratios in massive (and presumably shocked) molecular outflows driven by the powerful AGN in Mrk 231 would be consistent with this view (Aalto *et al.* 2012). A possible HCN maser amplification is also proposed (Matsushita *et al.* 2015) in the case of the HCN enhancement in M51 (Kohno *et al.* 1996).

Figure 5 displays the latest compilation of measurements of HCN(4-3)/HCO$^+$(4-3) and HCN(4-3)/CS(7-6) line ratios currently available, presented by Izumi *et al.* (2015b). This plot demonstrates that the HCN enhancement is still visible even in the $\sim$50 pc resolution ALMA measurements by García-Burillo *et al.* (2014) if we compare those values with star-forming and starburst galaxies. Although the origin of HCN enhancement is yet controversial (see also Viti 2015, this volume), more (spatially resolved) observational

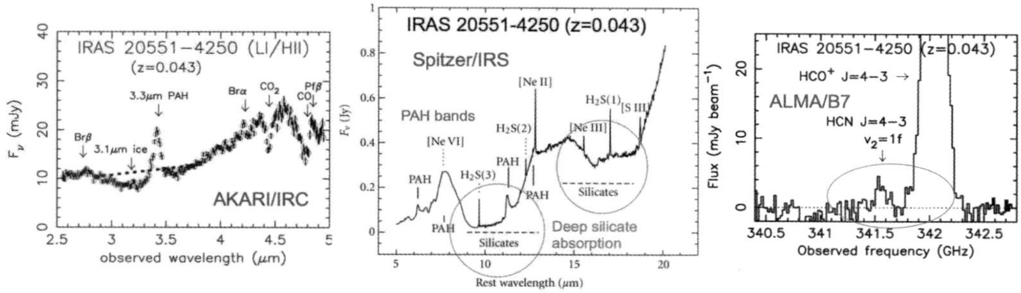

Figure 6. Mid-IR and submm spectra of IRAS 20551-4250, a heavily obscured ULIRG at $z = 0.043$. Vibrationally-excited HCN emission has been detected (right). Figures taken from Imanishi *et al.* (2010), Sani & Nardini (2012), and Imanishi & Nakanishi (2013).

work with ALMA will improve our understanding of such physical and chemical processes and result in identification of useful and reliable tracers of power sources in dusty galaxies.

Vibrationally excited HCN and synergy with mid-infrared spectroscopy: Regarding the nature of HCN enhancement, it is also important to consider the role of intense mid-IR radiation because HCN can be pumped by 14μm radiation, indicating the indispensable synergy between mm/submm and mid-IR facilities such as JWST and SPICA. Vibrationally excited HCN has been detected in IR-luminous galaxies such as NGC 4418, Mrk 231, and Arp 220 (e.g., Sakamoto *et al.* 2010; Costagliola *et al.* 2015; Aalto *et al.* 2015a,b). Mid-IR pumping seems to play a significant role in the enhanced high-J HCN emission in the $z = 3.91$ quasar APM 08279+5255 (e.g., Riechers *et al.* 2010).

Figure 6 presents a collection of submm (taken with ALMA) and mid-IR spectra (AKARI and Spitzer) toward the local ultraluminous IR galaxy IRAS 20551-4250, exhibiting vibrationally excited HCN emission (Imanishi & Nakanishi 2013). Because vibrationally excited HCN is preferentially found at heavily obscured nuclei with highly elevated mid-IR luminosity (e.g., Aalto *et al.* 2015b), deep silicate absorption features at 9.7 μm and 18 μm (indicating A_V >50 - 100 mag) are often seen in the mid-IR spectra (although their connection is not yet established). Furthermore, PAH band features (the brightest one being at 7.7 μm) are also useful to constrain the relative contribution of AGN to the total IR luminosity in such dusty nuclei (e.g., Menéndez-Delmestre *et al.* 2009 and references therein; see also Privon *et al.* (2015) for an example of joint analysis of mm-wave HCN(1-0) and mid-IR properties.). JWST will be a powerful tool for studying such mid-IR radiation for local to low-z galaxies, whereas SPICA, equipped with SMI (for 12 - 36 μm)† and SAFARI (for 35 - 210 μm), will be crucial for higher redshift galaxies to catch these redshifted major mid-IR features.

References

Aalto, S., García-Burillo, S., Muller, S., *et al.* 2012, *A&A*, 537, A44
Aalto, S., García-Burillo, S., Muller, S., *et al.* 2015, *A&A*, 574, A85
Aalto, S., Costagliola, S., Martín, S., *et al.* 2015, *A&A*, in press (arXiv:1504.06824)
Aladro, R., Viti, S., Bayet, E., *et al.* 2013, *A&A*, 549, A39

† SMI contains 3 spectrometer modules with different resolutions, including the High Resolution Spectrometer, HRS, allowing us to obtain R = 25,000 spectra at 12 - 17 μm. Note that JWST-MIRI does not have such high spectral resolution spectroscopic capability. See Sibthorpe *et al.* 2015 for details of SMI and the newly defined SPICA mission, a candidate to ESA's 5th medium class mission.

Ando, R., Kohno, K., Tamura, Y., Izumi, T., *et al.* 2015, *PASJ*, in press (arXiv:1510.05004)
Sibthorpe, B., Helmich, F., Roelfsema, P., Kaneda, H., Shibai, H., & the SPICA consortium, 2015, in *Conditions and Impact of Star Formation From Lab to Space* (EAS Publications Series), in press.
Combes, F. 2008, *Ap&SS*, 313, 321
Costagliola, F., Sakamoto, K., Muller, S., *et al.* 2015, *A&A*, 582, 91
Curran, S. J., Whiting, M. T., Combes, F., *et al.* 2011, *MNRAS*, 416, 2143
Downes, D. & Eckart, A. 2007, *A&A*, 468, L57
Espada, D., Peck, A. B., Matsushita, S., *et al.* 2010, *ApJ*, 720, 666
García-Burillo, S., *et al.* 2010, *A&A*, 519, A2
García-Burillo, S., *et al.* 2014, *A&A*, 567, A125
Garrod, R. T., Weaver, S. L. W., & Herbst, E. 2008, *ApJ*, 682, 283
Gerin, M., Goicoechea, J. R., Pety, J., & Hily-Blant, P. 2009, *A&A*, 494, 977
Harada, N., Herbst, E., & Wakelam, V. 2010, *ApJ*, 721, 1570
Harada, N., Thompson, T. A., & Herbst, E. 2013, *ApJ*, 765, 108
Henkel, C., Menten, K. M., Murphy, M. T., *et al.* 2009, *A&A*, 500, 725
Imanishi, M., Nakanishi, K., & Kohno, K. 2006, *AJ*, 131, 2888
Imanishi, M., Nakagawa, T., Shirahata, M., Ohyama, Y., & Onaka, T. 2010, *ApJ*, 721, 1233
Imanishi, M. & Nakanishi, K. 2013, *AJ*, 146, 91
Izumi, T., Kohno, K., Martín, S., *et al.* 2013, *PASJ*, 65, 100
Izumi, T., Kohno, K., Aalto, S., *et al.* 2015a, *ApJ*, 811, 39
Izumi, T., Kohno, K., Aalto, S., *et al.* 2015b, *ApJ*, in press (arXiv:1512.03438)
Jackson, J. M., Paglione, T. A. D., Ishizuki, S., & Nguyen-Q-Rieu 1993, *ApJ* (Letters), 418, L13
Croat, T. K., Stadermann, F. J., & Bernatowicz, T. J. 2005, *ApJ*, 631, 976
Kanekar, N., Ubachs, W., Menten, K. M. *et al.* 2015, *MNRAS*, 448, 104
Kohno, K., Kawabe, R., Tosaki, T., & Okumura, S. K. 1996, *ApJ* (Letters), 461, L29
Kohno, K., Matsushita, S., Vila-Vilaró, B., *et al.* 2001, in: J.H. Knapen, J.E. Beckman, I. Shlosman, & T.J. Mahoney (eds.), *The Central Kiloparsec of Starbursts and AGN: The La Palma Connection* (San Francisco: Astronomical Society of the Pacific), p. 672
Liszt, H. S., Pety, J., Gerin, M., & Lucas, R. 2014, *A&A*, 564, A64
Loenen, A. F., Spaans, M., Baan, W. A., & Meijerink, R. 2008, *A&A*, 488, L5
Lucas, R. & Liszt, H. 1998, *A&A*, 337, 246
Maloney, P. R., Hollenbach, D. J., & Tielens, A. G. G. M. 1996, *ApJ*, 466, 561
Martín, S. Martín-Pintado, & Mauersberger, R. 2009, *ApJ*, 694, 610
Martín, S., Kohno, K., Izumi, T., *et al.* 2015, *A&A*, 573, 116
Martín-Pintado, J., Bachiller, R., & Fuente, A. 1992, *A&A*, 254, 315
Matsushita, S., Trung, Dinh-V., Boone, F., *et al.* 2015, *ApJ*, 799, 26
Meier, D. S., Walter, F., Bolatto, A. D., *et al.* 2015, *ApJ*, 801, 63
Meijerink, R., Spaans, M., & Israel, F. P. 2007, *A&A*, 461, 793
Menéndez-Delmestre, K., Blain, A. W., Smail, I., *et al.* 2009, *ApJ*, 699, 667
Nakajima, T., Takano, S., Kohno, K., *et al.* 2015, *PASJ* 67, 8
Okuda, T., Iguchi, S., & Kohno, K. 2013, *ApJ*, 768, 19
Privon, G. C., Herrero-Illana, R., Evans, A. S., *et al.* 2015, *ApJ*, 814, 39
Riechers, D. A., Weiß, A., Walter, F., & Wagg, J. 2010, *ApJ*, 725, 1032
Sakamoto, K., Wang, J., Wiedner, M. C., *et al.* 2008, *ApJ* (Letters), 684, 957
Sakamoto, K., Aalso, S., Evans, A. S., *et al.* 2010, *ApJ* (Letters), 725, L228
Sani, E. & Nardini, E. 2012, *Advances in Astronomy*, vol. 2012, id. 783451
Schinnerer, E., Eckart, A., Tacconi, L. J., & Genzel, R. 2000, *ApJ*, 533, 85
Scoville, N. & Murchikova, L. 2013, *ApJ*, 779, 75
Tacconi, L. J., Genzel, R., Blietz, M., *et al.* 1994, *ApJ* (Letters), 426, L77
Takano, S., Nakajima, T., Kohno, K., *et al.* 2014, *PASJ*, 66, 75
Thompson, T. A., Quataert, E., & Murray, N. 2005, *ApJ*, 630, 167
Usero, A., García-Burillo, S., Fuente, A., *et al.* 2004, *A&A*, 419, 897
Viti, S., Jimenez-Serra, I., Yates, J. A., *et al.* 2011, *ApJ* (Letters), 740, L3

Viti, S., García-Burillo, S., Fuente, A., *et al.* 2014, *A&A*, 570, 28
Watanabe, N. & Kouchi, A. 2002, *ApJ* (Letters), 571, L173
Wiklind, T. & Combes, F. 1994, *A&A*, 286, L9
Wiklind, T. & Combes, F. 1996, *Nature*, 379, 139
Wilson, C. D., Rangwala, N., Glenn, J., *et al.* 2014, *ApJ* (Letters), 789, L36
Zhang, Z.-Y., Gao, Y., Henkel, C., *et al.* 2014, *ApJ* (Letters), 784, L31

From Interstellar Clouds to Star-Forming Galaxies:
Universal Processes?
Proceedings IAU Symposium No. 315, 2015
P. Jablonka, Ph. André & F. van der Tak, eds.

doi:10.1017/S1743921316007535

AGN feedback and star formation in ETGs: negative and positive feedback

Luca Ciotti[1], Jeremiah P. Ostriker[2,3], Andrea Negri[1,4], Silvia Pellegrini[1], Silvia Posacki[1] and Greg Novak[5]

[1]Dept. of Physics and Astronomy, University of Bologna, via Ranzani 1, I-40127 Bologna, Italy,
email: luca.ciotti@unibo.it

[2]Dept. of Astrophysical Sciences, Princeton University, NJ, USA

[3]Dept. of Astronomy and Astrophysics, Columbia University, NY, USA

[4] Institut d'Astrophysique de Paris, CNRS, UMR 7095 98bis bvd Arago, F-75014 Paris, France

5 Observatoire de Paris, LERMA, CNRS, 61 Av de l'Observatoire, F-75014 Paris, France

Abstract. AGN feedback from supermassive black holes (SMBHs) at the center of early type galaxies is commonly invoked as the explanation for the quenching of star formation in these systems. The situation is complicated by the significant amount of mass injected in the galaxy by the evolving stellar population over cosmological times. In absence of feedback, this mass would lead to unobserved galactic cooling flows, and to SMBHs two orders of magnitude more massive than observed. By using high-resolution 2D hydrodynamical simulations with radiative transport and star formation in state-of-the-art galaxy models, we show how the intermittent AGN feedback is highly structured on spatial and temporal scales, and how its effects are not only negative (shutting down the recurrent cooling episodes of the ISM), but also positive, inducing star formation in the inner regions of the host galaxy.

Keywords. ISM: kinematics and dynamics, galaxies: active, galaxies: cooling flows, galaxies: elliptical and lenticular, cD, galaxies: starburst, galaxies: evolution

1. Introduction

In a quite widespread view, merging is considered the main cause of QSO activity, because it is believed to be, after the end of the galaxy formation epoch, the only mechanism to add fresh gas to the central SMBHs in early type galaxies (ETGs). However, it has also been shown (e.g., Ciotti & Ostriker 1997) that the stellar mass losses produced during stellar evolution cyclically feed a central gas inflow, and then trigger the QSO activity in isolated ETGs (sometimes considered "red and dead"). Indeed, recent observations indicate that the QSO activity can be independent of galaxy merging. QSO activity is also invoked as the explanation of star formation quenching in ETGs (negative feedback) but, as stressed in Ciotti & Ostriker (2007, hereafter CO07)), AGN feedback can induce star formation (positive feedback) during "cooling flow" episodes fueled by stellar mass losses. In fact, also in isolation, significant amounts of fresh gas ($\simeq 20-30\%$ of the initial stellar mass M_* of the galaxy, depending on the IMF) are injected over the galaxy body by stellar evolution, at a rate approximately given by $\dot{M}_*(t) \propto M_*(0)\, t^{-1.3}$ (Ciotti *et al.* 1991, see also Pellegrini 2012). These losses interact with the pre-existing hot ISM, and mix with it, due to thermalization of the stellar velocity dispersion. In a galaxy of total B-band luminosity $L_{\rm B}$, SNIa's explosions at a rate $R_{\rm SN}(t) \propto L_{\rm B}\, t^{-s}$ provide additional mass and heat to the ISM, with rates $\dot{M}_{\rm SN}(t) = 1.4 M_\odot\, R_{\rm SN}(t)$ and $L_{\rm SN}(t) = 10^{51}{\rm erg}\, R_{\rm SN}(t)$. Since recent estimates suggest $s \simeq 1$, then the SNIa's specific heating $L_{\rm SN}/\dot{M}_*$ increases with time. The resulting hot atmosphere however cannot be static. The interstellar mass keeps increasing, its cooling time decreases, until a cooling catastrophe takes place in the central galactic region and the SMBH feedback takes place.

In a series of papers, Ciotti, Ostriker and co-workers, with the aid of high-resolution hydrodynamical simulations, and a physically based feedback implementation, studied in detail the AGN activity cycle induced by accretion of stellar mass losses, also considering star formation. In particular, it was shown that AGN feedback is able to avoid the overgrowth of the SMBH that would be caused by accretion of the mass lost by stars if unimpeded, and also that the same feedback can induce significant star formation at the center of ETGs (CO07). By using state-of-the-art galaxy dynamical models, and an updated version of the 2D code of Novak *et al.* (2011, 2012, hereafter N12, and references therein), we are now investigating in detail the ISM behavior in realistic ETG models with evolving input ingredients from the stellar population (Ciotti *et al.*, in preparation).

2. The models

We consider axisymmetric two-component galaxy models with adjustable flattening embedded in a spherically symmetric NFW dark matter halo and with a central SMBH. The stellar component is described by the ellipsoidal deprojection of the de Vaucouleurs law. The total gravitational field, the solution of the two-integrals Jeans equations, and the projection of the resulting kinematical field on the plane of the sky are computed as described elsewhere (Posacki et al. 2013). The velocity dispersion and rotational fields of the stellar populations, needed to compute energy and momentum injection in the ISM due to stellar evolution, are computed by using a generalized Satoh k-decomposition (Ciotti & Pellegrini 1996). The E4 and E7 galaxy models are obtained by flattening (at fixed stellar mass M_*) spherical "progenitors" with central (aperture) velocity dispersion of 180, 210, 250 and 300 km/s, and reproduce observed scaling laws (Negri *et al.* 2014). The initial mass of the SMBH is taken to be $M_{\rm BH} = 10^{-3} M_*$, near to the Magorrian relation, as expected at the end of the period of galaxy formation.

The hydrodynamical equations and the input physics are given in Ciotti & Ostriker (2012), and integrated by an improved version of the N12 code. Mass, momentum and energy sources and sinks associated with evolution of the passively evolving initial stellar population, and with the new stars added by star formation, are treated as described in Negri *et al.* (2015, hereafter N15). In particular the star formation rate is given by $\dot{\rho}_{\rm SF} = 0.1\rho/\max(\tau_{\rm cool}, \tau_{\rm dyn})$, where ρ is the local ISM density, and $\tau_{\rm cool}$ and $\tau_{\rm dyn} = \min(\tau_{\rm Jeans}, \tau_{\rm rot})$ are the cooling and dynamical times, respectively. In N15 we showed that this prescription reproduces the observed Kennicutt-Schmidt law remarkably well. Radiative and mechanical feedback from the accreting SMBH (both depending on the hydrodynamically self-consistently determined mass accretion rate on the SMBH, $\dot{M}_{\rm BH}$), are finally considered. In the energy equation, photoionization and Compton heating and cooling ($T_{\rm C} \simeq 2\times 10^7$ K), bremsstrahlung and line cooling are taken into account, with modifications due to the ionization effects of AGN radiation. In the momentum equation, radiation pressure due to AGN activity is computed by solving the radiative transport (in spherical symmetry) for the accretion luminosity $L_{\rm BH}$, considering photoionization+Compton opacity, and electron scattering. For simplicity at this stage we ignore the detailed treatment of dust formation and destruction, and the associated radiation pressure effects (Hansley *et al.* 2014, N12), as well as the radiation pressure due to star light in different bands (e.g., see Ciotti & Ostriker 2012). For given $\dot{M}_{\rm BH}$, then,

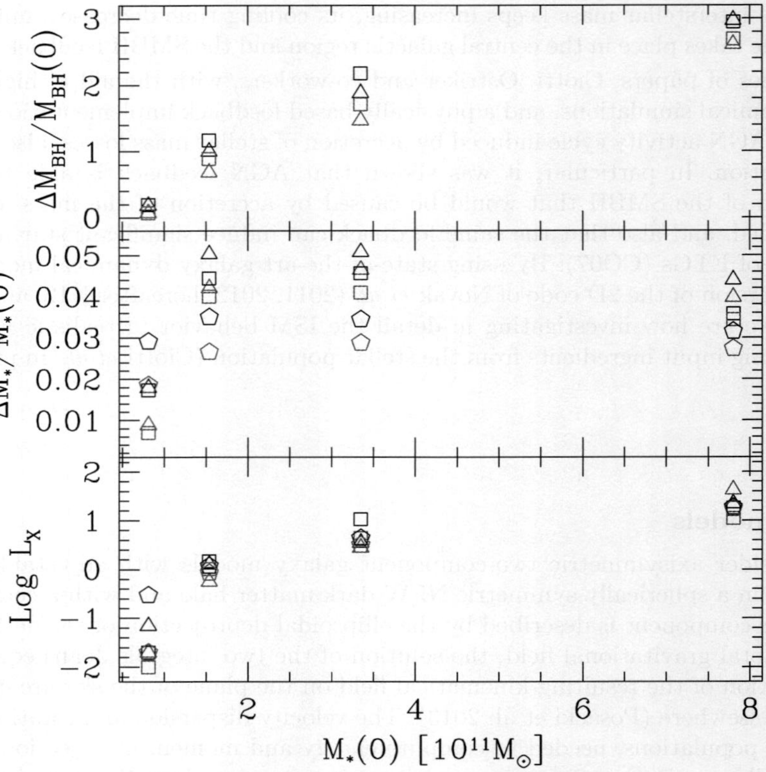

Figure 1. Global properties of the models at the end of the simulations (13 Gyr), as a function of their initial stellar mass, for E4 (black) and E7 (red) models. Triangles, squares, and pentagons refer to models with radiative plus mechanical AGN feedback, mechanical AGN feedback only, and no AGN feedback, respectively. *A color version of this figure is available online.*

one has:

$$L_{\rm BH} = \epsilon_{\rm EM} \dot{M}_{\rm BH} c^2, \quad \epsilon_{\rm EM} = \frac{\epsilon_0 A \dot{m}}{1 + A\dot{m}}, \quad \dot{m} \equiv \frac{\dot{M}_{\rm BH}}{\dot{M}_{\rm Edd}} = \frac{\epsilon_0 \dot{M}_{\rm BH} c^2}{L_{\rm Edd}}, \tag{2.1}$$

where $A = 100$, $\epsilon_0 = 0.125$ and $L_{\rm Edd}$ is the Eddington luminosity.

Mechanical feedback is due to mass, momentum, and energy injection of a conical-shaped nuclear wind, with half-opening angle of $\simeq 45°$. In terms of solid angle, this means that the wind is visible from $\sim 1/4$ of the available viewing angles. The relevant relations here are $\dot{M}_{\rm BH} = \dot{M}_{\rm in}/(1+\eta)$, $\dot{M}_{\rm out} = \eta \dot{M}_{\rm BH}$, $\dot{p}_{\rm w} = \dot{M}_{\rm out} v_{\rm w}$, $L_{\rm w} = \epsilon_{\rm w} \dot{M}_{\rm BH} c^2$, where $\eta \equiv 2\epsilon_{\rm w} c^2/v_{\rm w}^2$, $\dot{M}_{\rm in}$ and $\dot{M}_{\rm out}$ are the mass inflow and outflow rates at the first grid point (a few pc from the SMBH). In the current simulations, $\epsilon_{\rm w}$ follows a prescription similar to that adopted for $\epsilon_{\rm EM}$, with maximum value of 10^{-4}, while the wind velocity $v_{\rm w}$ is 10^4 km/s.

3. Results

As illustrated in Fig. 1 (top panel), the central SMBH grows more, in absolute terms ($\Delta M_{\rm BH}$) and also in percentage ($\Delta M_{\rm BH}/M_{\rm BH}(0)$), for increasing stellar mass M_* of the host galaxy; the growth reaches a factor of $\simeq 3$ in the biggest galaxies with AGN feedback (to be compared with a factor $\simeq 30$ in models without AGN feedback). Moreover, at any fixed M_*, more mass is accreted by the SMBH in E4 galaxies than in E7 galaxies. This is due to the fact that edge-on flattened galaxies are less bound (see N15) than more spherical systems. As expected, at fixed M_* and galaxy shape, the accreted mass increases from models with full AGN feedback (triangles), to models with mechanical AGN feedback only (squares), and finally to models without AGN feedback (pentagons). Therefore, since we start with a SMBH mass $\sim$ on the Magorrian relation, this relation is quite well preserved thanks to the effect of AGN feedback, which is able, in combination with large-scale SNIa heating, to maintain small the SMBH masses, avoiding the accretion of $\simeq 99\%$ of the mass injected by the stars over the galaxy body.

The stellar mass added to the galaxy due to star formation in the gas supplied by evolving stars (ΔM_*, middle panel), is larger in E4 galaxies than in E7 galaxies of same M_*. The total mass in new stars is of the order of $4-5\%$ of the initial M_*, indicating that an important fraction of the injected mass ($\simeq 0.2-0.3M_*$) escapes as a galactic wind, supported by SNIa heating with the additional contributions of the thermalization of stellar winds, of type SNII explosions in the new stellar populations, and of the AGN feedback. In general, as already found in the case of spherical models (CO07), models with AGN feedback tend to form *more* stars than models without (with the exception of very low mass galaxies, where AGN feedback is able to sustain global galactic winds). This positive feedback is explained by considering that the AGN prevents the rapid fall of the cooling material to the center, allowing for more time for star formation. A clear sign of this effect is the characteristic size of the region of star formation, which is confined to the very center of the galaxy in absence of AGN feedback (a small fraction of the optical effective radius), while in models with AGN feedback it reaches a size of the order of a kpc or more. The results of this work (in absence of AGN feedback) agree well with those extensively discussed in N15 including star formation.

An important observational feature of the models is of course the evolution of their luminosity, both of the central accreting SMBH ($L_{\rm BH}$), and of the galaxy hot ISM ($L_{\rm X}$). In Fig. 1 we show $L_{\rm X}$ in the X-ray $0.3-8$ keV band at the end of the simulations, with values in nice agreement with observations, and only moderately dependent on AGN feedback. The evolution of $L_{\rm BH}$ (not shown), is confirmed to be highly fluctuating, as already found in our previous works based on simpler galaxy models. The duty-cycle (phenomenologically defined as the fraction of time spent by the SMBH at $L_{\rm BH}$ higher than a few percents of $L_{\rm Edd}$), is of the order of $\simeq 3-5\%$.

References

Ciotti, L., D'Ercole, A., Pellegrini, S., & Renzini, A. 1991, *ApJ*, 376, 380

Ciotti, L. & Ostriker, J. P. 1997, *ApJ*, 487, L105

Ciotti, L. & Ostriker, J. P. 2007, *ApJ*, 665, 1038 (CO07)

Ciotti, L. & Ostriker, J. P. 2012, in D.W. Kim and S. Pellegrini (eds.), *Hot interstellar matter in elliptical galaxies*, (Springer, Heidelberg), ASSL, vol. 378, p. 83

Ciotti, L. & Pellegrini, S. 1996, *MNRAS*, 279, 240

Hensley, B., Ostriker, J. P., & Ciotti, L. 2014, *ApJ*, 789, 78

Negri, A., Posacki, S., Pellegrini, S., & Ciotti, L. 2014, *MNRAS*, 445, 1351

Negri, A., Pellegrini, S., & Ciotti, L. 2015, *MNRAS*, 451, 1212 (N15)

Novak, G., Ostriker, J. P., & Ciotti, L. 2011, *ApJ*, 737, 26

Novak, G., Ostriker, J. P., & Ciotti, L. 2012, *MNRAS*, 427, 2734 (N12)

Pellegrini, S. 2012, in D.W. Kim and S. Pellegrini (eds.), *Hot interstellar matter in elliptical galaxies*, (Springer, Heidelberg), ASSL, vol. 378, p. 21

Posacki, S., Pellegrini, S., & Ciotti, L. 2013, *MNRAS*, 433, 2259

From Interstellar Clouds to Star-Forming Galaxies: Universal Processes?
Proceedings IAU Symposium No. 315, 2015
P. Jablonka, Ph. André & F. van der Tak, eds.

doi:10.1017/S1743921316007547

Evolution of the ISM in main-sequence versus starburst galaxies: A motivation for molecular deep fields

Manuel Aravena

Núcleo de Astronomía, Facultad de Ingeniería, Universidad Diego Portales
Av. Ejército 441, Santiago Chile
email: manuel.aravenaa@mail.udp.cl

Abstract. In the last decade, significant progress has been made to understand the evolution with redshift of star formation processes in galaxies. Its is now clear that the majority of galaxies at $z < 3$ form a nearly linear correlation between their stellar mass and star formation rates and appear to create most of their stars in timescales of ~1 Gyr. At the highest luminosities, a significant fraction of galaxies deviate from this *main-sequence*, showing short duty cycles and thus producing most of their stars in a single burst of star formation within ~100 Myr, being likely driven by major merger activity. Despite the large luminosities of starbursts, main-sequence galaxies appear to dominate the star formation density of the Universe at its peak.

While progress has been impressive, a number of questions are still unanswered. In this paper, I briefly review our current observational understanding of this *main-sequence* vs *starburst* galaxy paradigm, and address how future observations will help us to have better insights into the fundamental properties of the interstellar medium of these galaxies. Finally, I show recent attempts to conduct molecular deep field observations and the motivation to perform molecular deep field spectroscopy with the Atacama Large Millimeter/submillimeter Array.

Keywords. galaxies: evolution, galaxies: high-redshift, galaxies: ISM

1. Introduction

One of the most important advances in the understanding of galaxy formation in the last few decades has been the determination of the evolution of the star formation rate (SFR) density of the Universe (e.g. Bouwens *et al.* 2014 and references therein). The current paradigm implies that galaxies were forming more actively ~8-10 billion years ago, and they are observed to have quenched their activity by an order of magnitude on average in the local Universe. Thus, the obvious question arises as to why the cosmic star formation density has this behavior, and what is the galaxy population that drives such evolution. Recent galaxy surveys have enabled a much more detailed view of the processes of star formation in large samples.

A major advance has been the determination that the bulk of star forming galaxies at all redshifts define a tight correlation between their stellar mass and their SFRs (Fig. 1; e.g., Brinchmann *et al.*2004; Elbaz *et al.*2007; Noeske *et al.* 2007; Daddi *et al.* 2007; Pannella *et al.* 2009; Karim *et al.* 2011; Rodighiero *et al.* 2011). Galaxies lying on top of this relation have been termed as "main-sequence" (MS), whereas galaxies lying above this relation on this plane have been called "starbursts". Consequentially, starburst galaxies have higher SFRs than MS galaxies for the same stellar mass (higher specific SFRs). At a fixed stellar mass, starburst galaxies will appear as outliers of the MS, however, as we go to higher masses both populations start to merge (e.g. Rodighiero *et al.* 2011).

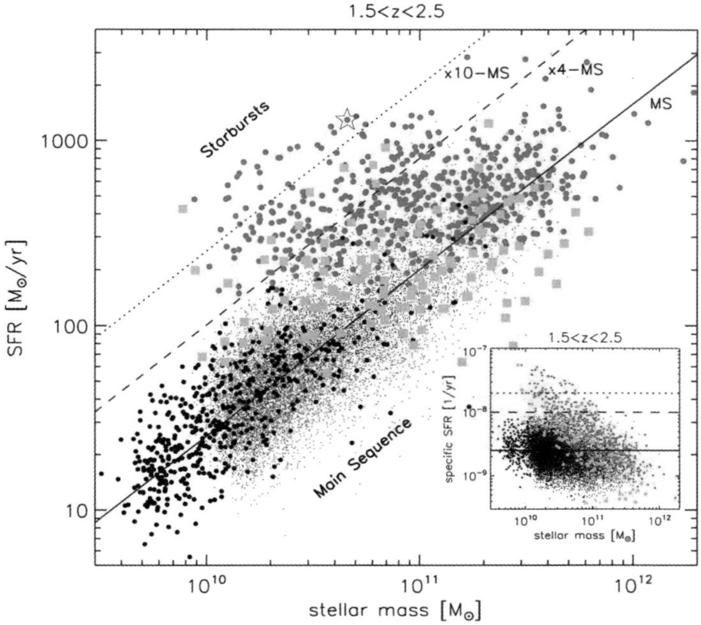

Figure 1. Stellar mass versus SFR for galaxies at $z = 1.5 - 2.5$ (from Rodighiero *et al.* 2011). Red circles and cyan circles represent IR selected samples from the GOODS South and COSMOS field surveys, respectively. Black filled circles and dots represent BzK color selected galaxies in the same fields, respectively. The solid line represents the MS relation at $z \sim 2$. *Color versions of the figures are available online.*

The MS has been observed to vary strongly with redshift and also to deviate from linearity at high stellar mass. The redshift evolution takes place mostly on the scaling of the MS, but not in the slope. This means that for the same stellar mass, MS galaxies at higher redshift tend to have higher SFRs. In other words, higher redshift galaxies have higher sSFRs. This increase in sSFRs is in hand with the increase of the cosmic SFR density from $z = 0$ to 3, but also as we will see later on, it goes together with the increase in gas fraction at higher redshifts. Another important result is the determination that the MS is not fully linear with stellar mass, with a flattening at the highest stellar masses. While this behavior is still under active discussion, it is possible that there are mechanisms that inhibit star formation at higher stellar masses (Whitaker *et al.* 2012; Panella *et al.* 2015; Lee *et al.* 2015).

With the advent of large galaxy surveys at infrared wavelengths, it has been possible to quantify the contribution of the different galaxy populations to the SFR density of the Universe. It appears that at $z = 2$, the most infrared luminous galaxies ($L_{\rm IR} > 10^{12}\ L_{\odot}$), typically associated with starburst galaxies, have an important contribution. This effect appears to be even more important at higher redshifts. However, more quiescent star forming galaxies tend to dominate at $z < 2$ (e.g. Magnelli *et al.* 2011, 2013; Sargent *et al.* 2012).

Despite the important progress made in the study of both MS and starburst galaxies with cosmic time, it is still hard to understand the fundamental differences between these populations and driving mechanisms for star formation. For this, we need to understand the star formation process in-situ by studying the raw material from which stars are formed. Namely, the cold molecular interstellar medium (ISM) in these galaxies.

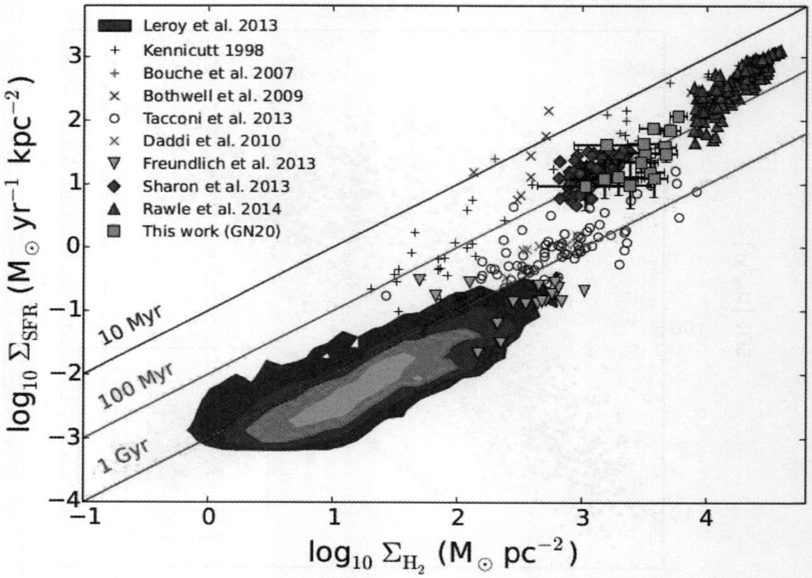

Figure 2. Gas and SFR surface density relationship for various samples of galaxies (from Hodge *et al.* 2015). Contours indicate the density of local galaxies. Magenta diamonds, cyan squares and green diamonds show the resolved densities for three different submillimeter galaxies (see legend). Notable is the clear separation between samples of MS galaxies and starbursts.

2. Studying the ISM

Before continuing, an obvious point should be noted. A typical, massive MS galaxy with a stellar mass of 3×10^{10} $M_\odot$, has a SFR of about a few tens solar masses per yr. The equivalently massive outlier of the MS will have several hundreds up to a thousand solar masses per yr. In this regime, galaxies will have large IR luminosities ($> 3\times10^{12}$ $L_\odot$), and thus will be routinely picked by large-area blank field far-IR and submillimeter surveys of the sky (see review by Casey *et al.* 2014). Most of the studies of the cold ISM (dust and molecular gas) have so far focused in the study of these "outliers", partly due to the easiness to detect such sources with available submillimeter/radio facilities (see review by Carilli & Walter 2013).

2.1. *The ISM of starburst galaxies*

A critical measurement to understand how the gas is converted into stars in galaxies is the relationship between their gas masses and SFRs (Fig. 2). Measurements of the dust and cold molecular gas for distant, luminous starburst galaxies (i.e. submillimeter galaxies and quasar host galaxies) indicated that, as observed in local galaxies, this "star-formation law" relationship also holds for these objects, spanning for several orders of magnitude (see review Solomon & Vanden Bout 2005). Such relation implies that these starburst galaxies have gas depletion timescales of $\sim$ 100 Myr, similar to that of local ULIRGs. Another important piece of information to gain further insight into the ISM properties of galaxies is the study of the excitation of the CO emission, i.e. the relative luminosity ratio of CO lines with different rotational quantum number (J). The specific distribution of this CO spectral line energy distribution (SLED) is sensitive to the density and temperature of the gas. Recently, it has been possible to obtain measurements of the typical CO excitation of SMGs (Bothwell *et al.* 2013; Spilker *et al.* 2014). The main conclusion is that starburst galaxies have typical CO SLEDs peaking at J = 5 or 6, being less extreme than what is found in local ULIRGs or high-redshift quasars. The main difference might come from the different sizes of the emitting regions, as local ULIRGs are shown to be fairly compact.

2.2. *The ISM of main-sequence galaxies*

Recently, numerous measurements of the gas content of MS galaxies have enabled us to determine that these objects follow a different star-formation law than starburst galaxies at similar redshifts (Fig. 2). This difference is visible even if only plotting observable quantities. This implies that MS galaxies have gas depletion timescales of $\sim 0.5 - 1$ Gyr, which is longer than that observed for starbursts. Another implication of these measurements is that there seems to be a trend for higher-redshift MS galaxies (which have higher SFRs) to have lower gas depletion timescales (e.g. Daddi *et al.* 2010; Genzel *et al.* 2010; Sargent *et al.* 2014). Also, there is a clear increase in the fraction of molecular gas measured from $z = 0$ to 2 (Tacconi *et al.* 2010, 2013; Scoville *et al.* 2014; Santini *et al.* 2014; Dessauges-Zavadsky *et al.* 2015). This rapid increase has been suggested to halt at the higher redshifts ($z > 3$; Saintonge *et al.* 2013), with indications that it may stay constant out to $z = 4$ (Bothwell *et al.* 2013).

Modeling of the ISM in MS galaxies based on line intensity ratios has been very slow and observationally expensive since it requires at least 2 detected CO lines. Currently the number of galaxies with more than one CO line transition detected is very modest (Aravena *et al.* 2010, 2014; Bolatto *et al.* 2015; Daddi *et al.* 2015). The main problem is that as we attempt to observe more typical, less massive galaxies, they become fainter in CO emission, and thus are hard to detect with current facilities. In particular, obtaining low-J CO measurements for these galaxies is extremely challenging.

Daddi *et al.* (2015) recently reported the detection of CO($J = 5 - 4$) line emission in 4 main sequence galaxies. All sources showed bright emission, suggesting that the turnover of the CO SLED is happening somewhere at $J = 6$ or 7. On average, these galaxies appear to have gas excitation comparable to that of starbursting submillimeter galaxies, at least out to $J = 5$.

2.3. *High resolution studies of molecular gas*

High resolution molecular gas studies hold the key to understand the fundamental differences between starbursts and main sequence galaxies at all redshifts. Models predict that starbursts galaxies should show features typical of galaxy mergers, with the molecular gas and dust concentrated within a few kpc and dispersion dominated kinematics. Main sequence galaxies should show an extended disk with rotation dominated kinematics and several gas clumps within this disk. In spite of that, the number of resolved studies are still modest, being mostly focused in single source studies. These show important differences from source to source, with disturbed dynamics in some cases (Sharon *et al.* 2013) and rotation dominated dynamics in others (Hodge *et al.* 2015).Current work shows that it is possible to dissect the high redshift galaxies and resolve the star formation law relations within these galaxies (Fig. 2). These studies indicate that starbursting submillimeter galaxies tend to follow a different scaling in the star formation law, with higher SFR surface densities compared to main sequence galaxies (e.g. Freundlich *et al.* 2013; Sharon *et al.* 2013; Rawle *et al.* 2013; Hodge *et al.* 2015).

Despite the substantial progress, resolved observations are limited by the faintness of the CO line emission. Resolving the star forming activity and molecular gas (CO) in distant starbursting galaxies requires significant amounts of observing time, even in the era of the Atacama Large Millimeter/submillimeter Array (ALMA). This is accentuated when attempting to observe more "normal" galaxies, which are fainter. Therefore,

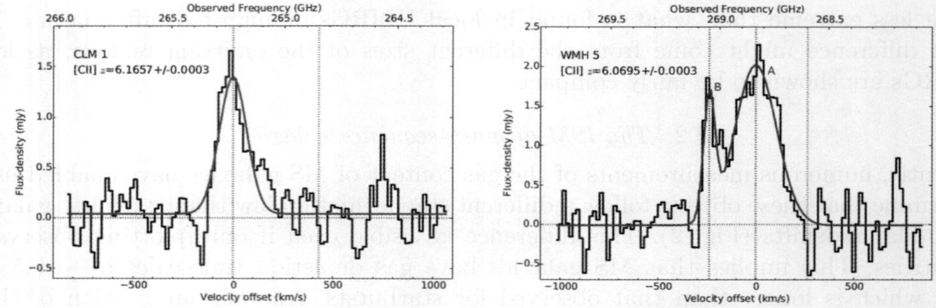

Figure 3. [CII] emission line spectra toward two typical galaxies at $z = 6$ (Willott *et al.* 2015). The recent detections of [CII] emission in several $z \sim 5 - 6$ galaxies has opened a remarkable opportunity to the study of star formation at the end of the cosmic reionization era.

acquisition of resolved dust and CO observations in statistical samples of distant star forming galaxies could be seriously limited. A solution used to overcome this limitation has been to take advantage of the amplification provided by gravitational lensing. However, such approach makes it complicated to interpret the emission given the necessity to model the foreground lens.

2.4. *How to access the z>4 Universe?*

As mentioned above, detecting CO line emission in normal galaxies is difficult even at $z = 2 - 3$ (e.g. Magdis *et al.* 2012). At $z > 4$, the current and most successful approach has been to target a brighter line, specifically the [CII] emission line. Observationally, this line has a rest-frame frequency of $\sim$ 1900.5 GHz, and thus is redshifted into the 380-210 GHz regime at $z = 4.0 - 8.0$. Most importantly, this line is a dominant coolant of the ISM in galaxies, typically making $0.1 - 1\%$ of the integrated IR luminosity. Furthermore, this line has been found to trace both photon dominated and HII regions in galaxies, being directly linked to the process of star formation (e.g. Carilli & Walter 2013).

Since the [CII] constitutes a critical tool for the study of the most distant galaxies, it has been widely used for this purpose. After several attempts to detect the [CII] line emission at $z > 4$ from normal galaxies (e.g. Lyman-α emitters and Lyman Break Galaxies), the unparalleled sensitivity of ALMA allowed us to detect this line in a number of galaxies at $z \sim 5 - 7$ (Fig. 3; e.g. Riechers *et al.* 2014; Capak *et al.* 2015; Willott *et al.* 2015; Maiolino *et al.* 2015).

Typically, galaxies at $z > 4$ are selected using the optical dropout technique, based on the identification of the Lyman break in the UV spectra. To confirm and obtain spectroscopic redshifts it is thus necessary that the Lyman-α 1216 $\AA$ emission line is sufficiently strong. However, it has been seen that the Lyman-α emission can be strongly suppressed by the intergalactic medium at $z > 6$. Furthermore, for objects at $z > 6.5$, this line is redshifted into a part of the spectrum that is contaminated by atmospheric emission, making it even more difficult to detect the Lyman-α from the ground. The recent detection of [CII] emission in distant galaxies suggest it may be possible to directly obtain spectroscopic redshifts based on this line, and also blindly identify such galaxies in deep field spectroscopic observations with ALMA.

3. Unbiased views of the cosmic evolution of the ISM

The star formation activity is necessarily linked to the molecular gas content at all redshifts (star formation law). The steep increase in the SFR density of the Universe

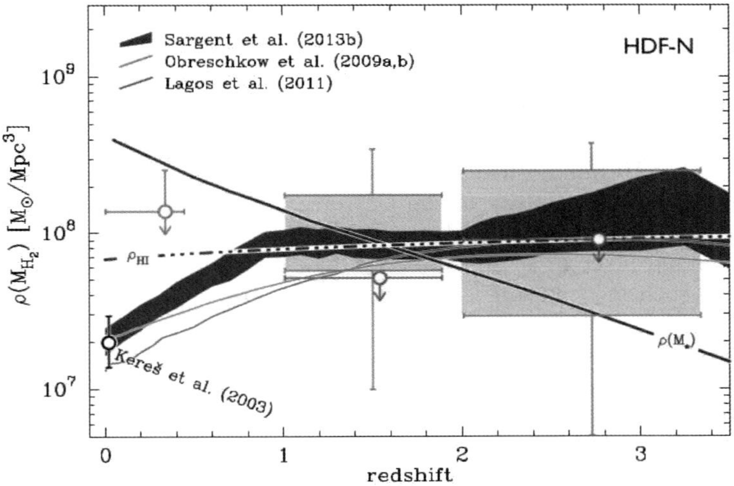

Figure 4. Evolution of the cosmic molecular gas mass density (from Walter *et al.* (2014)). The orange and green curves show the predictions from semi-analytical cosmological models. The grey shaded area show semi-empirical predictions from Sargent *et al.* (2014). The blue shaded area show the current limits based on PdBI observations of the Hubble Deep Field North. Deeper ALMA observations are needed to provide better constraints to this evolution.

between $z = 0$ to 2 appears to relate to the increase in molecular gas fractions in this redshift regime (i.e. the fraction of material that is available to sustain star formation). It thus follows that the evolution of the SFR density may be linked to a more fundamental quantity, the cosmic density of the molecular gas (the amount of molecular gas per comoving volume) across times.

It must be noted that most of what we know so far about the ISM properties of distant galaxies has been biased due to pre selection of sources in optical images which trace the stellar and star forming components of galaxies. What we need are deep wide-band molecular spectroscopic surveys over significant cosmic volumes, in order to directly pick up the cold gas phase of galaxies and select galaxies in this way.

A pilot study has been recently performed with the Plateau de Bureau Interferometer (PdBI) by Decarli *et al.* (2014) and Walter *et al.*(2014). See Figure 4. They conducted deep single pointing covering the full 3mm window, and thus covering the most relevant redshift ranges for CO line emission ($z < 0.5$, $z = 2 - 6$). Based on the number of detected sources at different CO luminosities, they were able to put constraints on the CO luminosity functions at three different redshifts, and thereby constrain the cosmic density of molecular gas with redshift. While this study could only provide rough constraints, their results provide the first insight into the relationship between the cosmic densities of star formation and molecular gas. To provide further constraints we obviously need to go deeper and cover larger areas. The only way to achieve this is using ALMA. Deep millimeter spectroscopic observations of large cosmological volumes with ALMA will be critical to provide an unbiased census of the cold gas phase of galaxies at all redshifts. As in Decarli *et al.* (2014), ALMA spectroscopy covering the full 3-mm window will enable us to target the CO line emission at $z > 2$. ALMA band 6 measurements, covering the full 1-mm window, will allow us to cover precisely the redshift range ($z < 2$) that is not constrained by the band 3 observations for CO line emission, making simultaneous observations of both bands and the same region of the sky highly complementary. Most importantly, ALMA band 6 observations will cover the [CII] line emission for galaxies at $z = 6 - 8$.

4. Concluding remarks

Despite the large IR luminosities of starburst galaxies, MS galaxies appear to dominate the evolution of the SFR density of the Universe out to $z = 2$. MS galaxies seem to be subject to a different star forming law, suggesting that the mechanisms in which star formation is taking place in these galaxies differs from that of starbursts. The best way to characterise such mechanisms is by studying the gas and dust that form the ISM of galaxies.

Despite the great advances made by studies of the ISM properties of both starburst and MS galaxies at high redshift, the number of detections of molecular gas is still scarce, and there is a particular need for resolved observations of gas and dust. The advent of ALMA brings forth the hope that such observations could become accessible in larger samples in the near future, even though it will still require significant investments of observing time. Finally, deep field molecular line spectroscopy with ALMA over considerable volumes promises to provide a critical census of the molecular gas content and properties in galaxies out to z=8.

References

Aravena, M. *et al.* 2010, *ApJ*, 718, 177
Aravena, M. *et al.* 2014, *MNRAS*, 442, 558
Bolatto, A. *et al.* 2015, *ApJ*, 809, 175
Bothwell, M. *et al.* 2013, *MNRAS*, 429, 3047
Bouwens, R. *et al.* 2014, *ApJ*, 795, 126
Brinchmann, J. *et al.* 2004, *MNRAS*, 351, 1151
Capak, P. *et al.* 2015, *Nature*, 522, 455
Carilli, C. L. & Walter, F. 2013, *ARA&A*, 51, 105
Daddi, E. *et al.* 2007, *ApJ*, 670, 156
Daddi, E. *et al.* 2010, *ApJ* (Letters), 714, 118
Daddi, E. *et al.* 2015, *A&A*, 577, 46
Decarli, R. *et al.* 2014, *ApJ*, 782, 78
Dessauges-Zavadsky, M. *et al.* 2015, *A&A*, 577, 50
Elbaz, D. *et al.* 2007, *A&A*, 468, 33
Freundich, J. *et al.* 2013, *A&A*, 553, 130
Genzel, R. *et al.* 2010, *MNRAS*, 407, 209
Karim, A. *et al.* 2011, *ApJ*, 730, 61
Magnelli, B. *et al.* 2011, *A&A*, 528, 35
Magnelli, B. *et al.* 2013, *A&A*, 553, 132
Maiolino, R. *et al.* 2015, *MNRAS*, 452, 54
Noeske, K. G. *et al.* 2007, *ApJ*, 660, 47
Pannella, M. *et al.* 2009, *ApJ*, 698, 116
Riechers, D. *et al.* 2014, *ApJ*, 796, 84
Rodighiero, G. *et al.* 2011, *ApJ*, 739, 40
Saintonge, A. *et al.* 2013, *ApJ*, 778, 2
Santini, P. *et al.* 2014, *A&A*, 562, 30
Sargent, M. T. *et al.* 2014, *ApJ*, 793, 19
Sargent, M. *et al.* 2012, *ApJ*, 747, 31
Scoville, N. *et al.* 2014, *ApJ*, 783, 84
Solomon, P. & Vanden Bout 2005, *ARA&A*, 43, 677

Spilker, J. S. *et al.* 2014, *ApJ*, 785, 149
Tacconi, L. *et al.* 2013, *ApJ*, 768, 74
Tacconi, L. *et al.* 2010, *Nature*, 463, 781
Walter, F. *et al.* 2014, *ApJ*, 782, 79
Willott, C., Carilli, C. L. & Wagg, J. 2015, *ApJ*, 807, 180

From Interstellar Clouds to Star-Forming Galaxies: Universal Processes?
Proceedings IAU Symposium No. 315, 2015
P. Jablonka, Ph. André & F. van der Tak, eds.

doi:10.1017/S1743921316007559

Interactions and star formation

Johan H. Knapen[1,2], **Mauricio Cisternas**[1,2] **and Miguel Querejeta**[3]

[1]Instituto de Astrofísica de Canarias, E-38205 La Laguna, Tenerife, Spain
[2]Departamento de Astrofísica, Universidad de La Laguna, E-38205 La Laguna, Tenerife, Spain
[3]Max Planck Institute for Astronomy, Königstuhl 17, 69117, Heidelberg, Germany

Abstract. We investigate the influence of interactions on the star formation by studying a sample of almost 1500 of the nearest galaxies, all within a distance of ∼45 Mpc. We define the massive star formation rate (SFR), as measured from far-IR emission, and the specific star formation rate (SSFR), which is the former quantity normalized by the stellar mass of the galaxy, and explore their distribution with morphological type and with stellar mass. We then calculate the relative enhancement of these quantities for each galaxy by normalizing them by the median SFR and SSFR values of individual control populations of similar non-interacting galaxies. We find that both SFR and SSFR are enhanced in interacting galaxies, and more so as the degree of interaction is higher. The increase is, however, moderate, reaching a maximum of a factor of 1.9 for the highest degree of interaction (mergers). The SFR and SSFR are enhanced statistically in the population, but in most individual interacting galaxies they are not enhanced at all. We discuss how those galaxies with the largest SFR and/or SSFR enhancement can be defined as starbursts. We argue that this study, based on a representative sample of nearby galaxies, should be used to place constraints on studies based on samples of galaxies at larger distances.

Keywords. Galaxies: general, Galaxies: interactions, Galaxies: starbursts

1. Introduction

Galaxy-galaxy interactions and mergers are an important factor in the dominant cosmological model of galaxy evolution, in which mergers between dark matter haloes and/or luminous galaxies occur frequently, increasing their mass and ultimately shaping them into the galaxies that we observe at the current epoch. A frequent assumption is that interactions and mergers lead to star formation rates (SFRs) that are temporarily increased, often by very large amounts. We know from observations that this can and does happen, in the most extreme form as Ultra-Luminous InfraRed Galaxies (ULIRGs), and numerical simulations can also produce much enhanced SFRs (e.g., Joseph & Wright 1985; Barnes & Hernquist 1991).

A question that is vital but has not been answered is whether any galaxy-galaxy interaction or merger is accompanied by a significant increase in the SFR, and whether this increase is always so large that is can be classified as a so-called 'starburst'. From various observational and modeling studies we know that indeed there is such a general increase, but also that this is limited to around a factor of two in the SFR (e.g., Larson & Tinsley 1978; Hernández-Toledo *et al.* 2005; Di Matteo *et al.* 2007; Knapen & James 2009; Ellison *et al.* 2013; Knapen, Cisternas & Querejeta 2015 and references therein).

We summarise here the results of our statistical study of the SFR and the specific SFR (SSFR; SFR normalized by stellar mass) of a sample of some 1500 galaxies in the very local Universe, within a distance of ∼45 Mpc (Knapen *et al.* 2015; Knapen & Cisternas 2015). This sample is unique because it covers galaxies down to stellar masses of around $10^8\ M_\odot$ and of all morphological types, and should be considered as a local analogue for studies of galaxies at higher redshifts.

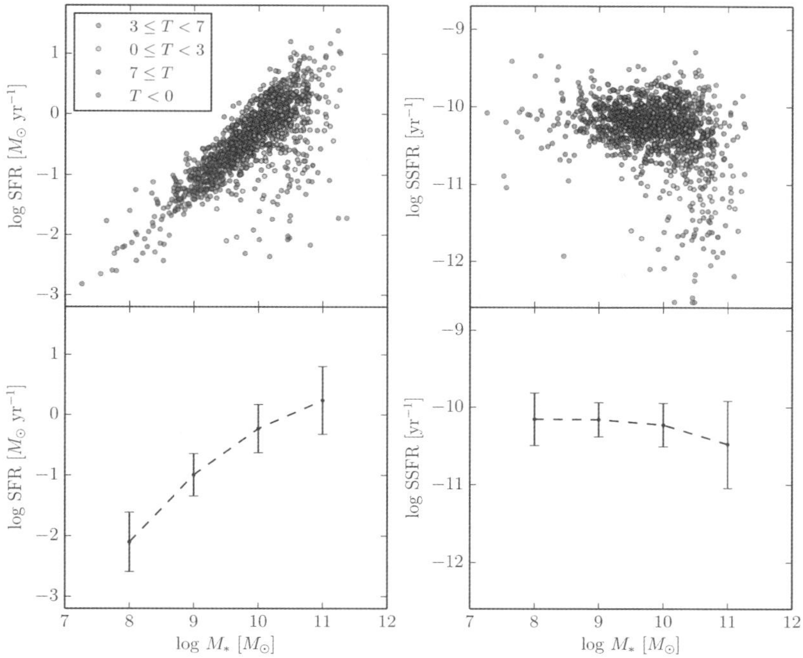

Figure 1. Log of the SFR (left panels) and the SSFR (right panels) as a function of stellar mass. Top panels show the individual galaxies, colour-coded by morphological type (for reference, red roughly corresponds to Hubble types E and S0; yellow, to early-type spirals; green, to late-type spirals; blue, to irregulars), while the lower panels show the median values, with the $1\,\sigma$ spread indicated by the error bars. The dashed curve connects the points. From Knapen *et al.* (2015).

2. Data and Analysis

For a sample of 1478 nearby galaxies ($D < 45$ Mpc) selected from the *Spitzer* Survey of Stellar Structure in Galaxies (S^{4}G; Sheth *et al.* 2010) we use the SFR (from a combination of IRAS 60 and 100 μm fluxes; from Querejeta *et al.* 2015), the SSFR (SFR divided by stellar mass, the latter from the dust emission-corrected *Spitzer* 3.6 μm images from the S^{4}G; also from Querejeta *et al.* 2015), and whether the galaxies are interacting (classes A—merging, B—highly distorted, C—with minor distortions; from Knapen *et al.* 2014).

We then calculate, for each sample galaxy, the enhancement in its SFR and SSFR, E(SFR) and E(SSFR), by dividing these parameters by the median values for a control sample. The control sample is created for each galaxy individually, and consists of all those galaxies which are not interacting and which are close in morphological type (within ± 1 numerical class) and stellar mass (within ± 0.2 in $\log(M/M_\odot)$; see Knapen *et al.* 2015).

We can use the values of E(SFR) and E(SSFR) to define which galaxies are starbursts: namely those which have values of E above a certain level, in particular $E(\text{SFR}) > 5$ and $E(\text{SSFR}) > 4$. As we show in Knapen *et al.* (2015; see also Knapen & James 2009), these choices are reasonable, but in any case the main results obtained do not depend on the exact values of the cut-off used to define starbursts.

3. Results

In Figure 1 we show how the SFR and the SSFR vary with stellar mass and with morphological type (the latter colour-coded in the top panels). This is a well-known

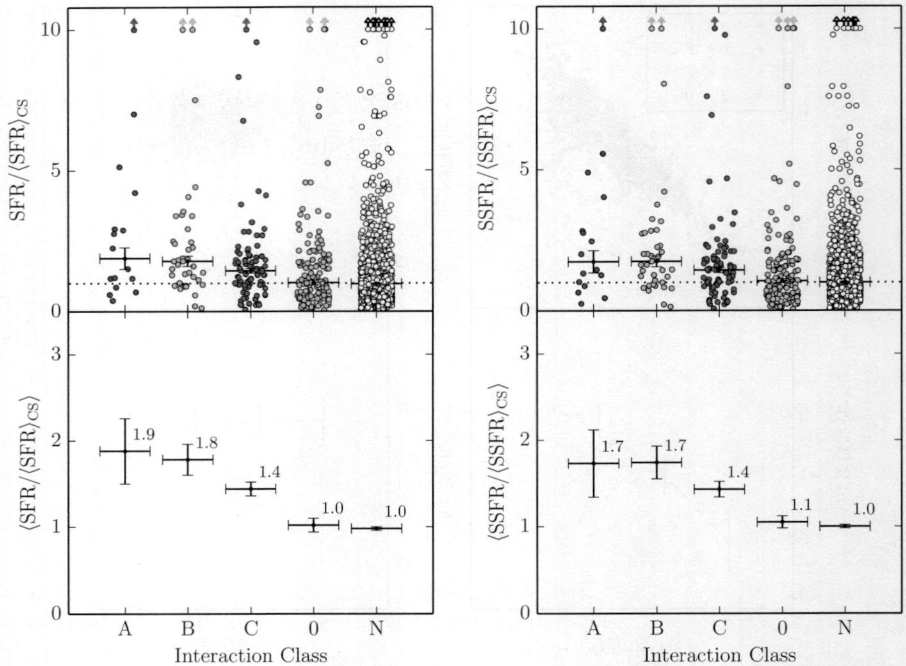

Figure 2. The SFR (left panels) and SSFR (right panels) enhancement, defined as the (S)SFR of a galaxy normalized by the median (S)SFR for its control population, separated by interaction class (where A is the most extreme class, of mergers, and N contains non-interacting galaxies). Values larger than 10 are indicated by lower limits. Median values per class are indicated with their $1\,\sigma$ uncertainty in both top and lower panels, but are amplified in the lower panels where the values are indicated next to the data points. From Knapen *et al.* (2015). *A color version of this figure is available online.*

plot (often referred to as the "galaxy main sequence") but nicely illustrates a number of important points, such as that our (S)SFR determinations seem correct, and that our sample spans a large range in both morphological type and mass. We claim that our sample is a much better representation of at least the nearby Universe than what a typical galaxy main-sequence plot from the literature might lead one to believe.

Figure 2 shows the enhancement of the SFR (left panels) and the SSFR (right panels) for each galaxy individually (top panels) and in the median (lower panels) for the sample galaxies as a function of their interaction class (see Sect. 2), where the most strongly interacting galaxies are category A on the left, and the non-interacting control sample galaxies are class N on the right. We can derive a number of main results from this figure.

(*a*) From the individual values (top panels) we see that extreme values of SFR and SSFR, of more than 10 times that of the control sample, occur across the board, in all categories. These are not limited to interacting galaxies.

(*b*) Values below unity also occur across the board, and in particular we note that a good number of interacting and merging galaxies (classes C, B and A) form stars at a lower rate than might be expected from their mass and type. Many interacting galaxies have no enhanced (S)SFR at all.

(*c*) From the median SFR and SSFR enhancement values (lower panels) we see that there is indeed an increase in (S)SFR in interacting as compared to non-interacting (class N) galaxies, but that this enhancement is modest, just below a factor of two.

(*d*) This increase in (S)SFR is more pronounced as the interaction gets stronger, towards class A or towards the left in the plots.

(*e*) Finally, we see that our control sample technique works: the median values for class N (the non-interacting galaxies, each of which has also been analyzed with its own control sample) are 1.0, as they should be.

4. Conclusions

Using a sample of 1500 nearby galaxies from the S^4G we quantify the role of galaxy-galaxy interactions in enhancing the star formation. To do this, we measure SFRs from *IRAS* fluxes, and use stellar masses determined from 3.6 μm S^4G images corrected for dust emission. We normalize the SFR by galaxy mass (yielding the SSFR), and explore the distribution of the SFR and SSFR with morphological type and galaxy stellar mass. For each galaxy we calculate the SFR and SSFR enhancement by normalizing them to the values obtained for a specific and individually constructed control sample, consisting of all galaxies within a morphological type range of ±1, and a stellar mass range of ±0.2 in $\log(M/M_{\odot})$.

We find that both SFR and SSFR are enhanced in interacting galaxies, and more so as the interaction becomes stronger. The increase is, however, moderate, of at most a factor of two. We can use the SFR and SSFR enhancements to define galaxies as starbursts. The largest values of (S)SFR enhancement, indicative of the most extreme starbursts, occur in all classes, not just among the interacting galaxies. Many interacting galaxies have SFRs and SSFRs that are not enhanced at all with respect to a control population, and in several cases their (S)SFR is even lower. As this study is based on a representative sample of nearby galaxies, we argue that it should be used to place constraints on studies based on samples of galaxies at larger distances.

5. Acknowledgments

We acknowledge support to the DAGAL network from the People Programme (Marie Curie Actions) of the European Union's Seventh Framework Programme FP7/2007-2013/ under REA grant agreement PITN-GA-2011-289313, and from the Spanish Ministry of Economy and Competitiveness (MINECO) under grant number AYA2013-41243-P.

References

Barnes, J. E. & Hernquist, L. E. 1991, *ApJ*, 370, L65
Di Matteo, P., Combes, F., Melchior, A.-L., & Semelin, B. 2007, *A&A*, 468, 61
Ellison, S. L., Mendel, J. T., Patton, D. R., & Scudder, J. M. 2013, *MNRAS*, 435, 3627
Hernández-Toledo, H. M., Avila-Reese, V., Conselice, C. J., & Puerari, I. 2005, *AJ*, 129, 682
Joseph, R. D. & Wright, G. S. 1985, *MNRAS*, 214, 87
Knapen, J. H. & Cisternas, M. 2015, it ApJL, 807, L16
Knapen, J. H., Cisternas, M., & Querejeta, M. 2015, *MNRAS*, 454, 1742
Knapen, J. H., Erroz-Ferrer, S., Roa, J., *et al.* 2014, *A&A*, 569, AA91
Knapen, J. H. & James, P. A. 2009, *ApJ*, 698, 1437 (KJ09)
Larson, R. B. & Tinsley, B. M. 1978, *ApJ*, 219, 46
Querejeta, M., Meidt, S. E., Schinnerer, E., *et al.* 2015, *ApJS*, 219, 5
Sheth, K., Regan, M., Hinz, J. L., *et al.* 2010, PASP, 122, 1397

From Interstellar Clouds to Star-Forming Galaxies: Universal Processes?
Proceedings IAU Symposium No. 315, 2015
P. Jablonka, Ph. André & F. van der Tak, eds.

doi:10.1017/S1743921316007560

The Evolution of Gas Content and Star Formation from z=3 to z=0

Francoise Combes[1] and the PHIBSS collaboration
[1]Observatoire de Paris, LERMA, College de France, CNRS, PSL, Sorbonne Univ. UPMC, F-75014, Paris, France
email: francoise.combes@obspm.fr

Abstract. The cosmic star formation rate density first increases with time towards a pronounced peak 10 Gyrs ago (or z=1-2) and then slows down, dropping by more than a factor 10 since z=1. The processes at the origin of the star formation quenching are not yet well identified: either the gas is expelled by supernovae and AGN feedback, or prevented to inflow. Morphological transformation or environment effects are also invoked. Recent IRAM/NOEMA and ALMA results are reviewed about the molecular content of galaxies and its dynamics, as a function of redshift. Along the main sequence of massive star forming galaxies, the gas fraction was higher in the past (up to 80%), and galaxy disks were more unstable and more turbulent. The star formation efficiency increases with redshift, or equivalently the depletion time decreases, whatever the position of galaxies, either on the main sequence or above. Attempts have been made to determine the cosmic evolution of the H_2 density, but deeper ALMA observations are needed to effectively compare with models.

Keywords. Galaxies, Star formation, AGN, Interstellar medium, Molecular gas

1. Introduction

The cosmic star formation and its evolution through the Hubble time is now well established (e.g. Madau & Dickinson 2014). One of the main issue is to understand the rather fast winding down of star formation after z=2, both globally and also in individual galaxies. The suppression of star formation in galaxies appears rather sudden and therefore is called quenching; this has been revealed by their bimodal distribution, between a blue cloud of galaxies actively forming stars, and a red sequence of dead objects. The paucity of galaxies in an intermediate sequence (e.g. Baldry *et al.* 2006) means that the quenching time-scale is typically smaller than 1 Gyr. However the quenching mechanism is hard to identify. In a recent study of 56 GOODS galaxies at z=1.7, Mancini *et al.* (2015) found an equal proportion of AGN in both star forming and quenched galaxies, and their morphology, traced by the bulge to disk ratio or their Sersic index, was also comparable over a large range of specific star formation rate (SFR).

Since star formation is directly linked to the amount of molecular gas, it is primordial to determine the gas fraction as a function of redshift, and also the star formation efficiency (SFE), defined by the ratio between SFR and gas mass. In the recent years, a large variety of data have been obtained, with somewhat discrepant results, revealing either low or high SFE at high redshift. In sub-millimeter galaxies (SMG), and ultra-luminous starbursts (ULIRGs) the SFE increases with redshift, and the depletion time (the inverse of SFE) which is of the order of 0.5-1Gyr for ULIRGs at z$\sim$ 0 becomes as low as 10-100 Myr at z=1-2 (Greve *et al.* 2005, Combes *et al.* 2011, 2013). But there exists a population of massive BzK galaxies, selected from their near-IR and optical colors, which are also actively forming stars (they are ULIRGs), but with a lower efficiency, with a depletion time-scale of the order of 0.3 Gyr at z$\sim$ 1.5 (Daddi *et al.* 2008), i.e. similar to local

ULIRGs. This might be due to their extended molecular component (10 kpc scales), and their low density state, revealed by their low excitation: the CO emission as a function of the upper level J peaks at $J = 3$, as in the Milky Way (Dannerbauer *et al.*2009), justifying the adoption of the standard CO-to-H_2 conversion factor (5 times that adopted for ULIRGs). Let us recall that local main-sequence galaxies, such as the Milky Way, have a depletion time-scale of the order of 2 Gyr (Bigiel *et al.* 2011).

2. Starburst galaxies

In a recent work with IRAM and ALMA, Silverman *et al.* (2015) detected CO emission in 7 galaxies from the COSMOS survey, with SFR$\sim$ 300-800 $M_\odot$/yr, at $1.4 < z < 1.7$. Their gas fraction was determined to be $\sim$ 30-50%, and their global SFE is enhanced, although the objects are unresolved, and the starbursting regions cannot be separated.

To trace redshift evolution and in particular the winding down of star formation after z=1, the intermediate redshift epoch $0.2 < z < 1$ is a key region to explore. We have observed the gas content of ULIRGs at these intermediate z, to make the link with local starbursts (Combes *et al.* 2011, 2013). Out of 69 ULIRG, 33 were detected in CO emission, with variable excitation. Some of the objects have a lower H_2 density, which can be explained in terms of at least two components. The objects for which an interferometric map is available show the separation of the molecular emission in a nuclear starburst and an extended gas disk of scale $\sim$20 kpc. The presence of such an extended component may explain lower SFE, in particular when low gas excitation suggests a higher CO-to-H_2 conversion ratio.

This study at intermediate z, compared to all other starburst data, showed that both gas fraction and SFE increase with redshift, by a factor 3$\pm$1, between z=0 and 1, with or without taking into account upper limits.

Recently, the Bzk galaxies at z=1.5, which show globally low excitation of the CO lines at low J were observed to have quite high CO(5-4) fluxes, revealing a second component of more excited, denser and warmer molecular gas (Daddi *et al.* 2015). For this excited component, the objects have the same correlations than for ULIRGs, confirming that the starburst occurs either in a separate nuclear region, or in hot and dense star forming clumps. In local starbursts, the total CO SLEDs observed with Herschel reveal a very large variety of shapes, with several molecular components (Mashian *et al.*2015), suggesting that the overall CO-to-H_2 conversion ratio could explore even wider ranges than expected, from α =0.4 to 5 $M_\odot$/(K km/s pc^2) for the M(H_2) to L'CO ratio.

3. Main sequence galaxies

Most of the cosmic star formation ($\sim$ 90%) occurs in the main sequence galaxies, and only about 10% in starbursts. The main sequence is clearly defined in an SFR versus stellar mass diagram as a power-law with slope slightly lower than 1. This power-law is similar at all redshift ranges, but the zero point is increasing with redshift, following the evolution of the cosmic star formation rate density described in the introduction. Large surveys of hundred thousands of galaxies locally (SDSS) or at high redshift (GOODS, COSMOS, ..) have shown a correlation between morphology and stellar populations since z$\sim$2.5 (Wuyts *et al.* 2011): blue star forming galaxies on the main sequence are exponential disks (Sersic index near 1), while quiescent red systems are of de Vaucouleurs type (Sersic index more near 4).

With the goal to explore the molecular gas content of main sequence galaxies, we undertook the PHIBSS project (Plateau de Bure HIgh z Blue Sequence Survey, Tacconi

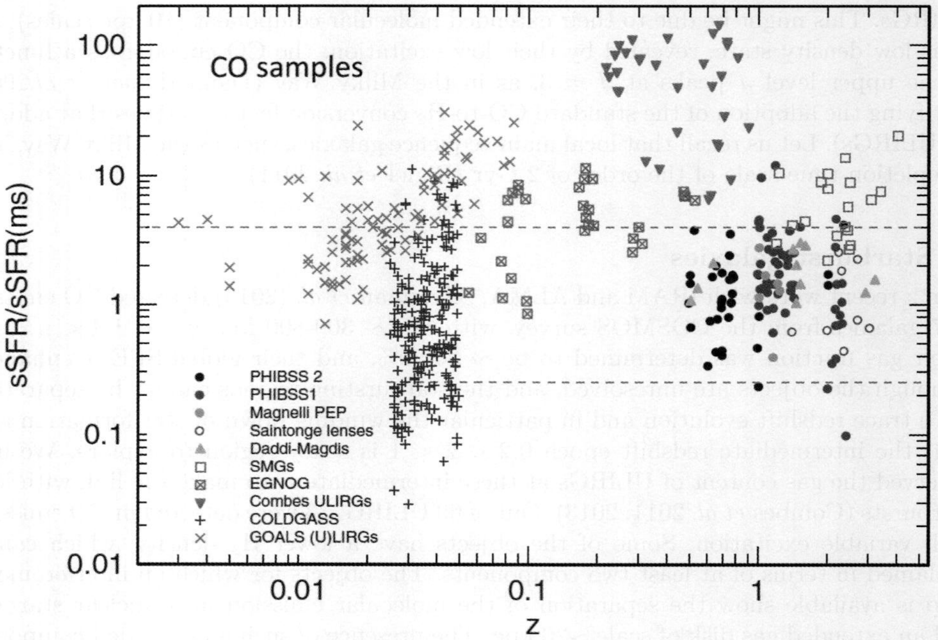

Figure 1. Distribution in the redshift versus specific star-formation rate plane of all the star forming galaxies with CO flux measurements found in the literature, compared with the PHIBSS data by Genzel *et al.* (2015). The sSFR on the vertical axis is normalized to the main sequence (ms) value of sSFR at a given mass and redshift, according to the scaling relation given by Whitaker *et al.* (2012). The membership to the main sequence is defined by the two horizontal dashed lines, situated at ±0.6 dex from the mid MS line. *A color version is available online.*

et al. 2010, 2013). In the first part of the project, 52 galaxies were detected at IRAM in the CO(3-2) line at z=1.2 and 2.3. The targets were selected to be massive ($M_* >$ 2.5 10^{10} $M_\odot$) star forming galaxies (SFR > 30 $M_\odot$/yr). Adopting a standard CO-to-H_2 conversion ratio for these main sequence objects, molecular masses were found between 10^{10} and 3 10^{11} $M_\odot$, corresponding to gas fraction in average of 33% at z=1.2 and 47% at z=2.3. The SFE was found to increase slightly with z, and the depletion time scale is in average 0.7 Gyr at z=1.2.

The CO detection rate was quite high (>85%), in these normal massive Star Forming Galaxies (SFG). Some were mapped at high spatial resolution, and a rather regular velocity field was found, confirming the absence of major mergers. At z=1.2, it was possible to resolve four galaxy disks in clumps with the help of the velocity information, both with CO and [OII] lines (gas content and SFR), since there is a good correlation between molecular and ionised gas. This allowed us to draw a resolved Kennicutt-Schmidt (KS) relation (Freundlich *et al.* 2013). The high-z points extend the local KS relation towards high gas and SFR surface densities, with the same slope. For one galaxy at z=1.5, it was possible to observe the Hα line at high resolution from the ground (Genzel *et al.* 2013). The KS slope depends strongly on the dust extinction model adopted, but falls around 1.

The evolution of specific SFR with redshift was compatible with the results of optical surveys, provided that the depletion time is varying on the main sequence as t_{dep}=1.5/(1+z) Gyr. The PHIBSS project is now being extended with the goal of observing CO lines in about 150 galaxies, exploring also intermediate redshifts (z=0.5-0.7), and some galaxies departing from the main sequence, either starbursts above the MS or

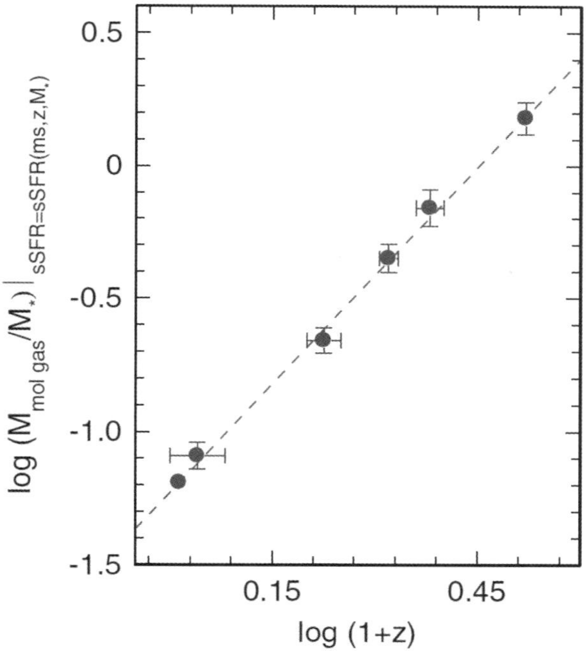

Figure 2. Ratio of the molecular to stellar mass versus redshift, for galaxies on the main-sequence (Genzel *et al.* 2015). The best linear fit has a slope of 2.71 (dashed line).

quiescent galaxies below (PHIBSS2). Detecting smaller masses, and bright galaxies with more spatial resolution and/or more molecular lines in addition will be attempted with ALMA.

The first results and the scaling relations obtained on the main sequence, by comparison with all other data with CO detections at high redshift (cf Fig 1), were presented in Genzel *et al.* (2015). On the MS, the gas fraction increases regularly with z, as shown in Fig 2.

Fig 3 displays the depletion time as a function of redshift, for the galaxies on the main sequence. There is a slight decrease, and the effect depends on the way the molecular gas content is estimated, either from the CO line, or from the dust emission.

A recent survey with ALMA of the continuum dust emission of 180 star forming galaxies between z=1 and 6.4 results in slightly different results (Scoville *et al.* 2015). The gas fraction estimated from the dust emission, assuming a constant dust temperature, is also highly increasing with redshift, reaching values as high as 50-80%. The depletion time-scale is found to be the same for starbursts and main sequence galaxies; it decreases strongly with redshift, to reach 200 Myr at z> 1, therefore 10 times lower than for local MS galaxies.

4. The cosmic H_2 density

One of the key issues to understand the cosmic star formation history is to observe the cosmic evolution of the H_2 density. Theoretical considerations and semi-analytical models predict that the molecular gas density must increase with redshift, and dominate over the atomic gas in galaxies (Obreschkow & Rawlings, 2009, Obreschkow *et al.* 2009). The phase transition from atomic to molecular hydrogen is favored by pressure (Blitz & Rosolowsky 2006), while surface density and consequently the pressure is higher in high-z

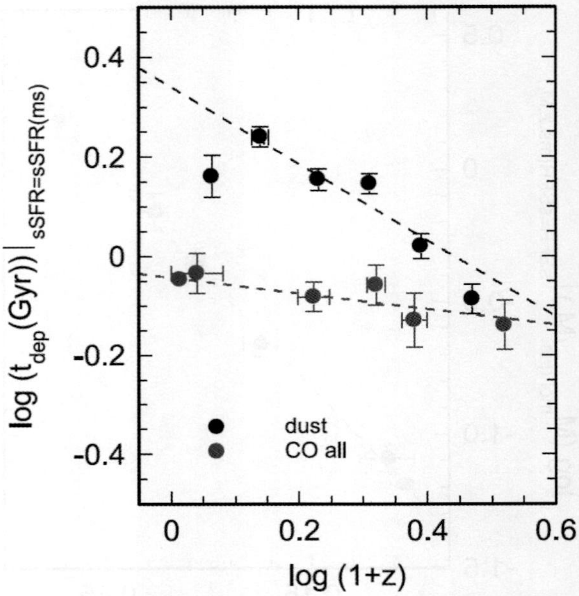

Figure 3. Depletion time versus redshift for galaxies on the main sequence, computed from the dust emission (black circles), and from the CO lines (blue circles). The best linear fit has a slope of –0.77 (black dashed line). *A color version is available online.*

galaxies. Models predict a molecular-to-atomic ratio H_2/HI varying as $(1+z)^m$, with m as high as 1.6. Some models however predict flatter evolutions of the H_2 density at high z (Lagos *et al.* 2011, Popping *et al.* 2014).

Decarli *et al.* (2014) and Walter *et al.* (2014) have attempted to constrain the H_2 density by observing a large cosmic volume of $\sim$7000 Mpc^3, in the Hubble deep field North with Plateau de Bure. They separate the results in three redshift bins: z<0.45, 1.01<z<1.89 and z>2. A blind molecular line survey has been carried out through scanning the whole 3mm band. The blind detection of 17 CO lines, together with the upper limits obtained by stacking the observations towards spectroscopically identified objects, constrain the CO luminosity functions at the corresponding redshifts. The results show that optical/mid-IR bright galaxies contribute less than 50% to the star formation rate density at $1 < z < 3$, and the normalised density $\Omega(H_2)$ at high z tends to be higher than the predictions.

It might be easier and certainly quicker to determine the evolution of the H_2 density from dust emission surveys. A recent 870μm continuum survey with ALMA of 99 SMG in the Extended Chandra Deep Field South (Swinbank *et al.* 2014) has discovered that the well detected sources ($S_{870} > 4.2$ mJy) are in average ULIRGs with SFR=300 $M_\odot$/yr, and dust temperatures of 32 K. They contribute to only 1-2% of SFR. The extrapolation of the counts down to $S_{870} > 1$ mJy through stacking shows that these sources contribute to 20% of the cosmic star formation density over z=1-4 (see Fig 4). Deriving H_2 masses from dust masses, the average SFE is found rather high, with depletion time-scale of 130 Myr.

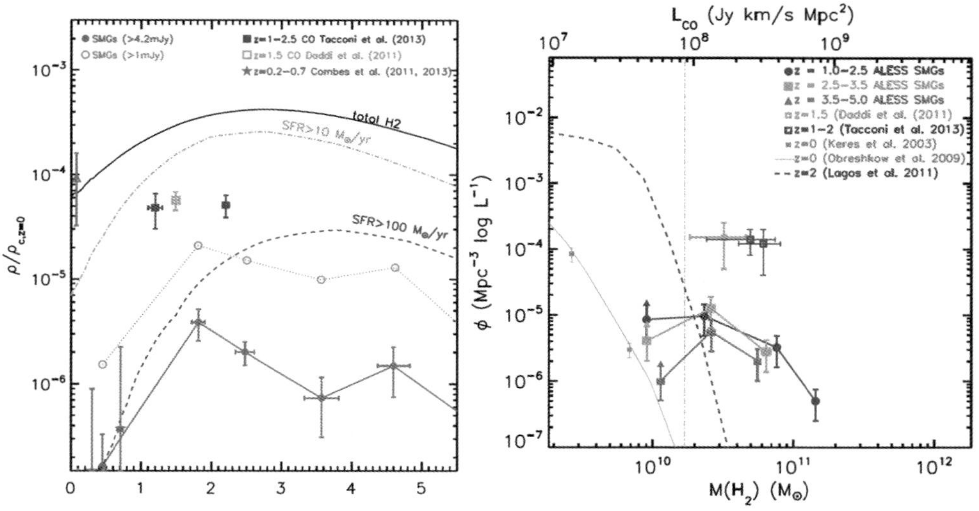

Figure 4. Left: The normalised H_2 density $\Omega(H_2)$ as a function of redshift, for SMG and other high-z starbursts (Swinbank *et al.*2014, Combes *et al.* 2013) in red, compared to main sequence galaxies in blue (Tacconi *et al.*2013). The dashed lines are the model predictions, for different SFR (Lagos *et al.*2011). **Right:** The H_2 mass function for SMG and main sequence galaxies compared with models. The main sequence galaxies at z=1.5 (green) are from Daddi *et al.* (2010) and at z=1-2 (blue) from Tacconi *et al.* (2013). They are well above the predictions at z=2 by Lagos *et al.*(2011) indicated by the blue dashed curve. From Swinbank *et al.* (2014). *A color version is available online.*

5. Conclusions

It is now well established that galaxies at high redshift have a larger gas fraction than local ones, whatever their position on the main sequence or above, in the starburst domain. The gas fraction can reach 50% and above.

There is not yet a consensus on the exact evolution of the star formation efficiency with redshift. The inverse of the SFE, the depletion time scale, is decreasing with redshift, however the amplitude of its variation with z is still debated. The various results depend on the way to estimate the total molecular gas amount, either from CO lines or from dust emission. The results may also depend on the definition of the Main Sequence (e.g. Renzini & Peng, 2015).

A higher SFE at high z might be explained by a higher surface density of molecular gas, if the Kennicutt-Schmidt relation is non-linear, which is not yet well known. Alternatively, a starburst can be triggered in nuclear regions when the gas is concentrated. Diagnostics could be searched for with CO excitation and the observation of several J lines, and also dense gas tracers (HCN, HCO^+).

ALMA observations begin to estimate the evolution of the molecular gas mass in galaxies, however we are still far from a total census of $\Omega(H_2)$ as a function of redshift.

References

Baldry, I. K., Balogh, M. L., Bower, R. G. *et al.* 2006, *MNRAS* 373, 469
Bigiel, F., Leroy, A. K., Walter, F. *et al.* 2011, *ApJ*, 730, L13
Blitz L., Rosolowsky E. 2006, *ApJ*, 650, 933
Combes, F., Garcia-Burillo, S., Braine, J. *et al.* 2011 *A&A* 528, A124
Combes, F., Garcia-Burillo, S., Braine, J. *et al.* 2013 *A&A* 550, A41

Daddi, E., Dannerbauer, H., Elbaz, D. *et al.* 2008, *ApJ*, 673, L21
Daddi, E., Bournaud, F., Walter, F. *et al.* 2010, *ApJ*, 713, 686
Daddi, E., Dannerbauer, H., Liu, D. *et al.* 2015, *A&A* 577, A46
Dannerbauer, H., Daddi, E., Riechers, D. A. *et al.* 2009, *ApJ*, 698, L178
Decarli, R., Walter, F., Carilli, C. *et al.* 2014, *ApJ*, 782, 78
Freundlich, J., Combes, F., Tacconi, L. J. *et al.* 2013, *A&A* 553, A130
Genzel, R., Tacconi, L. J., Kurk, J. *et al.* 2013, *ApJ*, 773, 68
Genzel, R., Tacconi, L. J., Lutz, D. *et al.* 2015, *ApJ*, 800, 20
Greve, T. R., Bertoldi, F., Smail, I. *et al.* 2005 *MNRAS* 359, 1165
Lagos C. d P., Baugh C. M., Lacey C. G. *et al.* 2011, *MNRAS* 418, 1649
Madau, P. & Dickinson, M. 2014, *ARAA* 52, 415
Mancini, C., Renzini, A., Daddi, E. *et al.* 2015, *MNRAS* 450, 763
Mashian, N., Sturm, E., Sternberg, A. *et al.* 2015, *ApJ*, 802, 81
Obreschkow D., Croton D., de Lucia G. *et al.* 2009, *ApJ*, 698, 1467
Obreschkow D., Rawlings S.: 2009, *ApJ*, 696, L129
Popping G., Somerville R. S., Trager S. C., 2014, *MNRAS*, 442, 2398
Renzini, A. & Peng, Y.-J. 2015, *ApJ*, 801, L29
Scoville, N., Sheth, K., Aussel, H. *et al.* 2015, *ApJ*, in press, arXiv1505-02159
Silverman, J. D., Daddi, E., Rodighiero, G. *et al.* 2015, *ApJ*, in press, arXiv1505-04977
Swinbank A. M., Simpson J. M., Smail I., *et al.* 2014, *MNRAS*, 438, 1267
Tacconi, L. J., Genzel, R., Neri, R. *et al.* 2010, *Nature*, 463, 781
Tacconi, L. J., Neri, R., Genzel, R. *et al.* 2013, *ApJ*, 768, 74
Walter, F., Decarli, R., Sargent, M. *et al.* 2014, *ApJ*, 782, 79
Whitaker, K. E., van Dokkum, P. G., Brammer, G., & Franx, M. 2012, *ApJ*, 754, L29
Wuyts, S., Förster Schreiber, N. M., van der Wel, A. *et al.* 2011 *ApJ*, 742, 96

From Interstellar Clouds to Star-Forming Galaxies: Universal Processes?
Proceedings IAU Symposium No. 315, 2015
P. Jablonka, Ph. André & F. van der Tak, eds.

doi:10.1017/S1743921316007572

Theoretical considerations for star formation at low and high redshift

Bruce G. Elmegreen

IBM T.J. Watson Research Center
1101 Kitchawan Road, Yorktown Heights, NY, USA
email: bge@us.ibm.com

Abstract. Star formation processes in strongly self-gravitating cloud cores should be similar at all redshifts, forming single or multiple stars with a range of masses determined by local magneto-hydrodynamics and gravity. The formation processes for these cores, however, as well as their structures, temperatures, Mach numbers, etc., and the boundedness and mass distribution functions of the resulting stars, should depend on environment, as should the characteristic mass, density, and column density at which cloud self-gravity dominates other forces. Because the environments for high and low redshift star formation differ significantly, we expect the resulting gas to stellar conversion details to differ also. At high redshift, the universe is denser and more gas-rich, so the active parts of galaxies are denser and more gas rich too, leading to slightly shorter gas consumption timescales, higher cloud pressures, and denser, more massive, bound stellar clusters at the high mass end. With shorter consumption times corresponding to higher relative cosmic accretion rates, and with the resulting higher star formation rates and their higher feedback powers, the ISM has greater turbulent speeds relative to the rotation speeds, thicker gas disks, and larger cloud and star complex sizes at the characteristic Jeans length. The result is a more chaotic appearance at high redshift, bridging the morphology gap between today's quiescent spirals and today's major-mergers, with neither spiral nor major-merger processes actually in play at that time. The result is also a thick disk at early times, and after in-plane accretion from relatively large clump torques, a classical bulge. Today's disks are thinner, and torque-driven accretion is slower outside of inner barred regions. This paper reviews the basic processes involved with star formation in order to illustrate its evolution over time and environment.

Keywords. stars: formation, galaxies: formation, galaxies: starburst, globular clusters: general

1. Introduction

Star formation (SF) in young L* galaxies looks very different from local SF. At a redshift of around 2, galaxies have a high gas fraction close to 50% (Tacconi *et al.* 2013; Duncan *et al.* 2014), a high ratio of gas velocity dispersion to rotation speed of 30% or more (Förster-Schreiber *et al.* 2009; Kassin *et al.* 2012), a high ratio of interstellar Jeans Length to galaxy radius of 30%, which follows from the high dispersion and gives big clumps of SF and a thick star-forming disk, a theoretically predicted high accretion rate (e.g., Dekel *et al.* 2013), which produces a high SF rate, and strong outflows from the high SF rate (Newman *et al.* 2012). High redshift galaxies also seem to produce more massive clusters than today's galaxies (see below), perhaps because of their larger ISM pressures.

SF today looks very different. L* galaxies have a low gas fraction, $\sim$ 5%, the ratio of dispersion to rotation speed is low, $\sim$ 10%, producing relatively small star-forming clumps in a thin disk, the accretion rate is low, which contributes to the low SF rate, outflows are relatively weak, and the maximum cluster mass is lower.

Even with these differences, the SF processes themselves appear to be uniform for young and old galaxies. This uniformity is manifest in several ways, as discussed in the next section.

2. Uniformity in star formation processes

We enumerate here the many ways in which star formation processes appear to be similar over a wide range of redshifts.

1. Most star-forming galaxies lie on a main sequence in a plot of SF rate versus galaxy mass. This main sequence has a constant slope with redshift, up to $z \sim 2.5$, with a change only in the absolute SF rate for a given galaxy mass, which increases at higher redshift (Whitaker *et al.* 2012). The constant slope suggests that the SF process is uniform on a galactic scale from galaxy to galaxy. More massive galaxies have more of the same processes of star formation.

The upward shift in the SF rate per unit stellar mass, or the specific SF rate, sSFR, satisfies approximately sSFR$\propto (1+z)^2$ for $\log M_{\rm star} \sim 9.7 \pm 0.3$ from SED and nebular line emission (Duncan et al. 2014). What drives this shift is probably not a change in process, but rather a change in the available fuel, i.e., the relative gas mass (Tacconi *et al.* 2013; Duncan *et al.* 2014). This increased gas mass follows from an increased gas accretion rate, according to cosmological models (Dekel *et al.* 2013).

2. The Kennicutt-Schmidt relation is nearly invariant with redshift for star-forming galaxies (Tacconi et al. 2013, see also Genzel *et al.* 2010; Daddi *et al.* 2010; Saintonge *et al.* 2012; Magdis *et al.* 2012). This implies that the gas depletion time is about constant with redshift, varying only by a factor of ~ 2 up to $z \sim 4$, considering observations of gas mass from either molecular or dust emissions (Genzel *et al.* 2015; Béthermin *et al.* 2015). This is a much slower variation than expected if the SF rate scales with the average internal rotation rate of a galaxy.

3. Fluctuations in the SF rate around the SF main sequence are small in the disk, corresponding to a factor of ~ 2 at both low (Abramson *et al.* 2014; Guo *et al.* 2015; Casado *et al.* 2015) and intermediate (Salmi *et al.* 2014) redshifts. According to Casado *et al.* (2015) who studied 82,500 SDSS galaxies, Hα and u-v colors show an aging sequence with internal evolution for field galaxies driven by accretion, and then a rapid quenching of SF in dense environments.

4. The normalized mass-metallicity relation is uniform over redshift. Zahid *et al.* (2014) showed that the redshift dependent mass-metallicity relation for galaxies can be scaled so that different redshifts lie on a common relationship. The scaling is set by the stellar mass where the gas and star masses are equal in a galaxy, and by a peak metallicity where the metal production rate by stars minus the loss rate in winds equals the metal lock-up rate in stars. With this scaling, the relationship between oxygen metallicity and the gas-to-star ratio is constant. Thus the apparent redshift evolution of the mass-metallicity relation is from the varying gas-to-star ratio, independent of SF processes.

5. There is uniformity in feedback over galaxy mass and cosmic time in the form of a nearly constant M* at the break of the galaxy mass function (Song *et al.* 2015), which is probably determined by feedback (similar results are in Duncan *et al.* 2014; Grazian *et al.* 2015).

6. There is uniformity over a wide redshift range in the dichotomy between SF galaxies and ULIRGs/mergers, which have higher efficiencies corresponding to lower depletion times (Tacconi *et al.* 2013; see also Genzel *et al.* 2010; Daddi *et al.* 2010; Saintonge *et al.* 2012; Magdis *et al.* 2012). This dichotomy was also observed by Silverman et al. (2015) using ALMA and IRAM/PdBI CO observations of z=1.6 starburst galaxies and

a molecular mass from the virial theorem. It was observed by Béthermin *et al.* (2015) using molecular mass from dust observations and by Sargent *et al.* (2014), who calibrated molecular mass using an H_2/CO conversion factor that depends on metallicity (which depends on SF rate and stellar mass).

The usual explanation for this dichotomy is that the star formation time scale in gas of a given mass depends on the average density. In ULIRGs and mergers, torques shrink the gas disk and increase the density. Shrinkage by 1/4 for example, from R_{25} to the disk scale length, causes the gas mass surface density to increase by a factor of 16 and the pressure to increase by a factor of 256 (scaling as the square of the surface density). Thus an increase in density by a factor of 100 is not unreasonable, and this would decrease the free fall time for a given column density by a factor of 10. In addition, mergers create large shocked regions where the ISM from the two galaxies crashes together, and this makes high densities too. Teyssier *et al.* (2010) showed this effect in numerical simulations: mergers make dense gas with a short consumption time, placing them higher on the KS relation for a given gas column density.

3. What are the unifying processes of star formation?

What is the unifying physics in the star formation main sequence? Krumholz, Dekel & McKee (2012) suggest it is feedback-controlled disk stability that makes the Toomre Q always about 1 or 2. If we write the star formation rate per unit gas mass as $\Sigma_{\rm SFR} = f(H_2)\epsilon_{\rm ff}\Sigma_{\rm gas}/\tau_{\rm ff}$ for molecular fraction $f(H_2)$, constant efficiency per unit free fall time $\epsilon_{\rm ff}$, gas mass surface density $\Sigma_{\rm gas}$, and free fall time $t_{\rm ff}$, and take for $t_{\rm ff}$ the minimum value of the free fall time in giant molecular clouds and the average free fall time for a $Q \sim 1$ disk, then the Kennicutt-Schmidt relation is a uniform linear function of the areal SF rate versus $\Sigma_{\rm gas}/t_{\rm ff}$. The Toomre free fall time comes from the average midplane density $\rho \sim \Sigma_{\rm gas}/2H$ for scale height $H \sim \sigma^2/(\pi G\Sigma_{\rm gas})$. When $Q \sim 1$ by self-regulation, then $\sigma \propto \Sigma_{\rm gas}$ at constant κ because $Q = \sigma\kappa/(\pi G\Sigma_{\rm gas})$ for velocity dispersion σ and epicyclic frequency κ. When σ is proportional to $\Sigma_{\rm gas}$, then H is also proportional to $\Sigma_{\rm gas}$ and the midplane density is about constant. Higher areal SF rates then correspond to thicker disks at constant κ.

The self-regulation of Q in disks is commonly seen in numerical simulations when spirals are active. Low Q makes spirals which stir the stars and gas and increase Q. The application of Q to star formation is more indirect and contains assumptions which may not be correct in some cases. Constant Q reflects a balance between self-gravity and centrifugal force in a radially contracting perturbation that was spun up by the Coriolis force. $Q = 1$ when these two forces balance on the scale of the fastest growing perturbation, which is twice the threshold Jeans length.

There are problems with applications of an instability parameter like Q to gas and SF. First, the gas is easily torqued by magnetic fields and viscosity, which removes the angular momentum and reduces spin-up. Second, gas turbulent energy always dissipates, removing the Jeans length as a minimum scale (Elmegreen 2011). Third, the instability condition was derived for rings (spirals have an additional parameter for instabilities in the azimuthal direction) and is applicable on a scale equal to $2\pi G\Sigma/\kappa^2$, which is comparable to a spiral arm spacing. It is 2-dimensional and not directly connected with 3D collapse in midplane gas. There are also thickness effects that should be considered explicitly in the stability condition even for two-dimensional instabilities (e.g., Romeo & Wiegert 2011).

In a study of 20 dwarf Irregular galaxies, Elmegreen & Hunter (2015) used the observed rotation speed to determine κ, and the observed values of $\Sigma_{\rm gas}$, $\sigma_{\rm gas}$ and $\Sigma_{\rm star}$, along with a scaling of $\sigma_{\rm star}$ with absolute galaxy magnitude, to determine Q for the gas and stars in radial annuli. The gas Q averaged about 4 and the stellar Q was about 10 or larger. These values are so high that even the combined Q for a 2-fluid instability (Romeo & Wiegert 2011) would be high. Yet SF is pervasive and normal-looking without $Q \sim 1-2$ as in spirals. Evidently, SF and Q do not self-regulate to make $Q \sim 1-2$ everywhere, and moderately high Q does not stop SF, even though it apparently stops spiral arms from forming (as in dwarfs).

The question about the relevance of Q to star formation comes up again for the central regions of star-forming galaxies where a high Q from the bulge is suggested to cause "morphological quenching" (Martig *et al.* 2009; Genzel *et al.* 2014). This quenching in a bulge is a clear effect in observations and simulations, but the cause of it could be something other than a high Q. For example, it could be an effect of a high critical tidal density for clouds to be self-bound, or a lack of spirals inside the inner Lindblad resonance. The tidal density for a co-rotating cloud is $\rho_{\rm tidal} = (3/2\pi G)\Omega^2$ for rotation rate Ω. That is a three-dimensional threshold for ambient gas to turn into self-gravitating clouds, independent of spiral arms. This should be high in bulge regions because Ω is high, regardless of Q. A condition for bulge quenching would then be $\rho << \rho_{\rm tidal}$ for ambient density ρ.

Analytical models assume that the SF rate is the collapse rate at all densities above some critical value in a density distribution function that is log-normal from turbulence. Federrath & Klessen (2012) compared several analytical models of SF to simulations and found that they agreed with each other to within a factor of 2. The models also agreed reasonably well with observations. This agreement suggests that various details in the models, such as the dependence of the critical density on model parameters, do not matter much for the overall SF rate. Federrath *et al.* also note that the power law form of the density pdf, which develops after the gas begins to be strongly self-gravitating, does not significantly affect the predicted rates. Power law forms for this pdf now seem preferred (Lombardi *et al.* 2015). Federrath & Klessen (2012) suggest that the log-normal density pdf is still appropriate for the initial conditions in the models.

The dual-track Kennicutt-Schmidt relation with one track for normal star forming galaxies and another for starbursts (see above), could have a contribution from metallicity changes. Tan *et al.* (2013) note that relatively low CO emission at a given SF rate for high redshifts ($z > 3$), which is another way of expressing the second track from starbursts, might be from low metallicity instead of high density.

Local observations show that CO emission drops significantly at low-metallicity for a given molecular mass or star formation rate (Bolatto *et al.* 2013; Leroy *et al.* 2011). CO emission in the local group galaxy WLM, which has a metallicity of 13% solar, is very weak compared to the mass of molecular hydrogen inferred from dust and HI emission (Elmegreen *et al.* 2013). ALMA observations of the individual CO clouds also show them to be very tiny compared to the extent of the H_2 (Rubio *et al.* 2015). Nevertheless, the clouds are like small CO clouds in the solar neighborhood: they have about the same pressure, density and column density, and they satisfy the usual size-linewidth relation for other dwarf galaxies, although at the low mass end. Morphologically, the CO in WLM is highly confined to the core regions of much larger H_2 clouds. This is unlike CO clouds in the Milky Way which share more of the H_2 volume. The conversion from CO emission to H_2 mass is $\sim$ 30 times higher in WLM than in the Milky Way because of this confinement.

The centralized location for CO seems to be necessary for low-pressure galaxies like WLM because the CO needs a certain column of dust and H_2 to shield it from background radiation, and it needs a high density for excitation which cannot come from the ambient

ISM pressure alone. What gives it this high density is the large weight of the overlying H_2 and HI layers, which both shield it and compress it (Rubio *et al.* 2015).

Galaxies at high redshift might show the same centralized structure for CO if the metallicity is low. The change would be in the direction of greater SF rate per CO luminosity, which is what also happens in ULIRGs. The reason for this change would be different in the two cases because in ULIRGS, the CO gas should be denser than usual, causing faster SF per unit CO mass, and at low metallicity, the CO should be more sparse than usual, causing normal SF rates but with less CO.

4. Massive Clusters at high redshift

A possible implication of the small CO cores in local metal-poor galaxies is that only small star clusters might form. This would be the case if SF happened only in CO-emitting regions. If SF also occurs in cold, shielded regions without CO (Glover & Clark 2012; Krumholz, Leroy & McKee 2011) then clusters more massive than the CO cloud could form. In either case, it seems possible that low-mass, low-metallicity galaxies left on their own will not be able to make bound star clusters as massive as the metal poor globular clusters currently seen in galactic halos. These clusters probably formed in dwarf galaxies (Searle & Zinn 1978; Elmegreen, Malhotra, & Rhoads 2012) when the metallicity was as low as we see it today in the clusters. The clusters could also have been much more massive when they formed, considering the evaporation of stars in the intervening time (McLaughlin & Fall 2008). Thus there could be a problem in understanding how low-mass and low-metallicity galaxies formed massive bound clusters. A possible solution is that the clusters formed during mergers between such galaxies, when a large mass of interstellar material would have been at high pressure (Bekki 2008; Rubio *et al.* 2015).

Cluster formation at high redshift seems to have made more massive bound clusters than are being produced today. Jordan *et al.* (2007) found that the Schechter cutoff mass for globular cluster systems decreases with galaxy mass but in all cases, this cutoff is about 10 times higher than the cutoff mass seen in local galaxies of the same mass, even including the Antenna galaxy merger. In Jordan *et al.* (2007), the cutoff mass for globular clusters associated with galaxies having an absolute blue magnitude of -21.6 is around $2 \times 10^6\ M_\odot$, whereas in M51 and the Antenna, which have the same magnitude, it is $10^5\ M_\odot$ and $4 \times 10^5\ M_\odot$, respectively (Gieles *et al.* 2006).

Zaritsky *et al.* (2015) identified globular clusters in the Spitzer Survey of Stellar Structure in Galaxies (S^4G). They calculated the number of globular clusters per $10^9\ M_\odot$ of galaxy mass. For dwarfs, approximately 2% of today's galaxy mass is in globular clusters. This means that at 10% of the universe's age, when the clusters formed, and for ten times the current cluster mass in the initial state (as constrained by the observation of dual populations – D'Ercole *et al.* 2012), the total globular cluster mass was comparable to the galaxy mass at the time of globular cluster formation. Then the dwarf galaxy would have been dominated by giant clusters. The same high fraction is observed for WLM (Elmegreen, Malhotra, & Rhoads 2012) and Fornax dSph (Larsen *et al.* 2012).

Cluster formation thus appears to have been more efficient at high redshift than it is today. If the formation efficiency of bound clusters is higher at higher pressure and higher star formation rate (Larsen 2009; Kruijssen 2012; Adamo *et al.* 2015) then perhaps the greater cutoff mass of clusters at high redshift also followed simply from the higher gas fraction, like the other changes with redshift discussed above.

5. Conclusions

SF processes look the same for a wide range of cosmic time. The SF Main Sequence and constant M* at the cutoff in the galaxy mass function suggest a uniformity of SF among different galaxies. The small dispersions in the SF main sequence, the Kennicutt-Schmidt relation, and the mass-metallicity relation all suggest a uniformity of SF within galaxies. There also appears to have been the same distinction between ULIRGs and normal star-forming galaxies for a wide range of cosmic time, leading to higher SF efficiencies and shorter gas consumption times in mergers, presumably from the higher densities in shocks and inflows in these systems.

Stars form in dwarf irregular galaxies even when the Toomre Q parameter is high. Thus the validity of a Q threshold for SF in dwarfs is questionable, and, by analogy, also questionable in the outer parts of spirals, which have similar properties. The applicability of a Q threshold to quenching of SF in the bulge regions of spirals might also be questioned. The utility of Q may be limited to spiral-scale instabilities in thin disks.

The microscopic SF mechanism is probably universal: turbulent and directed compressions of cold gas in self-gravitating clouds, followed by local collapse. Low metallicity at high redshift should change the large-scale correlations by changing the conversion factor between CO emission and cold gas mass, but the microscopic processes of SF could still be about the same.

Even with these similarities, the first clusters that formed appear to extend to much higher masses than the largest clusters forming today. Perhaps this reflects some differences in physical processes, such as less feedback at lower metallicities or more prevalent galaxy collisions in early times, or perhaps it is another indication, similar to what is starting to be observed locally, that cluster formation efficiency increases with pressure. Higher pressure are expected at high redshift because of the greater gas fractions and densities of disks.

References

Abramson, L. E., Kelson, D. D., Dressler, A., *et al.* 2014, *ApJ*, 785, L36
Adamo, A., Kruijssen, J. M. D., Bastian, N., Silva-Villa, E., & Ryon, J. 2015, *MNRAS*, 452, 246
Bekki, K. 2008, *MNRAS*, 388, L10
Béthermin, M., Daddi, E., Magdis, G., *et al.* 2015, *A&A*, 573A, 113
Bolatto, A. D., Wolfire, M., & Leroy, A. K. 2013, *ARA&A*, 51, 207
Casado, J., Ascasibar, Y., Gavilán, M., *et al.* 2015, *MNRAS*, 451, 888
Daddi, E., Elbaz, D., Walter, F., *et al.* 2010, *ApJ*, 714, L118
Dekel, A., Zolotov, A., Tweed, D., Cacciato M., Ceverino D., & Primack, J. R. 2013, *MNRAS*, 435, 999
DErcole A., DAntona F., Carini, R., Vesperini, E., & Ventura, P. 2012, *MNRAS*, 423, 1521
Duncan, K., Conselice, C. J., Mortlock, A., *et al.* 2014, *MNRAS*, 444, 2960
Elmegreen, B. G. 2011, *ApJ*, 737, 10
Elmegreen, B. G., Malhotra, S., & Rhoads, J. 2012, *ApJ*, 757, 9
Elmegreen, B. G., Rubio, M., Hunter, D. A., Verdugo, C., Brinks, E., & Schruba, A. 2013, *Nature*, 495, 487
Elmegreen, B. G. & Hunter, D. A. 2015, *ApJ*, 805, 145
Federrath, C. & Klessen, R. S. 2012, *ApJ*, 761, 156
Förster Schreiber, N. M., Genzel, R., Bouché, N., *et al.* 2009, *ApJ*, 706, 1364
Genzel, R., Tacconi, L. J., Gracia-Carpio, J., *et al.* 2010, *MNRAS*, 407, 2091
Genzel, R., Tacconi, L. J., Lutz, D., *et al.* 2015, *ApJ*, 800, 20
Gieles, M., Larsen, S. S., Bastian, N., & Stein, I. T. 2006, *A&A*, 450, 129
Glover, S. C. O. & Clark, P. C. 2012, *MNRAS*, 421, 9

Kruijssen, J. M. D. 2012, *MNRAS*, 426, 3008
Kruijssen, J. M. D. 2015, *MNRAS*, 454, 1658
Krumholz, M. R., Leroy, A. K., & McKee, C. F. 2011, *ApJ*, 731, 25
Genzel, R., Förster Schreiber, N. M., Lang, P., *et al.* 2014, *ApJ*, 785, 75
Grazian, A., Fontana, A., Santini, P., *et al.* 2015, *A&A*, 575, A96
Guo, K., Zheng, X. Z., Wang, T., & Fu, H. 2015, *ApJ*, 808, L49
Jordán, A., McLaughlin, D. E., Côté, P. *et al.* 2007, *ApJS*, 171, 101
Kassin, S. A., Weiner, B. J., Faber, S. M., *et al.* 2012, *ApJ*, 758, 106
Krumholz, M. R., Dekel, A., & McKee, C. F. 2012, *ApJ*, 745, 69
Larsen, S. S. 2009, *A&A*, 494, 539
Larsen, S. S., Strader, J., & Brodie, J. P. 2012, *A&A*, 544, 14
Leroy, A. K., Bolatto, A., Gordon, K., *et al.* 2011, *ApJ*, 737, 12
Lombardi, M., Alves, J., & Lada, C. J. 2015, *A&A*, 576, L1
Magdis, G. E., Daddi, E., Béthermin, M., *et al.* 2012, *ApJ*, 760, 6
Martig, M., Bournaud, F., Teyssier, R., & Dekel, A. 2009, *ApJ*, 707, 250
McLaughlin, D. E. & Fall, S. M. 2008, *ApJ*, 679, 1272
Newman, S. F., Genzel, R., Förster-Schreiber, N. M., *et al.* 2012, *ApJ*, 761, 43
Romeo, A. B. & Wiegert, J. 2011, *MNRAS*, 416, 1191
Rubio, M., Elmegreen, B. G., Hunter, D. A., Brinks, E., Cortés, J. R., & Cigan, P. 2015, *Nature*, 525, 218
Saintonge, A., Tacconi, L. J., Fabello, S., *et al.* 2012, *ApJ*, 758, 73
Salmi, F., Daddi, E., Elbaz, D., *et al.* 2012, *ApJ*, 754, L14
Silverman, J. D., Daddi, E., Rodighiero, G., *et al.* 2015, arXiv150504977
Sargent, M. T., Daddi, E., Béthermin, M., *et al.* 2014, *ApJ*, 793, 19
Searle L., Zinn R. 1978, *ApJ*, 225, 357
Song, M., Finkelstein, S. L., Ashby, M. L. N., *et al.* 2015, arXiv150705636
Tacconi, L. J., Neri, R., & Genzel, R. 2013, *ApJ*, 768, 74
Tan, Q., Daddi, E., Sargent, M., *et al.* 2013, *ApJ*, 776, L24
Teyssier, R., Chapon, D., Bournaud, F., *et al.* 2010, *ApJ*, 720, L149
Whitaker, K. E., van Dokkum, P. G., Brammer, G., & Franx, M. 2012, *ApJ* 754, L.29
Zahid, H. J., Dima, G. I., Kudritzki, R.-P., *et al.* 2014, *ApJ*, 791, 130
Zaritsky, D., McCabe, K., Aravena, M., *et al.* 2015, *ApJ*, submitted

From Interstellar Clouds to Star-Forming Galaxies: Universal Processes?
Proceedings IAU Symposium No. 315, 2015
P. Jablonka, Ph. André & F. van der Tak, eds.

doi:10.1017/S1743921316007584

Molecular gas, stars, and dust in sub-$L^\star$ star-forming galaxies at $z \sim 2$: Evidence for universal star formation and non-universal dust-to-gas ratio

Miroslava Dessauges-Zavadsky[1], Michel Zamojski[1], Daniel Schaerer[1], Françoise Combes[2], Eiichi Egami[3], Panos Sklias[1], Mark A. Swinbank[4], Johan Richard[5] and Tim Rawle[6]

[1]Observatoire de Genève, Université de Genève, 51 Chemin des Maillettes, 1290 Versoix, Switzerland

[2]Observatoire de Paris, LERMA, 61 Avenue de l'Observatoire, 75014 Paris, France

[3]Steward Observatory, University of Arizona, 933 North Cherry Avenue, Tucson, AZ 85721, United States

[4]Institute for Computational Cosmology, Durham University, South Road, Durham DH1 3LE, United Kingdom

[5]CRAL, Observatoire de Lyon, Université Lyon 1, 9 Avenue Ch. André, 69561 Saint Genis Laval Cedex, France

[6]ESAC, ESA, PO Box 78, Villanueva de la Canada, 28691 Madrid, Spain

Abstract. Current star-forming galaxies (SFGs) with CO measurements at $z \sim 2$ suffer from a bias toward high star formation rates (SFR) and high stellar masses (M_*). It is yet essential to extend the CO measurements to the more numerous $z \sim 2$ SFGs with $L_{\rm IR} < L^\star = 4 \times 10^{11}$ L$_\odot$ and $M_* < 2.5 \times 10^{10}$ M$_\odot$. We have achieved CO, stars, and dust measurements in 8 such sub-$L^\star$ SFGs with the help of gravitational lensing. Combined with CO-detected galaxies from the literature, we find that the $L_{\rm IR}$, $L'_{\rm CO(1-0)}$ data are best-fitted with a single relation that favours a universal star formation. This picture emerges because of the enlarged star formation efficiency spread of the current $z > 1$ SFGs sample. We show that this spread is mostly triggered by the combination of redshift, specific SFR, and M_*. Finally, we find evidence for a non-universal dust-to-gas ratio (DGR) with a clear trend for a lower DGR mean in $z > 1$ SFGs by a factor of 2 with respect to local galaxies and high-redshift sub-mm galaxies at fixed about solar metallicity.

Keywords. Gravitational lensing, galaxies: evolution, galaxies: high-redshift

---

1. Introduction

The study of cold gas, the fuel for star formation, across the cosmic time is crucial to understand how galaxies have converted their gas into stars. Thanks to the sensitivity improvement of the IRAM Plateau de Bure interferometer (PdBI) these last years, molecular gas measurements have become accessible in star-forming galaxies at $1 < z < 2.5$ which lie on the "main sequence" (MS). Galaxies following this sequence between stellar mass and star formation rate contribute $\sim 90\%$ to the cosmic SFR density at these redshifts (Daddi *et al.* 2007; Rodighiero *et al.* 2011). Unfortunately, the current sample of $z > 1$ CO-detected SFGs is still biased toward objects distributed at the high SFR and high M_* end. Extending the molecular gas measurements to the more common and

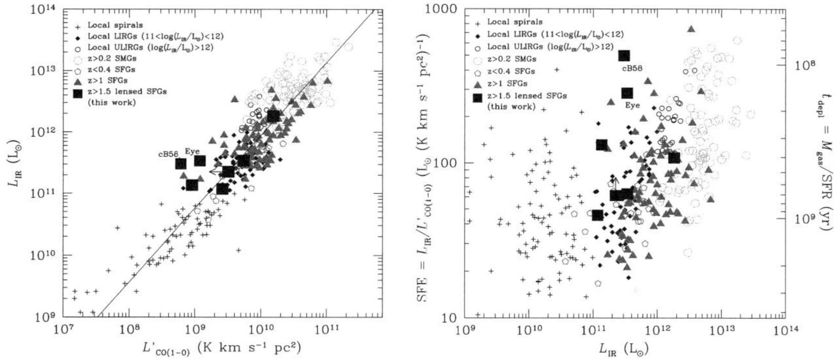

Figure 1. *Left:* IR luminosities as a function of CO(1–0) luminosities of our low-$L_{\rm IR}$-selected SFGs compared to our compilation of galaxies with CO measurements from the literature (see Dessauges-Zavadsky *et al.* 2015). A single linear $\log(L_{\rm IR})$–$\log(L'_{\rm CO(1-0)})$ relation seems to emerge, favouring a universal star formation. *Right:* Star formation efficiencies as a function of IR luminosities for the same galaxies. A large SFE spread is observed for the $z > 1$ SFGs. *Color versions of the figures are available online.*

numerous MS galaxies in the sub-$L^\star$ domain below the knee of the luminosity function is challenging, but feasible with the help of strong gravitational lensing.

2. Target selection

We have selected five targets for new IRAM/PdBI and IRAM/30m CO observations from the *Herschel* Lensing Survey (HLS; Egami *et al.* 2010), designed to observe 54 massive galaxy clusters to discover cluster-lensed, high-redshift background sources. The following selection criteria have been applied: high magnification factors $\mu > 10$; spectroscopic redshifts $z \sim 1.5-3$; lensing-corrected *Herschel* IR luminosities $L_{\rm IR} < 4 \times 10^{11}$ L$_\odot$, or equivalently SFR < 40 M$_\odot$ yr^{-1}, below the characteristic $L^\star$ of the IR luminosity function (Gruppioni *et al.* 2013); and complete photometry from optical, near-IR, IR to far-IR to enable to derive the stellar and dust properties from spectral energy distribution (SED) modeling. To these five targets, we have added the well known strongly-lensed galaxies, MS1512–cB58 and the Cosmic Eye, that also satisfy the above selection criteria.

When comparing our selected cluster-lensed, sub-$L^\star$ SFGs with our compilation of galaxies with CO measurements from the literature (see Dessauges-Zavadsky *et al.* 2015 for the references), we observe that: (1) they populate the domain with the lower $L_{\rm IR}$ at $z > 1.5$ and even fall below the *Herschel* blank field detection limit (Elbaz *et al.* 2011; Sklias *et al.* 2014); and (2) they lie on the MS at their respective redshifts, but probe one order of magnitude smaller M_* with $M_* < 2.5 \times 10^{10}$ M$_\odot$.

3. Evidence for a universal star formation

In Fig. 1 (left panel) we show the IR luminosities as a function of CO(1–0) luminosities of our compilation of CO-detected galaxies from the literature. We observe that the CO measurements achieved in our sample of sub-$L^\star$ SFGs nicely extend the $L_{\rm IR}$–$L'_{\rm CO(1-0)}$ distribution of $z > 1$ SFGs toward lower values. Moreover, evidence for a single linear $\log(L_{\rm IR})$–$\log(L'_{\rm CO(1-0)})$ relation with a slope of ~ 1.2 emerges over 5 orders of magnitude in $L_{\rm IR}$ and for galaxies gathering together spirals, MS SFGs, and starbursts at redshifts from $z = 0$ to 5.3. This strongly favours *a universal star formation*, which was questioned by Daddi *et al.* (2010), Genzel *et al.* (2010), and Sargent *et al.* (2014) who claimed the

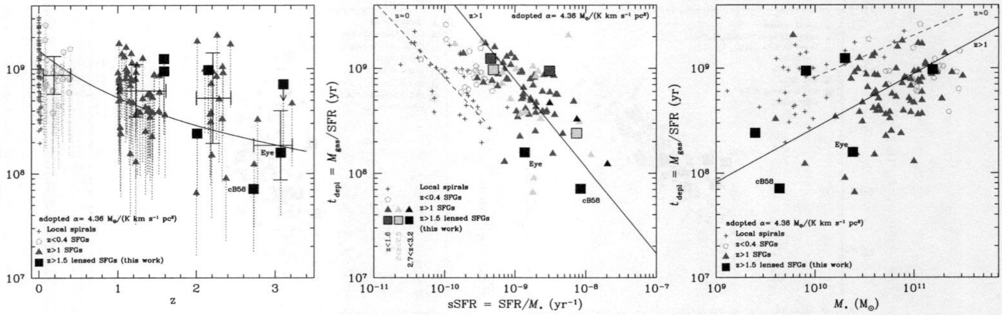

Figure 2. The dependence of the star formation efficiency on mainly three physical parameters: the redshift (*left*), the specific star formation rate (*middle*), and the stellar mass (*right*). All contribute to the large SFE spread observed in SFGs at $z > 1$ (see Fig. 1, right panel).

existence of a bimodal behaviour between the sequences of 'disks' and 'starbursts'. Their reported offset of 0.46−0.5 dex in the normalization between these two galaxy populations is now within the $1.5\,\sigma$ dispersion of the current sample of $z > 1$ SFGs. Indeed, when looking at the star formation efficiency defined as $\mathrm{SFE} = \mathrm{SFR}/M_{\mathrm{gas}} \equiv L_{\mathrm{IR}}/L'_{\mathrm{CO(1-0)}}$, we clearly see that $z > 1$ SFGs show an enlarged SFE spread and dispersion similar to that of high-redshift starbursts (or sub-mm galaxies) and do not have the SFE confined to the SFE of local spirals any more (Fig. 1, right panel).

The question which follows is: what drives such a large SFE spread in $z > 1$ SFGs? We have investigated the dependence of SFE (or molecular gas depletion timescale, $t_{\mathrm{depl}} = 1/\mathrm{SFE}$) in high-redshift SFGs on several physical parameters: the redshift, the specific star formation rate ($\mathrm{sSFR} = \mathrm{SFR}/M_*$), the M_*, the offset from the main sequence, and the compactness of the starburst. We find that the SFE spread is mostly triggered by the combination of redshift, sSFR, and M_*. The dependence of SFE on the offset from the MS is only valid for starburst galaxies with $\mathrm{sSFR}/\mathrm{sSFR}_{\mathrm{MS}}$ well above the MS limit.

In Fig. 2 we show from left to right the t_{depl}–z, t_{depl}–sSFR, and t_{depl}–M_* relations. While the first two relations have already been reported by Saintonge *et al.* (2011,2013) and Tacconi *et al.* (2013) although in a different context not linked to the SFE spread, for the first time one finds evidence for the last relation between SFE and M_* at high redshift. Both Saintonge *et al.* (2011) and Bothwell *et al.* (2014) reported a t_{depl} increase with M_* initially for massive galaxies with $10^{10} < M_*/\mathrm{M}_\odot < 10^{11.5}$ and then down to $M_* \sim 10^9$ $\mathrm{M}_\odot$. The few data points we have collected at the low-M_* end for the $z > 1$ SFGs seem to trigger the t_{depl}–M_* correlation at high redshift. If correct, it has several important implications: (1) it questions the constant t_{depl} of 0.7 Gyr observed by Tacconi *et al.* (2013); (2) it contradicts the bathtub model that assumes a constant t_{depl}; and (3) it refutes the linearity of the Kennicutt-Schmidt relation ($\Sigma_{\mathrm{SFR}} \propto \Sigma_{\mathrm{H_2}}^N$ with $N \neq 1$).

4. Evidence for a non-universal dust-to-gas ratio

Dust-to-gas ratio measurements from far-IR/sub-mm SED and CO luminosity have several caveats, because of a number of assumptions that have to be done. First, the dust mass estimates from SED are tributary to the adopted dust model, dust emissivity index, and dust mass absorption cross section. Second, a CO–H_2 conversion factor has to be assumed. And third, it is accepted that $M_{\mathrm{gas}} \approx M_{\mathrm{H_2}} \gg M_{\mathrm{HI}}$. To bypass at least the first caveat, Scoville *et al.* (2014) proposed to consider the rest-frame 850 μm continuum as the dust mass tracer. They then derived in a homogeneous way (same Galactic CO–H_2 conversion factor and β-slope = 1.8) the DGR in local galaxies (spirals and

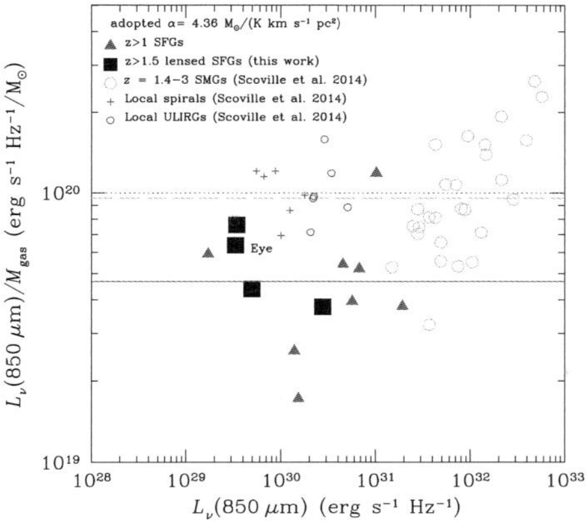

Figure 3. Dust-to-gas ratios as estimated from the 850 μm continuum over the molecular gas mass ratio following the prescriptions from Scoville *et al.* (2014). The DGR non-universality is favoured when comparing at fixed about solar metallicity local spirals, local ultra-luminous IR galaxies, high-redshift sub-mm galaxies, and high-redshift SFGs.

ultra-luminous IR galaxies) and $z > 1.4$ sub-mm galaxies and found the same DGR means. They interpreted the result as a strong evidence for a universal DGR.

We have done the same exercise for our sample of high-redshift, sub-$L^\star$ SFGs plus a few $z > 1$ SFGs from the literature for which the far-IR photometry has been published (Magdis *et al.* 2012; Saintonge *et al.* 2013). But, we have retained for the comparison with the Scoville *et al.* DGR measurements only the $z > 1$ SFGs with about solar metallicities ($Z/Z_\odot > 0.8$), as derived from nebular emission line metallicity estimators or the mass-metallicity relation. We were left with 12 SFGs in total. We find a clear trend for a lower DGR mean in $z > 1$ SFGs by ~ 0.33 dex at fixed about solar metallicity, namely a factor of 2 difference with the DGR measured by Scoville *et al.* in local galaxies and high-redshift sub-mm galaxies (see Fig. 3). This favours *a non-universal dust-to-gas ratio*. As a consequence, the use of the CO as a molecular gas mass estimator still remains highly recommended despite the uncertain CO–H_2 conversion factor, in comparison with the dust mass derived from the 850 μm continuum.

References

Bothwell, M. S., Wagg, J., Cicone, C., *et al.* 2014, *MNRAS*, 445, 2599
Daddi, E., Dickinson, M., Morrison, G., *et al.* 2007, *ApJ*, 670, 156
Daddi, E., Bournaud, F., Walter, F., *et al.* 2010, *ApJ*, 713, 686
Dessauges-Zavadsky, M., Zamojski, M., Schaerer, D., *et al.* 2015, *A&A*, 577, 50
Egami, E., Rex, M., Rawle, T. D., *et al.* 2010, *A&A*, 518, 12
Elbaz, D., Dickinson, M., Hwang, H. S., *et al.* 2011, *A&A*, 533, 119
Genzel, R., Tacconi, L. J., Gracia-Carpio, J., *et al.* 2010, *MNRAS*, 407, 2091
Gruppioni, C., Pozzi, F., Rodighiero, G., *et al.* 2013, *MNRAS*, 432, 23
Magdis, G. E., Daddi, E., Béthermin, M., *et al.* 2012, *ApJ*, 760, 6
Rodighiero, G., Daddi, E., Baronchelli, I., *et al.* 2011, *ApJL*, 739, L40
Saintonge, A., Kauffmann, G., Wang, J., *et al.* 2011, *MNRAS*, 415, 61
Saintonge, A., Lutz, D., Genzel, R., *et al.* 2013, *ApJ*, 778, 2
Sargent, M. T., Daddi, E., Béthermin, M., *et al.* 2014, *ApJ*, 793, 19
Scoville, N., Aussel, H., Sheth, K., *et al.* 2014, *ApJ*, 783, 84
Sklias, P., Zamojski, M., Schaerer, D., *et al.* 2014, *A&A*, 561, 149
Tacconi, L. J., Neri, R., Genzel, R., *et al.* 2013, *ApJ*, 768, 74

From Interstellar Clouds to Star-Forming Galaxies: Universal Processes?
Proceedings IAU Symposium No. 315, 2015
P. Jablonka, Ph. André & F. van der Tak, eds.

doi:10.1017/S1743921316007596

On the Variation of Gas Depletion Time

D. Utomo[1], L. Blitz[1], A. Bolatto[2], T. Wong[3], A. Leroy[4], S. Vogel[2] and EDGE-CALIFA Collaboration

[1]Department of Astronomy, University of California, Berkeley, CA 94720
[2]Department of Astronomy, University of Maryland, College Park, MD 20742
[3]Department of Astronomy, University of Illinois, Urbana, IL 61801
[4]Department of Astronomy, The Ohio State University, Columbus, OH 43210

Abstract. We present an updated status of the EDGE project, which is a survey of 125 local galaxies in the $^{12}CO(1-0)$ and $^{13}CO(1-0)$ lines. We combine the molecular data of the EDGE survey with the stellar and ionized gas maps of the CALIFA survey to give a comprehensive view of the dependence of the star formation efficiency, or equivalently, the molecular gas depletion time, on various local environments, such as the stellar surface density, metallicity, and radius from the galaxy center. This study will provide insight into the parameters that drive the star formation efficiency in galaxies at $z \sim 0$.

Keywords. Galaxies: star formation, galaxies: ISM, galaxies: stellar content.

1. Introduction

The present-day stellar mass of galaxies is an integral of the history of star formation and mergers over the life time of galaxies. Therefore, the rate at which galaxies convert their gas reservoir into stars is one of the important factors that drives galaxy evolution. We have seen in the Milky Way and other nearby galaxies that star formation occurs exclusively in molecular clouds (GMCs, Blitz 1993). In a simplified scheme, these clouds contract under their self-gravity and finally form stars. The contraction time-scale can be estimated to be proportional to the free-fall time, which is a function of gas volume density. Therefore, it is logical to connect the star formation rate (SFR) and the gas volume density (Schmidt 1959). However, the gas volume density is not an observable quantity in extragalactic studies. Instead, we can use the gas surface density. Previous studies showed that SFR surface density correlates with the molecular gas surface density as $\Sigma_{\rm SFR} \propto \Sigma_{\rm mol}^N$ with an exponent of $N \sim 1$ (Bigiel *et al.* 2008, Leroy *et al.* 2013).

A linear relation between $\Sigma_{\rm SFR}$ and $\Sigma_{\rm mol}$ implies that the *average* molecular gas consumption time-scale or gas depletion time, defined as $\tau_{\rm dep} \equiv \Sigma_{\rm mol}/\Sigma_{\rm SFR}$, is constant, with typical value of $\sim 2-3$ Gyr. The star formation efficiency (SFE) is the inverse of $\tau_{\rm dep}$ and measures the ability of gas to form stars per unit gas mass. Shorter $\tau_{\rm dep}$ means gas forms stars more efficiently. In this paper, we use both terms interchangeably.

The observed scatter in the Kennicutt–Schmidt diagram (~ 1 dex, Kennicutt 1998) means that for a given amount of molecular gas, some galaxies or parts of galaxies (on a ~ 1 kpc scale) are more efficient in forming stars than the others by a factor of ~ 10. The conditions or parameters that drive the variations of $\tau_{\rm dep}$ are the main issue that is addressed in this paper. In this respect, we need more than just the tracers of SFR (Hα+IR or UV+IR flux) and molecular gas density (CO flux). Therefore, in §2 we briefly describe the CALIFA and EDGE surveys that gather other physical properties of galaxies, such as stellar surface densities and metallicities. Then, we analyze if any of these parameters correlate with the SFE or $\tau_{\rm dep}$ in §3.

2. The CALIFA and EDGE Surveys

The Calar Alto Legacy Integral Field Area (CALIFA) is an optical Integral Field Unit (IFU) survey of ~ 600 local galaxies at $0.005 < z < 0.03$. The samples are selected from the SDSS database based on their diameter in r-band ($45'' < D_{25} < 80''$), so that they fit well within the IFU field-of-view. A detailed description of the survey is presented by Sanchez *et al.* (2012). The main data products of CALIFA survey are various properties and kinematic maps of the stellar populations and ionized gas.

We select 176 galaxies from the CALIFA samples, based on their brightness in the WISE 22μm band, to be observed in the $(J = 1-0)$ ^{12}CO and ^{13}CO spectral lines using the CARMA observatory in the E-array configuration ($8''$ resolution). Then, we chose 125 of them based on ^{12}CO detections to be observed in the D-array configuration to achieve higher angular resolution ($4''.5$ or $\sim 1 - 2$ kpc) that is closely matched to the CALIFA resolution ($2''$). This Extragalactic Database for Galaxy Evolution (EDGE) survey is the first major, resolved CO survey matched to an IFU survey. The survey descriptions will be presented elsewhere (Bolatto *et al.* in prep.). The main data products of the EDGE survey are the surface density and kinematic maps of the molecular gas.

We make use of the CALIFA and EDGE data products to investigate if there is any systematic trend between the depletion time and other galaxy physical properties. We derive the SFR surface densities using the Hα line, following the prescription of Calzetti *et al.* (2007). The Hα lines have been corrected for dust extinction by using the Balmer decrement method and the Cardelli *et al.* (1989) extinction curve. The CO integrated luminosities are converted to the molecular gas surface densities by applying a constant CO-to-H_2 conversion factor, α_{CO}, of 4.3 $M_{\odot}$ (K km s^{-1} pc^2)$^{-1}$. Additionally, we can derive the gas-phase metallicity by using the ratio of strong emission lines (e.g., Kewley & Dopita 2002), and the stellar mass surface density, Σ_*, by applying the stellar population synthesis technique to each line-of-sight spectrum.

3. Preliminary Results

We divide our studies into three parts: global (galaxy-by-galaxy), local ($\sim 1 - 2$ kpc scales), and radial analyses as a function of distance from the galaxy centers.

3.1. *Global Variations*

Globally, $\tau_{\rm dep}$ increases with the stellar mass, M_* (left panel of Figure 1). Here, we compute $\tau_{\rm dep}$ by adding all of the molecular gas mass and dividing it by the sum of all star formation within the galaxy. More massive galaxies tend to have longer $\tau_{\rm dep}$, in agreement with the COLD GASS result (solid red line in Figure 1, Saintonge *et al.* 2011). They argued that the galaxies that show reduced $\tau_{\rm dep}$ are those undergoing minor starbursts due to distant tidal encounters, variations in the intergalactic medium accretion rate, or secular processes within galactic discs, while morphological quenching and AGN feedback prevent the molecular gas from forming stars in the high mass galaxies.

By using HERACLES samples, Leroy *et al.* (2013) proposed that the correlation in the $\tau_{\rm dep} - M_*$ diagram can be explained *partially* by a variation in α_{CO} with respect to the dust-to-gas ratio, and hence, gas-phase metallicity, Z. According to the $M_* - Z$ relation (Tremonti *et al.* 2004), less massive galaxies have lower Z, and therefore, higher α_{CO} (Bolatto *et al.* 2013). Increasing α_{CO} will shift the low mass galaxies data points upward in the $\tau_{\rm dep} - M_*$ diagram to make a flatter slope. However, the correlation cannot be completely removed, so that a variation in α_{CO} is not the only explanation. A check of

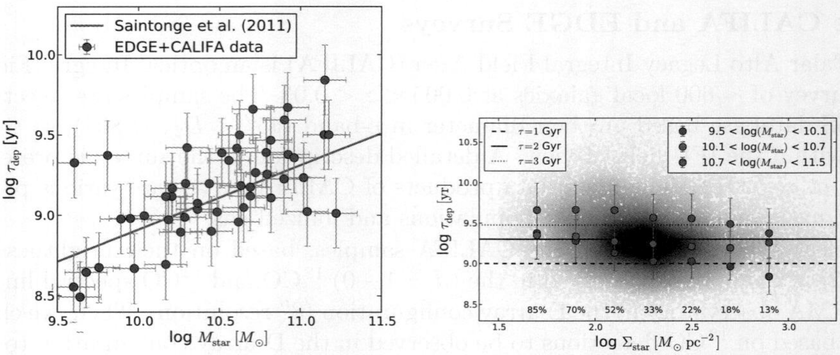

Figure 1. Plots of the depletion time vs. stellar mass (left) and stellar surface density (right). *Color versions of the figure are available online.*

how robust our result is with respect to the variation of α_{CO} will be done in a future work (Utomo *et al.* in prep.).

3.2. *Local Variations*

In order to understand the underlying physics that drive the global trend in τ_{dep} vs. M_*, we make detailed analyses on a scale of $\sim 1-2$ kpc. From many galaxy properties that can be resolved by the IFU observations, we investigate two of them: Σ_* and Z. These two properties are independent of the τ_{dep} measurements, but are physically linked to the star formation. Briefly, Σ_* is the parameter that dominates the mid-plane pressure, P_h (Elmegreen 1989), which correlates to the H_2-to-HI mass ratio, R_{mol}, as $R_{mol} \propto P_h^{0.92}$ (Blitz & Rosolowsky 2006). Higher Σ_* raises R_{mol}, which in turns produces more sites of star formation (if the molecular gas bound in GMCs). Similarly, higher Z means more shielding of the gas (Wolfire, Hollenbach & McKee 2010), so that the gas is more likely to produce molecular clouds, which are the sites of star formation. Based on this reasoning, it makes sense to investigate if there is a correlation between τ_{dep} and Σ_* or Z.

On the right panel of Figure 1, we plot τ_{dep} vs. Σ_*. The red, green, and blue dots are the median values of various ranges of M_*. At face value, it seems there is a weak correlation so that a higher Σ_* has a shorter τ_{dep}. This means star formation is more efficient in regions of high P_h, probably because P_h exerts an additional inward force to GMCs so that GMCs are more prone to contract and form stars. Recently, Huang & Kauffmann (2015) also found this correlation in the HERACLES, ATLAS 3D, and COLD GASS samples. However, they also mentioned the specific SFR (sSFR) as another parameter that drives τ_{dep}, which we do not consider in this paper because it is not an independent parameter with respect to τ_{dep}, i.e. sSFR $=$ SFR$/M_*$.

The scatter in τ_{dep} itself is driven by different M_* (right panel of Figure 1), so that for a given Σ_*, more massive galaxies have longer τ_{dep} than less massive galaxies. It is interesting that M_*, a global property of galaxies, still leaves its imprint on the local scales. Since the scatter in τ_{dep} is orthogonal to Σ_*, the local variations in τ_{dep} depend on *both* Σ_* and M_*. Note that Σ_* here is Σ_* on ~ 1 kpc scale, not its average value over a whole galaxy, $\bar{\Sigma}_* \propto M_* \, R_{gal}^{-2}$. Therefore, two galaxies with different M_* can have regions with the same Σ_*, but located at different radii from the centers due to different absolute scale of their Σ_* profile, i.e. it is closer to the center for lower M_* galaxy. This leads us to a question of whether τ_{dep} also varies with radius, which we discuss in §3.3.

The caveat in the local analyses is that the *non*-detection fractions of Σ_{mol} (labeled as percentages in Figure 1) rise steeply in the lower Σ_* regimes. In a future study, we

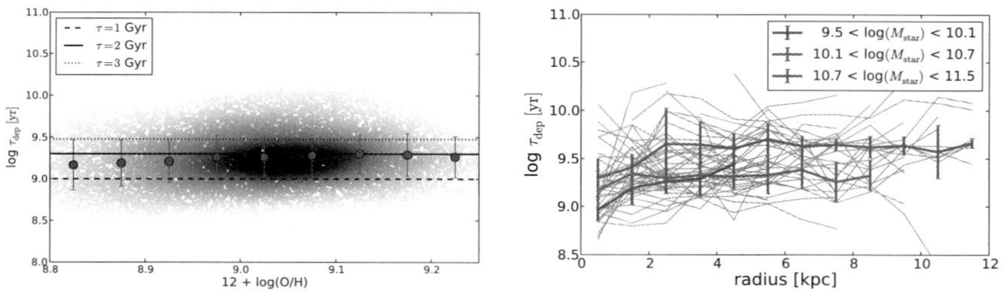

Figure 2. *Left:* depletion time vs. metallicities. *Right:* radial profile of depletion time.

will investigate the effect of non-detections by stacking the CO spectra to increase the sensitivity in the lower Σ_* regimes (Utomo *et al.* in prep.). Another caveat is the variation of $\alpha_{\rm CO}$ due to Z. Therefore, we also plot $\tau_{\rm dep}$ vs. Z, based on Kewley & Dopita (2002) prescriptions, in the left panel of Figure 2. The median values are shown as blue circles. We do not see a clear correlation, except for the low metallicity regime. This reassures us that our choice of $\alpha_{\rm CO} = 4.3\ M_\odot$ (K km s^{-1} pc^2)$^{-1}$ is reasonable.

3.3. *Radial Variations*

Lastly, we investigate the radial profile of $\tau_{\rm dep}$. The radial profiles for individual galaxies are generated by azimuthally averaging $\tau_{\rm dep}$ within a ring. These are shown as the grey lines on the right panel of Figure 2. Then, we group the radial profiles into three M_* bins, as in Figure 1, and take their median values. We find that the radial profiles of $\tau_{\rm dep}$ are roughly flat for $r \gtrsim 3$ kpc, with an exception for the lowest M_* bin (probably due to the lack of data samples). Interestingly, all three mass bin profiles drop toward the centers at $r \lesssim 3$ kpc, which means the star formation is more efficient in the galaxy centers.

The enhanced SFE at the centers has also been pointed out by Leroy *et al.* (2013) in the HERACLES samples. They mentioned that this enhanced efficiency coincides with the higher fraction of the ^{12}CO$(2-1)$ over ^{12}CO$(1-0)$ line intensities, which is an indicator of more excited gas and higher gas temperature at the centers. Furthermore, variations in $\alpha_{\rm CO}$ exaggerate the drop, so this enhancement must be genuine. In the future, we will investigate what causes the drop of $\tau_{\rm dep}$ at the centers, and whether this enhancement of SFE is linked to the morphology of the bulges (classical vs. pseudo-bulges and barred vs. unbarred) and the gas free-fall time due to the underlying weight of stars and gas.

References

Bigiel, F., Leroy, A., Walter, F., *et al.* 2008, *AJ*, 136, 2846
Blitz, L. 1993, in: E. H. Levy & J. I. Lunine (eds.), Protostars and Planets Vol. III, p. 125
Blitz, L. & Rosolowsky, E. 2006, *ApJ*, 650, 933
Bolatto, A. D., Wolfire, M., & Leroy, A. K. 2013, *ARAA*, 51, 207
Calzetti, D., Kennicutt, R. C., Engelbracht, C. W., *et al.* 1993, *ApJ*, 666, 870
Cardelli, J. A., Clayton, G. C. & Mathis, J. S. 1989, *ApJ*, 345, 245
Elmegreen, B. 1989, *ApJ*, 338, 178
Huang, M.-L. & Kauffmann, G. 2015, *MNRAS*, 450, 1375
Kennicutt, R. 1998, *ARAA*, 36, 189
Kewley, L. J. & Dopita, M. A. 2002, *ApJS*, 142, 35
Leroy, A. K., Walter, F., Sandstrom, K., *et al.* 2013, *AJ*, 146, 19
Saintonge, A., Kauffmann, G., Wang, J., *et al.* 2011, *MNRAS*, 415, 61
Sánchez, S. F., Kennicutt, R. C., Gil de Paz, A., *et al.* 2012, *A&A*, 538, 8
Schmidt, M. 1959, *ApJ*, 129, 243
Tremonti, C. A., Heckman, T. M., Kauffmann, G., *et al.* 2004, *ApJ*, 613, 898
Wolfire, M. G., Hollenbach, D., & McKee, C. F. 2010, *ApJ*, 716, 1191

Author index

Poster index

The posters are available online at http://journals.cambridge.org/IAU in usual journal style.

Aikawa et al: Evaporation of grain-surface species by shock waves onto a forming protoplanetary disk

Akhter et al: High Mass Star Formation: Properties of NH_3 clumps

Aller et al: Connecting the Interstellar Gas and Dust Properties in Distant Galaxies Using Quasar Absorption Systems

Bagoly et al: Apparent brightness distribution of GRB host galaxies

Barnes et al: ThrUMMS: A New View of the Molecular Milky Way

Beck et al.: A supernova scenario for magnetic fields and rotation measures in galaxies

Beckman et al.: Detection and measurement of expanding ionized shells around young star clusters

Beckman et al.: A Statistical Comparison of the Near Infrared Emission with the Hα Emission from the HII regions of NGC4321, NGC4736 and NGC4254

Berczik et al: Time evolution of chemistry with fixed physical parameters in TMC-1

Burton et al.: Extended Carbon Emission in the Galaxy: Dark Gas along the G328 Sightline

Catalán-Torrecilla, C. et al.: SFR bulge-to-disk ratios from the CALIFA IFS nearby galaxies survey

Chengpeng et al: The feedback of Herbig Ae/Be stars

Clark et al.: Quantifying the Magnetic Alignment of HI and Dust in the Diffuse ISM

Colombo et al.: The road toward a full, high resolution molecular Cloud catalog of the Galaxy

Crider et al.: Detecting H II Regions in "Pure" Starburst Galaxies with SDSS Data

Datta et al: Universal Accretion Growth Using Sandpile Models

Dawson et al.: First results from SPLASH

Falstad et al: Herschel spectroscopic observations of Zw 049.057 and Arp 299a

Ferreras et al.: The IMF-SFH connection in ETGs

Fischer et al.: The WISE Census of Young Stellar Objects and Clusters in Canis Major

Ford et al.: A Green Bank Telescope Search for Highly Extended HI Disks Around Spiral Galaxies

Freundlich et al.: Resolved star formation relations at high redshift from the IRAM PHIBSS program

Fu et al.: HI Gas in Disk and Dwarf Galaxies in the Semi-analytic Models of Galaxy Formation

Fujimoto et al.: Birthing star forming clouds in the grand design

Gao et al.: Dust Extinction toward SN 2014J in M82

Gezari et al.: Mid-IR Imaging of Orion BN/KL: Modeling of Physical Conditions and Energy Balance

Gibson et al.: A Global View of Molecule-Forming Clouds in the Galaxy

Gonçalves et al.: Properties of the ISM in UV-luminous galaxies: clues from the low-redshift universe

Gorski et al.: SWAN: NGC 253's Nucleated Star Bursting Environment

Green et al.: Intensity ratios for XDR/PDR identification

Hanawa et al.: Structure and Stability of Filamentary Clouds Supported by Lateral Magnetic Field

Harada et al.: Investigating formation of isolated intermediate/massive YSOs in the LMC

Heiderman et al: The Real Solar Neighborhood Protostars

Hojaev: All-sky catalogs and analysis of interstellar clouds

Holdship et al.: Chemical Tracers of Pre-Brown Dwarf Cores Formed Through Turbulent Fragmentation

Hsieh et al.: IRAS 16253–2429: the First Proto-Brown Dwarf Binary Candidate Identified through Dynamics of Jets

Jaskot et al.: ALFALFA Hα Reveals How Galaxies Use Their H I Fuel

Jiang et al.: Hydrocarbons in Massive Star Forming Regions: C_2H Observations

Kalinova et al.: Bridging Galaxy Dynamics and Baryon Efficiency of 40 EDGE-CALIFA galaxies

Kavila et al.: Possible density dependent local variations in the IMF

Kim et al: Star Formation in Nuclear Rings of Barred-Spiral Galaxies

Kong et al.: Spectroscopic Observations of Hundreds of Star Formation Regions in M33

Könyves et al.: The connection between prestellar cores and filaments in the Aquila molecular cloud complex

Kusune et al.: A Study of Magnetic Fields on Bright-Rimmed Clouds

Kwon et al.: Warm dust and gas of massive young stellar objects revealed by Herschel PACS spectroscopy

Ladjelate et al.: Results from the Herschel Gould Belt Survey in the Ophiuchus Main Cloud

Lawther et al.: FeLoBAL quasars: evolution, orientation or just odd-BALs?

Lehmann et al.: Dissipation of turbulence by magnetohydrodynamic shock waves

Leung et al.: Nature of highly extended CS($J = 7-6$) emission around low-mass protostar L483

Levrier et al.: Polarized thermal dust emission as seen by *Planck* : A comparison with MHD simulations and lessons from a toy model

Lindberg et al.: Protostellar chemistry dominated by external irradiation

Marchetti et al.: The HerMES Local Luminosity Function

Marton et al.: The Herschel Point Source Catalogue

Menéndez-Delmestre et al.: High-Redshift Protoclusters Traced by Submillimeter Galaxies. Tracing Star Formation Activity out to $z > 4$

Miyamoto et al.: Hot gas in the center of the Seyfert galaxy NGC 3079

Mok et al.: Environmental Effects on the ISM and Star Formation Properties of Nearby Spiral Galaxies

Montier et al.: The Planck Catalogue of Galactic Cold Clumps : PGCC

Nishimura et al.: Nobeyama 45m CO Galactic Plane Survey: Filament properties and star formation in M17

Oey et al.: Massive Star Formation in the Sparsest Environments

Oi et al.: Mass–metallicity relation for AKARI-FMOS infrared luminous galaxies at $z \sim 0.9$

Oliveira et al.: Young Stellar Objects in the Magellanic Clouds: Herschel Spectroscopy First Results

Onishi et al.: Large-Scale Molecular Gas Survey in ^{12}CO, ^{13}CO and $C^{18}O$ ($J = 2–1$) with the Osaka 1.85m mm-submm Telescope

Pan et al.: What is a GMC? Are observers and simulators discussing the same star-forming clouds?

Pattle et al.: A Full Virial Analysis of the Prestellar Cores in the Ophiuchus Molecular Cloud

Pellegrini et al.: Star formation in early-type galaxies: the role of stellar winds and kinematics

Pettitt et al.: Galactic spiral generation in tidal encounters

Ramirez-Tannus et al.: Massive pre-main sequence stars in M17

Ren et al.: Core mass function: a comparison study between Orion A and other massive filamentary clouds

Riguccini et al.: Mid-Infrared Enhanced Lenticulars caught in the dusty

midst of transformation in the Coma Cluster

Sahu et al.: Multiphase ISM in low luminosity radio galaxies: a case study of NGC 708
Saito et al.: Molecular Gas Excitation and Chemistry in VV 114 and NGC 1614 with ALMA
Shima et al.: From cloud crash to star birth:
star formation in cloud collisions

Takamiya et al.: SFR and Abundances of Nearby Galaxies
Tokuda et al: ALMA Observations of a High-density Core in Taurus: Dynamical Gas Interaction at the Possible Site of a Multiple Star Formation
Tóth et al.: Star formation in Taurus Auriga Perseus and California nebulae
Turnshek et al.: CaII Quasar Absorbers: Statistics, Metals, Dust, and Associated Galaxies

Varga et al.: Atlas of low-mass young stellar object disks from mid-infrared interferometry
Varga-Verebélyi et al.: Herschel Footprint Database and Service
Varosi et al.: Dust Temperatures and Opacities in the Central Parsec of the Galactic Center Modeled from Analysis of Multi-wavelength Mid-infrared Images

Wang et al.: Probing the magnetic field structure in the filamentary cloud IC5146
Ward et al.: Characterising the Protostellar Population of the Magellanic Clouds with VLT/ SINFONI
Williams et al.: Diffuse Interstellar Bands in Emission
Wu et al.: How surface density of galaxy disks affects metallicity? Outflow and Accretion

Zaragoza-Cardiel et al.: Gravity Binding and Pressure Bounding of HII regions and Molecular Clouds in Interacting Galaxies
Zhang et al.: WISE properties of OH megamaser galaxies: Guide for future FAST OHM survey?
Zhang et al.: Binary stars on the galaxy SFH